Distributed Data Services

Zichen Xu

Distributed Data Services

Fundamentals and Practices

Springer

Zichen Xu
College of Mathematics and Computer Science
Nanchang University
Nanchang, Jiangxi, China

ISBN 978-981-96-9391-7 ISBN 978-981-96-9392-4 (eBook)
https://doi.org/10.1007/978-981-96-9392-4

Jointly published with China Machine Press Co., Ltd.
The print edition is not for sale in Mainland China. Customers from Mainland China please order the print book from: China Machine Press Co., Ltd.

This Springer imprint is published by the registered company Springer Nature Singapore Pte Ltd.
The registered company address is: 152 Beach Road, #21-01/04 Gateway East, Singapore 189721, Singapore

Contents

Part I
Fundamentals and Theory of Distributed Systems

Chapter 1
Fundamentals of Distributed Systems

1.1 Definition of Distributed Systems

With the advancement of the global digitalization process, global data is growing explosively. According to the latest forecast by the International Data Corporation (IDC), the global data volume will continue to grow rapidly and is expected to reach approximately 384.6 ZB by 2028 (1 ZB $= 2^{10}$ EB $= 2^{20}$ PB $= 2^{30}$ TB), with a compound annual growth rate (CAGR) of 24.4%. Traditional standalone systems currently face severe performance bottlenecks. Moreover, despite their high cost, mainframe systems often lack robust disaster recovery capabilities, resulting in service interruptions and unavailability in case of failures. There is a growing awareness that traditional standalone systems can no longer meet the demands for large-scale data storage and computation in the technological field. Given the disparity between the volume of data in this era and standalone systems' storage and processing capabilities, distributed systems are gaining increasing attention. Efforts to address the explosive growth of data have explored two main approaches: scaling up individual machines and interconnecting multiple machines. The high cost and inflexibility associated with the former have made the latter approach increasingly preferable. However, according to the cost conservation principle, costs do not just disappear; software design costs often go up when hardware costs go down. Therefore, the principles of distributed systems can help lower these software costs.

A distributed system consists of independent computing devices interconnected through a network, presenting users with the appearance of a unified system. This architecture has two key dimensions: physically separate hardware components and integrated software that creates a seamless user experience. For example, in video rendering, the system can distribute individual frames across multiple nodes for parallel processing, demonstrating how complex tasks can be partitioned and processed simultaneously. Once the rendering, the managing application will assign a new phase of work to the nodes, namely, the aggregation phase. This phase will continue until all segments of the video are recombined. This task might be distributed across hundreds or even thousands of nodes, turning a job that might

Z. Xu, *Distributed Data Services*, https://doi.org/10.1007/978-981-96-9392-4_1

take a single computer several days to complete into one that can be done in minutes. The advantages of distributing and integrating computing devices (interconnecting machines) can be summarized as follows:

- **Greater Flexibility**. As the demand for services grows, it is easier to increase computing capacity. In most cases, servers can be added instantly to a distributed system.
- **Reliability**. A well-designed distributed system can tolerate the failure of one or more nodes without significantly affecting performance. In contrast, a failure in a monolithic system can cause the entire application to stop functioning.
- **Increased Speed**. When traffic becomes heavy, a single server may struggle, affecting the performance of each user. The scalability of distributed databases and other distributed systems makes them easier to maintain while sustaining high-performance levels.
- **Geographical Distribution**. Distributed content delivery is intuitive for any Internet user. Distributed systems mask geographical differences, which is crucial for organizing content from around the globe.

Distributed systems are a significant advancement in IT and computer science, with usage scenarios becoming increasingly large and complex. A single model and architecture for distributed systems cannot meet the demands of these varied and dynamic scenarios. Over years of development, various distributed system models and architectures have been proposed. We summarize some of distributed systems' more commonly used models and architectures:

- **Client-Server Systems**. This is the most traditional and simplest type of distributed system, organized as a master-worker architecture. It involves numerous networked computers interacting with a central server to achieve data storage, processing, or other shared goals. The primary drawbacks are single-point bottlenecks and failures. In simpler terms, it can be described as one machine gives commands, while the others carry them out.
- **Mobile Networks**. This represents a more advanced distributed system that distributes the workload among mobile phones, switching systems, and Internet-based devices. The structure of this system is multilayered, with each layer designed to handle different characteristics and tasks.
- **Peer-to-Peer Networks**. This type of distributed system follows a peer-to-peer architecture. The workload is distributed across hundreds or thousands of computers running the same software, with all computers being logically equivalent. This is another example of a distributed system architecture.
- **Web-based Systems**. The most common distributed systems in modern enterprises run through the web, handing off workloads to dozens of cloud-based virtual server instances created as needed and terminated once the task is completed.

Several monitoring methods have been proposed to monitor these various distributed systems. One method suitable for modern cloud and microservices architecture is distributed tracing. This is a method of monitoring applications that essentially involves distributed computing to observe the operations of applications running on a distributed system.

Although various distributed systems are different in their organizational structures and models, we can still identify some fundamental similarities. The following key features and characteristics typically define distributed systems:

- **Scalability**. The ability to grow with increasing workloads is a fundamental characteristic of distributed systems. This is achieved by adding additional processing units or nodes to the network as needed.
- **Concurrency**. Various components in a distributed system operate concurrently. Another characteristic is the absence of a "global clock," as tasks are performed at different rates and may not be synchronized.
- **Availability/Fault Tolerance**. If one node fails, the remaining nodes can continue to operate without interrupting the overall computation.
- **Transparency**. External programmers or end users perceive the distributed system as a single computing unit rather than its underlying parts. This allows users to interact with a single logical device without worrying about the system's architecture.
- **Heterogeneity**. In most distributed systems, nodes and components are typically asynchronous, featuring different hardware, middleware, software, and operating systems. This allows the distributed system to be expanded by adding new components.
- **Replication**. Distributed systems support information sharing and message passing, ensuring consistency among redundant resources (such as software or hardware components). This enhances fault tolerance, reliability, and availability.

These features and characteristics serve as fundamental starting points for the theoretical research and design of distributed systems. In Sect. 1.2, we will provide an overview of these concepts. Although these topics may not be explicitly highlighted in the subsequent chapters, readers will sense their underlying influence as a foundation for thinking about distributed challenges throughout the book.

The mismatch between single-machine capacity and data volume is becoming increasingly evident in various scenarios. In the foreseeable future, distributed systems are well-positioned to dominate computing, with almost every application or service incorporating some form of distributed computing. As users increasingly rely on mobile devices for their daily tasks, the demand for always-online, readily available computing is driving this trend. As enterprise developers continue to turn to distributed tools to simplify development, deploy systems and infrastructure, and streamline application operation and management, the significant role of distributed systems in global computing will be further solidified.

1.2 Design Goals of Distributed Systems

We need to design a distributed system to address problems that a single-machine system cannot solve. Different individuals have their own definitions of the objectives of distributed system design. Tanenbaum and Van Steen [130], for example, list connectivity of users and resources, transparency, openness, and scalability as goals of a distributed system. Meanwhile, Coulouris, Dollimore, and Kindberg [32] identify heterogeneity, openness, security, scalability, failure handling, concurrency, and transparency as design goals and challenges of distributed systems. In this book, we focus on how to design distributed systems from the perspective of distributed data services. We believe that the goals of a distributed system can be balanced among the following four points:

- Consistency.
- Availability.
- Partition tolerance.
- Scalability.

In this section, we will provide a brief overview of these four objectives.

1.2.1 Consistency

To better understand what consistency is, let's first imagine a simple scenario: suppose some data needs to be stored long-term and cannot be lost. How should we handle this? A natural idea is to prepare multiple hard drives and create several backup copies on different drives. In reality, making multiple copies is obviously easy to achieve. Although the hard drives are isolated and cannot communicate with each other, the backup data is static, and once initialized, the state does not change. Manual file copying operations can easily ensure consistency across the backup drives. However, in a distributed system facing diverse data services, we must consider how to replicate dynamic data across nodes under unreliable network conditions correctly.

Let's extend the scenario: suppose we have some constantly changing data that needs to be accurately stored on several different computers in a distributed network. How should we handle this? Such a scenario is not encountered in large monolithic machines. In a distributed system, read and write operations on machines cannot simultaneously reach different data sources due to physical limitations (geographical location, network transmission). Moreover, machine failures and rotations are common in distributed systems. We need consistency algorithms specifically designed for distributed systems to ensure smooth replication.

In the context of distributed systems, consistency refers to the outcomes of performing read and write operations on multiple physical copies of a logical data entity within the system. The "read/write" operations mentioned here are similar to the "data changes" in the previous example. In fact, the prerequisite for discussing data consistency is the simultaneous existence of both read and write operations; otherwise, it is meaningless. Combining these two factors results in a set of read/write strategies for multiple copies of data, known as a consistency model. This section will briefly introduce the concepts of data synchronization, levels of data consistency, and common data consistency algorithms. In subsequent chapters, we will provide detailed explanations of these concepts.

When discussing data synchronization, we usually classify synchronization methods into two categories: asynchronous and synchronous. Based on these two types of data synchronization methods, the following five points are common data synchronization mechanisms:

- **Backup**. Periodic backups have minimal impact on the performance of the existing system but can barely enable recovery during node failures.
- **Master-slave**. Asynchronous replication of each command can be seen as a more granular form of periodic backup.
- **Multi-master**. Also known as master-master replication, it is an enhanced version of master-slave replication, allowing writes on multiple nodes and synchronizing them later.
- **Two-Phase Commit**. Ensures all nodes are notified before writing, with performance bottlenecks easily occurring on the "master" node.
- **Paxos**. Similar to 2PC, multiple nodes can write simultaneously and only need to notify most nodes, resulting in higher throughput.

Data replication methods represented by synchronization are known as state transfer, and provide a reliability assurance approach that aligns more closely with human expectations, but often at the cost of reduced availability. These methods are challenging to implement and typically suffer from poor performance. When building distributed systems, these drawbacks are often unacceptable. Some critical systems must ensure both data correctness and reliability while also requiring high availability, for instance, systems that need to guarantee 99.999999% data reliability and 99.999% system availability. To mitigate the conflict between reliability and availability as the number of machines in a distributed system grows, today's prevailing replication method is based on operation transfer. The core idea of operation transfer is to ensure that the final state of multiple machines is consistent. If their initial states are consistent and the sequence of received operation commands is the same, then—whether the operations involve adding, modifying, deleting, or any other possible program action—the method effectively broadcasts a series of operation logs to each distributed node correctly. Figure 1.1 summarizes these data synchronization mechanisms.

With the above background, let's rethink what consistency means. Essentially, ensuring consistency means avoiding contradictions between multiple nodes in a distributed system. Many consistency models exist, making it difficult to differentiate. For ease of understanding, we temporarily classify consistency into the following two types, which will be further refined in subsequent chapters:

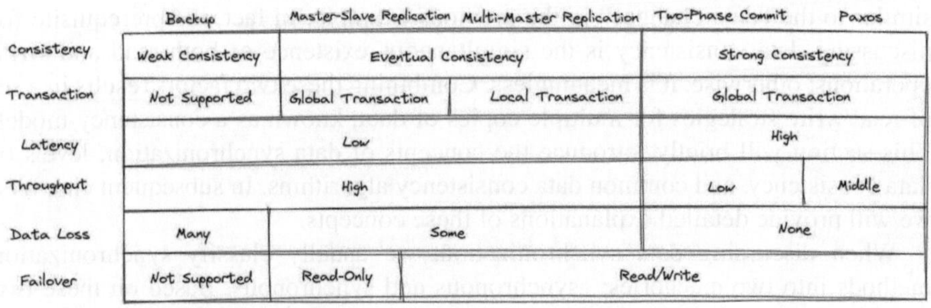

Fig. 1.1 Summary of data synchronization mechanisms

- **Strong Consistency**. At any time, all users or processes query the most recently successfully updated data, ensuring that updated data can be seen by subsequent accesses.
- **Weak Consistency**. Users or processes may query different data at a certain time, but eventually, all users or processes will see the successfully updated data.

Readers should note that the consistency mentioned here is different from the distributed consensus commonly seen. Consistency refers to the differences between different copies of data, while consensus refers to the methods and processes for achieving consistency. We will delve into the differences between these two concepts in subsequent chapters.

In distributed systems designed for strong consistency, we must mention the Paxos algorithm, which is so important that Mike Burrows, the inventor of Google Chubby service, once said, "There is only one consensus protocol, and that is Paxos. All other approaches are just broken versions of Paxos." This means that other consistency consensus algorithms are merely variants of Paxos. Although this statement may be somewhat exaggerated, common strong consistency algorithms such as Raft and ZAB can be considered variants of Paxos. These algorithms will be detailed in subsequent chapters.

In distributed systems, all consistency models other than strong consistency can be classified as weak consistency. Since network partitioning in a distributed environment cannot be eliminated, data synchronization must be balanced. When designing distributed systems, the goal is no longer to ensure that the data state of all nodes is consistent under all circumstances but to achieve eventual consistency using the principle of "light process, heavy result." A classic weak consistency algorithm is the Quorum algorithm, a voting algorithm used to ensure data redundancy and eventual consistency. We will elaborate on this algorithm in subsequent chapters.

1.2.2 *Availability*

Before diving into the availability details, we would like to formally introduce the well-known CAP theorem, which will underpin the consistency, availability, and partition tolerance sections. The CAP theorem was proposed by Professor Eric Brewer from the University of California, Berkeley, in 2000. Its core idea is that any network-based data-sharing system can satisfy at most two of the following three properties: consistency, availability, and partition tolerance. The definitions of these three properties are as follows:

- Consistency. This refers to all nodes having the latest version of the data. Note that in the context of the CAP theorem, consistency emphasizes that the data seen at any moment and by any distributed node is as excepted, representing strong consistency.
- Availability. This means that the data is highly available. A distributed system does not stop serving due to the failure of a single machine. When a machine fails, the system can quickly redirect traffic to the healthy machines and continue providing services.
- Partition Tolerance. This refers to the ability to tolerate network partitions. Even when communication between partitions is disrupted, the system can still provide services correctly.

The relationship among the three properties of CAP can be summarized as shown in Fig. 1.2. The CAP theorem tells us that we can achieve systems that favor CA, CP, or AP, but it is impossible to satisfy all three simultaneously. Since the CAP theorem has already been rigorously proven, we will not discuss why these three properties cannot all be achieved together.

The CAP theorem seems to define a pessimistic outcome for distributed systems. Here, we want to clarify a common misunderstanding that arises from treating the CAP theorem as an absolute law: classifying popular distributed systems based solely on CAP theory has often been inaccurate. It is acknowledged that in large-scale distributed environments, network partitioning is an unavoidable reality that must be tolerated, forcing a choice between availability and consistency. For example, HBase is considered a CP system, while Cassandra is viewed as an AP system.

Fig. 1.2 CAP theorem

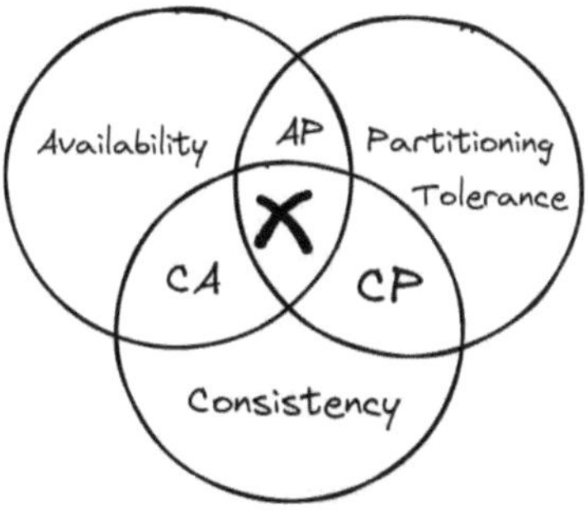

However, the CAP theorem asserts that a single piece of data can't achieve both availability and consistency simultaneously. In modern distributed systems, there are always various types of data existing, and the trade-off between consistency and availability can occur at very granular levels within the same system. Each decision may vary based on specific operations or involve particular data or users. Through the diversification of data types and the measures to optimize the balance between consistency and availability, we see that Cassandra can maintain eventual consistency even at the expense of certain performance metrics. Thus, delineating distributed systems solely from the CAP perspective is not rigorous.

Availability represents the system's ability to provide uninterrupted service. To understand the availability, it's important to first grasp two closely related metrics: reliability and maintainability. Reliability is measured by Mean Time Between Failures (MTBF), while Maintainability is measured by Mean Time To Repair (MTTR). Availability is determined by the ratio of operational time to total time, expressed by the formula: Availability = MTBF/(MTBF + MTTR). This indicates that availability is derived from the relationship between reliability and maintainability. For availability, the term "six nines" (99.9999%) is commonly used to describe a high level of uptime, which implies an average annual downtime of only 32 seconds.

Achieving availability differs from consistency, which has many mathematically proven theories. In distributed systems, availability often relies on practical design principles. First, redundancy design is essential. Single points of failure are inevitable in distributed systems, and the best way to reduce them is through redundancy. By deploying nodes in different physical locations, the system can avoid multiple failures occurring in a single data center. Currently, common redundancy designs include master-slave and peer-to-peer architectures. Then, the master-slave design can be further categorized into configurations with one master and multiple slaves or multiple masters and multiple slaves. After redundancy design, rate limiting can be implemented on the server side by taking into account the load-handling capacity of the distributed system itself. Generally, rate limiting can be designed in terms of service denial, service degradation, prioritized requests, and elastic scaling. Specifically, when the system experiences a surge in traffic, it may temporarily reject requests from clients that exceed a certain threshold during a defined period. There are also ways to sacrifice strong data consistency to achieve higher performance throughput. Resources can be allocated to users with higher priority, and performance monitoring systems can detect the busiest services and automatically scale them as needed.

1.2.3 Partition Tolerance

Distributed systems generally use networks as bridges to interconnect multiple machines. Networks are not always reliable, and network failures can cause one or more machines to experience network partitioning. The distributed systems

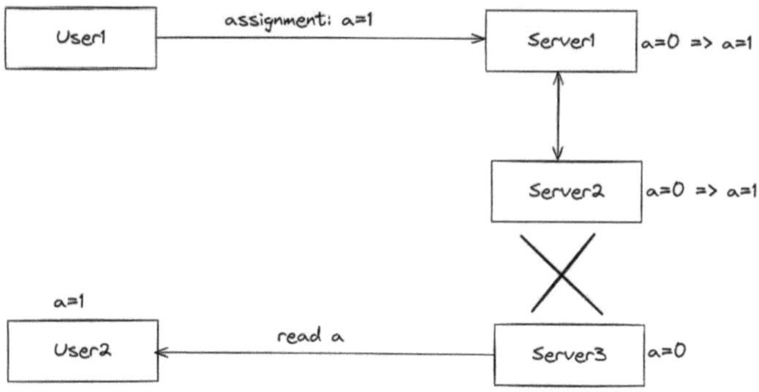

Fig. 1.3 Server partitioning

we design must be able to correctly handle network partitions and ensure data consistency in such cases.

As shown in Fig. 1.3, in the case of partitioning, a data item is stored only in the nodes of one partition, making it inaccessible to parts of the system that are not connected to that partition. In this scenario, the partition is not fault-tolerant. One way to improve partition tolerance is to replicate a data item across multiple nodes, so that in the event of a partition, the data item can be distributed across different partitions. This increases fault tolerance. However, replicating data across multiple nodes leads to consistency issues, as the data on different nodes may become inconsistent. To ensure consistency, each write operation must wait for all nodes to successfully write, which in turn affects availability. In general, the more nodes a piece of data exists on, the higher the partition tolerance, but the more updates need to be replicated, making consistency harder to maintain. To ensure consistency, the time required to update data across all nodes increases, which reduces availability.

However, with the advancement of technology, people have found that even in the presence of network failures or partitions, it is possible to create systems that can automatically handle failover correctly. When building multi-replica systems that can recover automatically while avoiding split-brain scenarios, it has been discovered that the key point is the majority vote, a fundamental concept used in the Raft algorithm.

As shown in Fig. 1.4, the first step of a majority voting system lies in having an odd number of servers rather than an even number. If the number of servers is odd, then the two partitions will no longer be symmetrical in the event of a network partition. Assuming a network partition occurs, one partition will have two servers while the other will have only one server, thus breaking the symmetry. This is the appealing aspect of majority voting. Therefore, you first need an odd number of servers. Then, to complete any operation, such as leader election in Raft or committing a log entry, you must gather more than half of the servers to approve the corresponding operation at any time. Here, "more than half" refers to exceeding half of the total number of servers. Intuitively, if there are three servers, you need

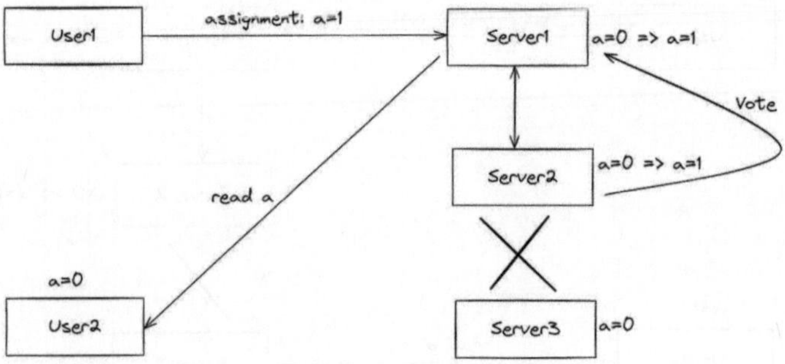

Fig. 1.4 Partition tolerance

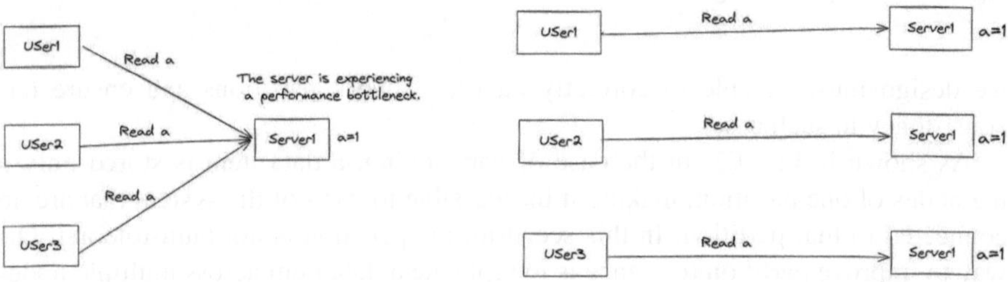

Fig. 1.5 Scalability

the approval of two servers to complete any operation. From another perspective, this system can tolerate the failure of one server, as any two servers are sufficient to complete the operation. If you need to build a more reliable system, you can add more servers to the system.

1.2.4 Scalability

The distributed systems we design must be scalable. Scalability means that we can achieve higher overall system throughput and better performance by using more machines. However, adding more machines does not always improve performance, especially in some complex computing scenarios where having more nodes may not necessarily result in better performance.

As shown in Fig. 1.5(left), its performance may degrade when a single server is accessed by many users simultaneously. We can improve performance by increasing the server's configuration, which is known as vertical scalability. However, a high-performance server is expensive, and the entire system becomes inaccessible if it fails. Scalability is a powerful feature. As shown in Fig. 1.5(right), if you build a system that can improve performance or throughput by simply adding more computers, it would be a significant achievement because you can just buy more computers to achieve this, which is known as horizontal scalability. Therefore, when constructing large websites using an entire data center of computers, scalability must always be considered to achieve the corresponding performance.

In summary, scalability provides the ability to handle ever-changing demands. It can add resources and scale up to handle increased customer demands and larger workloads seamlessly; conversely, resources can be seamlessly removed when demand and workloads decrease to save server resources. At the same time, scalability can improve system availability to some extent, ensuring continuous access to database resources during unplanned outages. Unplanned outages refer to unexpected system failures such as power outages, network interruptions, hardware failures, operating system or other software errors. Additionally, scalability can bring consistency and fault tolerance issues, so owning more machines does not always mean better performance.

1.3 Data Models

Data models may be the most important part of software development, as they profoundly influence not only how software is written but also how we think about the problems to be solved [64]. A data model is a collection of conceptual work that describes data, data relationships, data semantics, and consistency constraints. The choice of a data model often determines the system architecture design as well. The historical development of data models has been quite remarkable, evolving continuously under the influence of advancements in mathematical theory and the needs of application programs. This section will introduce three commonly used data models, the relational model, the document model, and the graph data model, and analyze their advantages and disadvantages in different use cases.

The relational model was first proposed by Edgar Codd in 1970 [31]. Initially, the relational model was merely a theoretical research concept, but it later evolved into SQL for relational databases and gained widespread acceptance. At that time, few people could have foreseen its immense success in dominating the development of data models for the following two to three decades. In the twenty-first century, the diverse forms of data representation have led to the emergence of NoSQL and document databases, reigniting debates over the best way to represent data relationships: is the strong data relational model truly the optimal choice for various scenarios? The document model has once again garnered significant attention. Nowadays, with the rise of social networks, many-to-many data structures are ubiquitous, and complex data associations among massive datasets are increasingly emerging. People are once again hopeful for the optimization of data models. A very natural idea is to model data as graph models. In fact, the document and graph data models, which have gained popularity in recent years, are not new concepts. After reading this section, readers should reflect on the reasons behind the development and choice of data models.

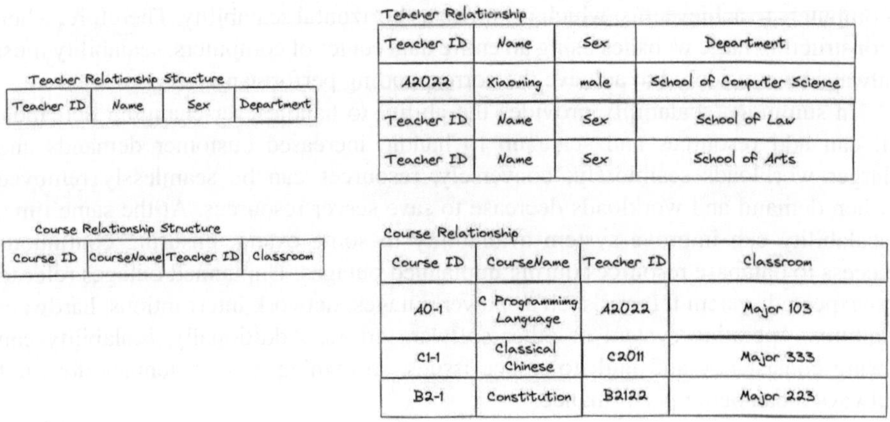

Fig. 1.6 Examples of relational model

1.3.1 Relational Model

The relational model used for database management is a data model based on predicate logic and set theory, aimed at hiding implementation details behind a simpler interface. SQL, which developed from the relational model, is now the most widely recognized data model. In Fig. 1.6, we use examples of a teacher relation and a course relation to illustrate the relational model: it succinctly represents the relationship between teachers and courses.

In SQL, which is based on the relational model, a relation refers to a table, a tuple refers to a row, and an attribute refers to a column. A relation instance represents a specific instance of a relation, which is the particular set of rows it contains. Since a relation is an unordered set of tuples, the order in which tuples appear in a relation is irrelevant. For each attribute of a relation, there exists a set of permissible values, known as the domain of that attribute.

In database management systems that use the relational model, the concept of a relation corresponds to the concept of a variable in programming languages, while the concept of a relational schema corresponds to the concept of a type definition in programming languages. A relational schema consists of a sequence of attributes and their corresponding domains. We can consider that, for data formatting, the relational model only stipulates that a relation (table) is merely a collection of tuples (rows) and nothing more. There are no complex nested structures, and any row or all rows in the table can be read. Figure 1.7 describes the process of transforming a conceptual model (the language used for communication between database designers and users) into a relational model.

Relational Database Management Systems (RDBMS) and SQL have been widely used as tools for storing and querying structured data since the 1980s. Along the path of being the primary technology for data storage and querying, the relational

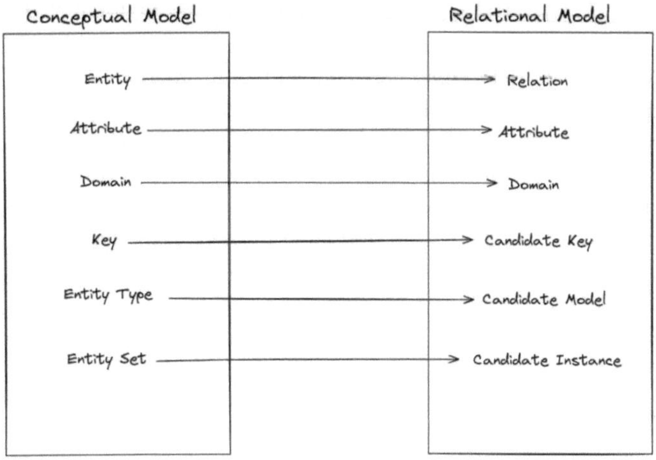

Fig. 1.7 Transformation of the conceptual model to the relational model

model has also encountered challenges from many competing technologies, with the network model and hierarchical model being two major ones.

A typical application of the hierarchical model is IBM's Information Management System (IMS), released for the Apollo space program [85]. The hierarchical model can be considered a precursor to the document model that will be introduced later, where all data is represented as records nested within records. However, this structure has significant limitations when it comes to modeling many-to-many relationships and associations between relationships. As an extension of the hierarchical model, the network model allows a record to have multiple parent nodes, thereby supporting the modeling of both one-to-many and many-to-many relationships. In the network model, relationships between records are not connected using foreign keys but rather through something akin to pointers in the C programming language. The only way to access a particular record is by selecting a path that starts from a root record. This traversed path is also known as the access path. It is akin to traversing a linked list, but in the context of many-to-many relationships, the multitude of access paths can become burdensome.

Compared to the two models mentioned above, the relational model stands out with three distinct advantages:

- In the relational data model, the access paths are completely hidden from the user, which results in a high degree of independence between programs and data. Furthermore, the data language exhibits a high level of non-procedurality.
- The structure is simple, and the relationships between entities are represented through common attributes of tables, making the structure straightforward and easy to understand.
- The operation is convenient, as the basic object of manipulation in the relational data model is a set rather than an individual tuple.

As the saying goes "There's no such thing as a free lunch," the advantages of the relational model are also the root of its disadvantages. The relational data model provides high data independence and non-procedural query capabilities (during a query, one only needs to specify the tables where the data exists and the columns where the desired data is located, without specifying the specific search path), which increases the burden on the system and results in lower query efficiency. Furthermore, the query capabilities of relational databases come from query optimizers, which can be considered complex beasts that require continuous investment and significant resources from research and development personnel over the years to update and iterate.

1.3.2 Document Model

Entering the twenty-first century, with the rapid development of Internet technology, data volumes have expanded swiftly, and humanity has gradually entered the era of big data. Big data poses significant challenges to traditional data management methods, and relational databases struggle to meet the demands of big data management in terms of capacity, performance, and cost. By compromising on the strict data consistency management of the relational model in relational databases, NoSQL databases have demonstrated significant advantages in scalability, model flexibility, economy, and accessibility, especially when using document models to construct databases for specific data structures.

In the document model, data is stored in the form of documents. A document is composed of records that describe the actual situation and data of the file. Nested documents can be used to provide information about subcategories of data. Additionally, documents can also be utilized to represent real-world objects. MongoDB [9] is a typical NoSQL database based on the document model. MongoDB supports a very loose data structure, storing semi-structured data as documents instead of normalizing data across multiple tables with unique and fixed structures, as in relational databases. The logical structure of storage is hierarchical, with documents in document databases using nested key-value pairs to provide the structure or schema of the document. Table 1.1 compares the terminology used in document databases (MongoDB) with that used in SQL databases.

A pertinent question arises regarding the advantages of the relational data model versus the document data model. Below are two general summaries of their respective strengths:

- The main argument favoring the document data model is centered around schema flexibility, which leads to better performance due to locality and the fact that it more closely resembles the data structures used by certain applications.
- The main argument supporting the relational data model lies in its support for join operations, allowing for the concise representation of one-to-many and many-to-many relationships.

Table 1.1 Terminology comparison

SQL	MongoDB
Table	Collection
Row	Document
Column	Field
Primary key	Objectid
Index	Index
View	View
Nested object	Nested document
Array	Array

Most document databases, as well as relational databases that support JSON, do not enforce a schema on the data within the documents. Support for XML in relational databases typically comes with optional schema validation. Schemalessness means that arbitrary keys and values can be added to documents, and when reading, clients have no guarantees about which fields a document might contain. The difference in flexibility between document databases and relational databases is most evident in their approach to data schemas. Specifically, the specific flexibility of document databases lies in their data structures following a schema-on-read approach (where the structure of data is implicit and only interpreted when reading the data), whereas the traditional approach in relational databases is schema-on-write (where the schema is explicit and the database ensures that all written data conforms to it). This different design paradigm remains a contentious topic, with no universally correct or incorrect answer. Strong schemas imply lower flexibility, whereas weak schemas provide greater flexibility but often result in data redundancy and a less concise representation of relationships. Below, we present two examples that demonstrate the flexibility of document database schemas. Please note that these are merely potential applications:

- There are many different types of objects, and it is impractical to place each type of object in its own table.
- The structure of the data is determined by an external system beyond your control and can change at any time.

In both of these scenarios, having a fixed schema can potentially do more harm than good, and a schema-less document might be a more natural data model. However, we should acknowledge that in cases where all records have the same structure, a schema is a useful mechanism for documenting and enforcing that structure.

Due to their lack of a strict data schema, databases that operate under the document model allow for arbitrary key-value pairs to be added to documents, making it easy to represent one-to-many relationships in the data. However, as the interconnectedness of data increases, particularly when dealing with many-to-many relationships, the document model becomes less appealing. While it is possible to reduce the need for joins by denormalizing the data, the application code must undertake additional work to maintain the consistency of this denormalized data.

Table 1.2 Summary of relational and document models

Relational model	Document model
Row-based	Document-based
Not ideal for hierarchical data	Commonly used for hierarchical data
Defined by a predefined schema	Contains a dynamic schema
Follows ACID properties (Atomicity, Consistency, Isolation, Durability)	Follows CAP theorem (Consistency, Availability, Partition Tolerance)
Supports complex joins	Does not support complex joins
Column-oriented	Field-oriented
Vertically scalable	Horizontally scalable
Limited dynamic replication support	Provides simplified replication support

By issuing multiple requests to the database, joins can be simulated within the application code, but this shifts the complexity to the application and often results in slower performance compared to joins executed by specialized code within the database. In such scenarios, the use of the document model may lead to more complex application code and poorer performance.

The primary justification for supporting the document data model lies in its exceptional schema flexibility, which results in better performance due to locality and, for certain applications, a closer alignment with the data structures used by those applications. The relational model, on the other hand, excels in scenarios involving joins, one-to-many, and many-to-many relationships. Actually, relational and document databases have become increasingly similar over time, which is a positive development: data models are complementary. If a database can handle data akin to documents and execute relational queries on them, applications can leverage the combination of features that best suit their needs. We predict that a hybrid of relational and document models represents a promising path forward for databases. Table 1.2 summarizes our comparison of the relational and document models.

1.3.3 Graph Data Model

As we saw in the previous two sections, many-to-many relationships are an important distinguishing feature among different data models. Suppose the data primarily consists of one-to-many relationships (which can be represented as treelike data) or there are no relationships between records. In that case, the document model is the most suitable choice. When there are only a few many-to-many relationships, the relational model can also try to handle them through simple transformations.

However, what if many-to-many relationships are common in the data? Imagine data structures like today's social networks, where the interconnectedness of data becomes increasingly complex. In such cases, modeling the data as a graph model would be a more natural fit.

Graphs consist of two types of objects: vertices (also known as nodes or entities) and edges (also known as relationships or arcs). In the real world, many types of data can be modeled as graphs. Some typical examples include:

- Social Networks. The vertices are people, and the edges represent the relationships between people.
- Web Graphs. The vertices are web pages, and the edges represent HTML links to other pages. Google's famous PageRank algorithm operates on web graphs to determine the search ranking of web pages.

In the examples mentioned above, the vertices of the graph represent the same type of entities. However, graphs are not limited to such homogeneous data. Their more practical use lies in providing a consistent way to store completely different types of objects within a single data store. Taking Weibo as an example, the vertices in its data can be people, locations, events, user comments, and so on, while the edges can represent relationships such as who follows whom (or one-way followership), who commented on which post, who participated in which event, and so forth.

There are various but related methods for constructing and querying data within graphs. This section will introduce the property graph model and the triple store model as a lead-in to subsequent chapters. Among the property graph models, the most notable graph database, neo4j [88], and its declarative query language, Cypher, will be discussed in detail in later chapters.

A property graph adds additional attribute information to graph data. In a property graph, both nodes and relationships have labels (indicating the type of node or relationship) and properties (providing additional descriptive information about the node or relationship). For example, a node with the label "User" could have properties such as "Name," "Registration Time," and "Registration Address." The property graph is one of the most common industrial-grade representations of graph data, widely applicable to data expression across various business scenarios. A property graph consists of a directed graph made up of vertices (also known as nodes), edges (also called relationships), labels, relationship types, and properties. Nodes and relationships are the most crucial entities in the graph. Specifically, nodes include:

- A unique identifier.
- A collection of outgoing edges.
- A collection of incoming edges.
- A collection of properties.

Each relationship includes:

- A unique identifier.
- The starting node of the edge (tail node).
- The ending node of the edge (head node).
- A label describing the type of relationship between the two vertices.
- A collection of properties (key-value pairs).

This design allows for efficient retrieval of all incoming and outgoing edges for a given node, enabling traversal of the graph, i.e., moving forward or backward along these vertex chains. By using different labels for different types of relationships, it is possible to store multiple types of information within a single graph while maintaining a clean and organized data model.

The triple store model is almost equivalent to the property graph model, using different terminology to describe the same concept. In a triple store, data is stored in a very concise form (subject, predicate, object). An example of this is the triple (Xiaoming, loves, Xiaoli), which segments the subject, predicate, and object into three parts to achieve the purpose of information recording. In practice, the subject of a triple corresponds to a node in a graph, and the object can either be the value of an attribute of the subject (in which case, the predicate is the key of the subject's attribute) or another node in the graph (in which case, the predicate represents an edge in the graph).

So far in this chapter, we have discussed the relational model, the document model, and the graph data model. Both the document model and the graph data model share a similar characteristic, which is that they are "schema-on-read," meaning they do not enforce validation on stored data, allowing for greater flexibility in adapting to changing requirements during application. The advantage of the relational model lies in its ability to somewhat clumsily emulate the document and graph models, making it relatively more versatile in its range of applications. Each of these models excels in their respective target domains and possesses unique strengths. As data representation is diverse, we will not conclude that any particular model is definitely better than another. Instead, our genuine concern lies in whether we can correctly select or combine these models in different business scenarios.

1.4 Data Storage

With the development of emerging technologies such as cloud computing, big data, and artificial intelligence, as well as the increasing complexity and demands of Internet businesses, the core value of data lies not merely in simple storage but in the processing, data mining, and analytical forecasting of massive amounts of data. Traditional storage technologies face numerous challenges, including handling large data volumes, supporting high concurrency, and achieving low latency. Distributed databases have become a widely adopted and effective solution in the industry to meet the requirements of modern applications for high performance, high availability, scalability, and cost-effectiveness. Distributed data storage employs a scalable architecture, distributing data across multiple devices and transparently handling data partitioning, fault tolerance, and load balancing to address the limitations of traditional databases in terms of storage capacity, high concurrency, and high availability. Furthermore, as the volume of data and concurrent operations in the system continue to grow, issues such as consistency, performance, and fault tolerance in distributed databases also require our attention and resolution.

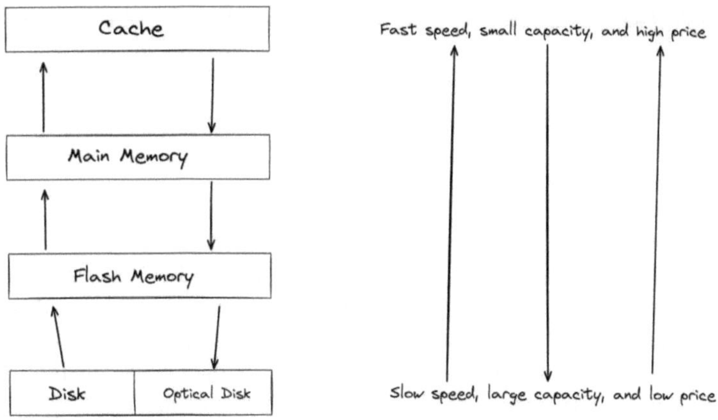

Fig. 1.8 Storage hierarchy architecture

Generally, there are various data storage media in computer systems. As shown in Fig. 1.8, they are organized in a hierarchical structure based on different speeds, capacities, and costs. The primary memory and cache, which are the most expensive and fastest storage media, are referred to as the main memory; the secondary storage (or online storage), such as disks, is located in the next level of the hierarchy; and the tertiary storage (or offline storage), such as optical disks, is at the bottom of the hierarchy.

In addition to differences in speed, capacity, and cost, the storage hierarchy also involves the issue of volatility. In Fig. 1.8, storage above the main memory is volatile, meaning that the data will be lost if certain conditions for data preservation are not met. For example, data stored in DRAM is volatile because as soon as the power supply to the DRAM is cut off, the data saved within it will be lost. Storage below flash memory is nonvolatile, and to ensure data security, data must be written to nonvolatile storage.

To enhance reliability and availability, a Redundant Array of Independent Disks (RAID) is employed, which enables concurrent read and write operations across multiple hard drives to improve program execution efficiency. Since data is redundantly stored across multiple disks, the failure of a single disk does not result in the loss of all data. Thus, this organization method also improves data reliability.

To strike a balance between cost and performance, RAID is categorized into different levels. RAID 0 involves striping data at the block level with no redundancy; RAID 1 involves striping data at the block level and using mirroring to store redundant data; RAID 5 involves striping data at the block level, with data blocks and parity blocks distributed alternately across disks. RAID 5 is capable of maintaining normal data access even when one disk fails. In a RAID 5 storage configuration consisting of four disks, the parity block Pk is stored on the (k mod 4)th disk, while the other three disks store the 3k, 3k + 1, and 3k + 2 data blocks. Table 1.3 demonstrates the storage details of some data blocks, with subsequent blocks repeating according to this pattern.

Table 1.3 RAID 5 disk array storage configuration

Disk 1	Disk 2	Disk 3	Disk 4
p_0	0	1	2
3	p_1	4	5
6	7	p_2	8
9	10	11	p_3
…	…	…	…

When choosing a RAID level, several factors should be considered: economic cost, performance requirements for system read/write operations, and the impact of disk failure on the system. The concept of RAID is applicable not only in disk storage but also in flash devices such as SSDs and in data broadcasting in wireless devices. Below is a summary of the characteristics of various RAID levels and references for their selection:

- **RAID 0** combines two or more hard drives to improve performance and capacity but does not provide fault tolerance. The failure of a single hard drive will result in the loss of all data in the array. RAID 0 is useful for noncritical systems that require a high price/performance balance.
- **RAID 1** is typically implemented with two hard drives. Data is mirrored between the drives, providing fault tolerance in the event of a hard drive failure. Read performance is improved, while write performance is similar to that of a single hard drive. It can tolerate the failure of a single hard drive without data loss. RAID 1 is often used when fault tolerance is critical, but space and performance are less important.
- **RAID 5** provides fault tolerance protection and can improve read performance. At least three hard drives are required to form a RAID 5 array. RAID 5 can tolerate the failure of a single hard drive. When a hard drive fails, the data on the failed hard drive is reconstructed from the parity information spread across the remaining hard drives. Therefore, when a RAID 5 array is in a degraded state, both read and write performance can be significantly impacted. RAID 5 is ideal when storage space and cost are more important than performance.
- **RAID 6** is similar to RAID 5, but it provides an additional layer of block extension capabilities, allowing it to withstand the failure of two hard drives. At least four hard drives are required to form a RAID 6 array. RAID 6 performs less than RAID 5 because of its additional fault tolerance mechanisms. RAID 6 is ideal when storage space and cost are important, and the ability to withstand multiple hard drive failures is required.
- **RAID 10** combines the advantages of RAID 1 and RAID 0. It improves both read and write performance, but only half of the total space is available for data storage. Four or more hard drives are required, making the cost relatively high, but it offers high performance while providing fault tolerance protection. RAID 10 can withstand multiple hard drive failures as long as the failures do not occur within the same subgroup. RAID 10 is an ideal solution for applications requiring high I/O, such as database servers.

1.4.1 *Internal Data Structures in Databases*

The underlying storage of databases utilizes various tree structures, among which binary trees offer the highest search efficiency (fewer comparison operations). However, since database file indexes are stored on disks, we must consider search efficiency and the number of disk seeks and loads. As a result, traditional relational databases store data on disks in the form of B-trees. These databases maintain indexes of this data in RAM using B-trees to ensure faster access speeds. Data operations in memory are significantly more efficient than loading data from disks, and the comparisons are performed in memory. Although B-trees may require more comparisons than binary search trees, they involve fewer disk operations. In fact, the number of disk loads generally correlates with the tree's height: the taller the tree, the more loads; the shorter, the fewer loads. Therefore, we generally prefer shorter and wider tree structures for storing such file indexes.

In distributed systems centered around customer-facing services such as messaging, chatting, real-time communication, and the Internet of Things (IoT), which involve abundant unstructured data, millions of write operations occur every hour. As a result, these systems are primarily write-intensive, and to meet their needs, databases must be capable of rapidly inserting data. Traditional databases often struggle to adequately fulfill such scenarios due to their inability to handle requirements like high availability, best-effort eventual consistency, flexibility for unstructured data, and low latency.

Hence, the Log-Structured Merge tree (LSM tree) emerged. The LSM tree is a disk-based data structure designed to provide low-cost indexing for files that experience high record insertion (and deletion) rates over extended periods. The LSM tree employs an algorithm that reduces latency and performs bulk index changes, cascading changes from memory-based components to one or more disk components in an efficient manner similar to merge sort. Throughout this process, all index values can be continuously accessed through either the memory component or one of the disk components (with the exception of very brief lock periods). Compared to traditional access methods like B-trees, this algorithm significantly reduces disk arm movements and improves cost-effectiveness in domains where the cost of disk arm movements in traditional access methods exceeds the cost of the storage medium.

The Log-Structured Merge tree (LSM tree) concept is quite simple. The core idea is that, assuming there is enough memory, writing data to disk with every update is unnecessary. Instead, the latest data can reside in memory first, and once enough data has accumulated, it can be merged and appended to the end of the disk structure using a merge sort approach. Since all the trees waiting to be sorted are already ordered, the data can be quickly merged together through merge sort.

The LSM-tree index structure is shown in Fig. 1.9. The process of exporting the memory component to form an ordered data file is called a flush. To avoid the flush impacting write performance, the current MemStore being written to is first marked as a Snapshot, and no new write operations are allowed to this Snapshot's

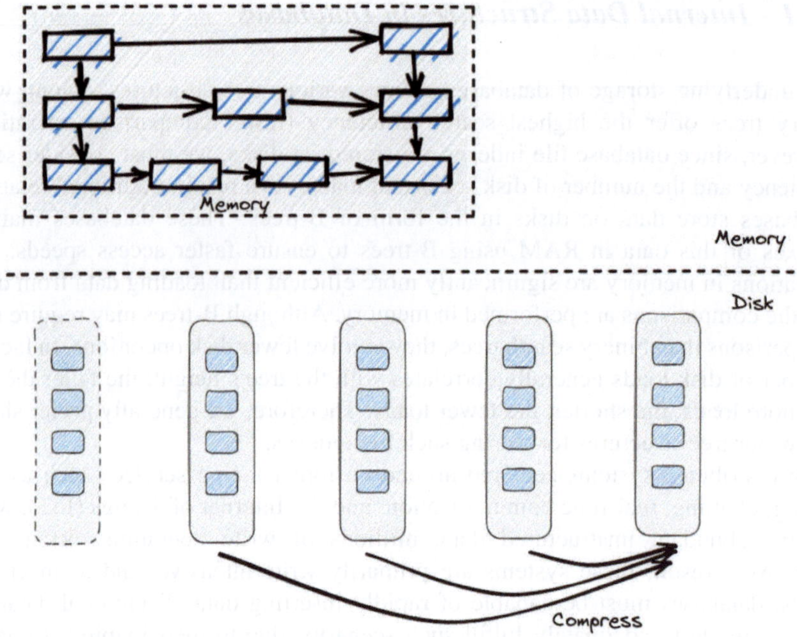

Fig. 1.9 LSM-tree index

MemStore. A new memory space is then allocated as a MemStore for subsequent data writes. The corresponding memory space can be released once the Snapshot's MemStore has completed its write process. In this way, stable write performance is achieved by using two MemStores.

In operating systems, databases are stored on disks in the form of files, and a database can be mapped to multiple files. Each file contains a sequence of records, where the size of each record does not exceed the size of a file block, and each record is stored within the same block. When a record is deleted from a block, shifting all subsequent blocks forward to reclaim space can result in significant resource waste. To address this, a pointer is maintained in the file header, pointing to the address of the deleted space, thus forming a free list. If the free list has an address, newly added data is inserted at that location. When the free list is empty, new records are appended to the end of the file.

Variable-length records refer to records where the length of each record can vary, such as string data types. Typically, 2 bytes are used to store the offset and length of variable-length attributes. A null bitmap is also used to indicate which attributes have null values. As shown in Fig. 1.10, a slotted page structure is used to store these variable-length records. The slotted page structure is commonly used to organize records within a block.

Distributed storage architecture is an essential component of distributed file systems (including object storage and block storage) and distributed databases.

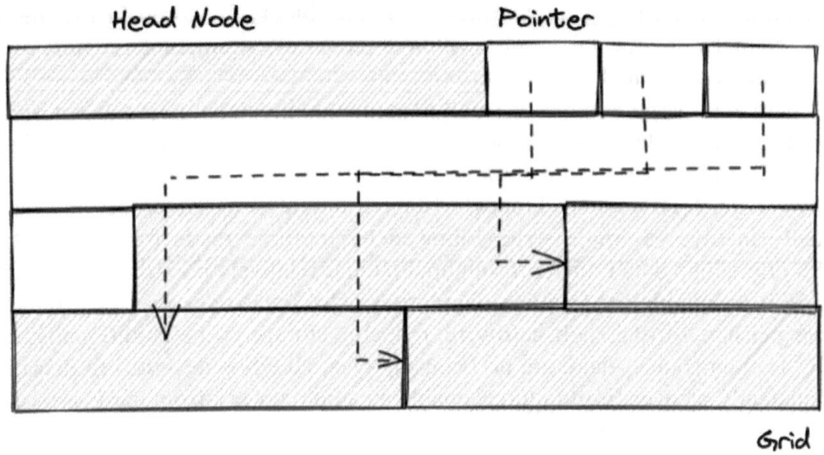

Fig. 1.10 Slotted pages structure

Typically, a distributed storage architecture comprises three parts: clients, metadata servers, and data servers:

- Client: Send read and write requests to an automatically routed or centralized server.
- Metadata Server: Metadata is separated from data management, and extensive metadata requires independent management. Therefore, the metadata server is responsible for managing metadata and processing client requests for metadata.
- Data Server: Responsible for storing user files and other data, ensuring data availability, integrity, and consistency.

Designing the architecture with the separation of metadata and user data offers two significant benefits: performance and capacity can be scaled simultaneously, providing strong system scalability.

Comparison of Distributed Storage

Distributed storage technologies have emerged in the market to meet the storage requirements of massive amounts of data. The rise of distributed storage is closely tied to the development of the Internet, and Internet companies, with their big data and asset-light characteristics, often utilize large-scale distributed storage systems. We have compared mainstream distributed storage architectures, each with its unique advantages and application scenarios.

HDFS (Hadoop Distributed File System) is primarily used in big data storage scenarios and is the storage component in the Hadoop big data architecture. From the outset of its design, HDFS has had a clearly defined application scenario: big data services. Its main application scenarios include the high-performance storage of large files, such as those measuring hundreds of megabytes to several gigabytes. This is because HDFS manages files through a metadata approach, where information related to metadata, directories, and blocks is stored in the memory of the NameNode. As the number of files increases, significant memory in the NameNode is consumed.

If there are a large number of small files, they will occupy a significant amount of memory space, leading to a decline in the overall performance of distributed storage. Therefore, it is more suitable to use HDFS to store large files. It is ideal for low-write, high-read workloads. In big data analytics, the processing pattern typically involves a single write followed by multiple reads for data analysis. HDFS offers high data transfer throughput but has relatively poor data read latency, making it unsuitable for frequent data writes. The DataNodes that store data rely on inexpensive, off-the-shelf hardware, reducing storage costs. Additionally, since HDFS is open-source, there are no licensing fees; HDFS is designed to detect and automatically recover from failures, enabling rapid recovery from hardware issues; HDFS is portable across all hardware platforms and compatible with multiple operating systems, including Windows, Linux, and Mac OS/X, making it versatile. HDFS is built specifically for high data throughput and is ideally suited for accessing streaming data.

Ceph is a scalable distributed storage system that is currently the most widely used among open-source options and is supported by numerous vendors. In many hyper-converged systems, the distributed storage is deeply customized based on Ceph. Furthermore, Ceph has become a "standard" for both Linux and OpenStack, used to support their respective storage systems. Ceph can provide object storage, block device storage, and file system storage services. The feature of supporting all three different types of storage services simultaneously is rare in distributed storage systems.

Ceph does not adopt the metadata addressing scheme of HDFS but instead utilizes the CRUSH algorithm, which ensures balanced data distribution and high parallelism. Ceph provides strong data consistency regarding block storage support, offering a user experience similar to traditional centralized storage. For object storage services, Ceph supports Swift and S3 API interfaces. In block storage, it supports thin provisioning, snapshots, and cloning. For file system storage services, Ceph supports the Posix interface and snapshots. However, compared to other distributed storage systems, Ceph's file performance is slightly weaker, and its deployment is more complex. Therefore, Ceph is generally used for block and object storage. Ceph is a decentralized distributed solution that requires careful planning and design in advance, and poses higher requirements on technical teams. Especially during Ceph expansion, its balanced data distribution characteristic can lead to a decrease in the overall storage system performance.

Swift is a highly available, distributed, and eventually consistent object storage solution. Organizations can use Swift to store massive amounts of data efficiently, securely, and cost-effectively. Swift is primarily oriented toward object storage, similar to the object storage service provided by Ceph. It is mainly used to address unstructured data storage problems.

The main differences between Swift and Ceph's object storage service are as follows: when clients access the object storage system, Swift requires them to retrieve data through the Swift gateway. In contrast, Ceph utilizes OSDs (Object Storage Devices) running on each storage node to access data information without a single entry point, making it somewhat more flexible than Swift. Regarding data consistency, Swift employs eventual consistency. This approach can be more efficient when processing large volumes of data. It is primarily suited for object storage cases requiring high data processing efficiency rather than scenarios demanding strict data consistency. Ceph, on the other hand, maintains strong consistency across the cluster at all times. Regarding primary application scenarios, OpenStack utilizes Swift for its object storage service rather than Ceph.

1.4.2 Column-Oriented Storage

There are two ways to organize relational databases: row-oriented and column-oriented. Row-oriented databases organize data by records, keeping all data associated with a record adjacent to each other in memory. Row-oriented databases are the traditional way of organizing data and still provide some key advantages for fast data storage. They are optimized for efficient reading and writing of rows. Common row-oriented databases include MySQL and PostgreSQL. Column-oriented databases organize data by fields, keeping all data associated with a field adjacent to each other in memory. Columnar databases are becoming increasingly popular and offer performance advantages for querying data. They are optimized for efficient reading and computing of columns. Common column-oriented databases include Redshift, BigQuery, and Snowflake.

Column-oriented storage is not a new technology, and it can be traced back to the 1983 paper by Cantor. However, due to limitations in early hardware conditions and application scenarios, most mainstream Online Transaction Processing (OLTP) databases adopted row-oriented storage. The concept of column-oriented storage regained popularity only with the recent rise of Online Analytical Processing (OLAP) databases.

Row-oriented databases have established the traditional approach to database management systems for storing data. They are designed to optimize reading and writing individual rows of data, which leads to design choices like using a row-based storage architecture. In row-oriented databases, data is stored row by row, with each row's first column placed immediately after the last column of the previous row. This is illustrated in Table 1.4.

Table 1.4 Example of row-oriented data

Name	City	Age
Matt	Hangzhou	27
Dave	Shanghai	30
Jim	Guangzhou	33

Table 1.5 Diagram of row-oriented data storage

Matt	Hangzhou	27	Dave	Shanghai	30	Jim	Guangzhou	33

Table 1.6 Adding a new row of data

Jenny	Shenzhen	30

Table 1.7 Changes after adding a row

Matt	Hangzhou	27	Dave	Shanghai	30	Jim	Guangzhou	33	Jenny	Shenzhen	30

These data will be stored on the disk of a row-oriented database by rows, as shown in Table 1.5.

This enables the database to quickly write a row by simply appending it to the end of the data. As shown in Table 1.6, if we want to add a new record.

As shown in Table 1.7, we can append it to the end of the current data.

Row-oriented databases are still widely used for OLTP applications, as they excel at handling write operations. However, databases are also used for data analysis. In OLAP scenarios, databases need to support ad hoc queries, whereas row-oriented databases can be slower than columnar (C-store) databases.

When performing ad hoc queries, various data sorting methods can enhance performance. For example, we might sort data by date, either in ascending or descending order. Additionally, if we are searching for a large volume of data related to a specific customer, sorting by customer can also improve performance.

In row-oriented databases, indexes can be created, but data is rarely stored in multiple sorting orders. In contrast, column-oriented databases allow data to be stored in various sorting orders. Beyond improving query performance, this also offers additional benefits. These differently sorted columns are known as projections, and they enhance the system's fault tolerance because the data is stored multiple times.

Fetching from Row-Oriented Databases Row-oriented databases are efficient when retrieving a single row or a small set of rows, but they perform less efficiently when executing aggregations. This is because they load additional data (columns) into memory, which is slower than selecting only the columns needed for the aggregation. Furthermore, row-oriented databases typically require accessing a larger number of disks.

Additional Data Loaded into Memory Suppose we want to obtain the sum of ages from the data in the table. To do this, we would need to load all nine data blocks into memory and then extract the relevant data for aggregation, which would waste computation time.

Number of Disks Accessed Let's assume that a disk can only hold enough data bytes to store three columns per disk. In a row-oriented database, as shown in Table 1.8, the above table would be stored as follows.

Table 1.8 Schematic of storing three columns per disk

Disk1			Disk2			Disk3		
Name	City	Age	Name	City	Age	Name	City	Age
Matt	Hangzhou	27	Dave	Shanghai	30	Jim	Guangzhou	33

Table 1.9 Schematic of column-oriented data storage

Name	City	Age
Matt	Hangzhou	27
Dave	Shanghai	30
Jim	Guangzhou	33

Table 1.10 Column store example

Matt	Dave	Jim	Hangzhou	Shanghai	Guangzhou	27	30	33

Table 1.11 Adding a new record to a columnar database

Jenny	Shenzhen	30

Table 1.12 Columnar storage insertion

Matt	Dave	Jim	Jenny	Hangzhou	Shanghai	Guangzhou	Shenzhen	27	30	33	30

To obtain the sum of all the ages, the computer needs to look at all three disks and all three columns within each disk to perform this query. Therefore, we can see that while adding data to a row-oriented database is fast and easy, retrieving data from it may require the use of additional memory and accessing multiple disks.

Column-oriented DBMS, or columnar DBMS, is a type of DBMS that stores data tables by columns rather than by rows. The benefits include more efficient access to data when querying only a subset of columns (by eliminating the need to read irrelevant columns) and increased options for data compression. However, they typically have lower efficiency in inserting new data. Relational Database Management Systems provide data represented in two-dimensional tables with columns and rows. Data warehouses are created to support the analysis of data, and these types of databases are optimized for reading. In a column-oriented or column-facing database, data is stored in a way that each row of a column is adjacent to other rows within the same column. Let's take a look again at the same dataset and see how it would be stored in a column-oriented database, as shown in Table 1.9.

As shown in Table 1.10, a table stores one column at a time in a row-wise manner.

Writing to a Columnar Database As shown in Table 1.11, if we want to add a new record.

As shown in Table 1.12, we must navigate through the data to insert each column into its appropriate position.

If the data is stored on a single disk, it will face the same additional memory issues as a row-oriented database because it needs to load everything into memory.

Table 1.13 Column store underlying diagram

Disk1			Disk2			Disk3		
Name			City			Age		
Matt	Dave	Jim	Hangzhou	Shanghai	Guangzhou	27	30	33

However, when stored on separate disks, a column-oriented database will have significant advantages.

As shown in Table 1.13, if we place the above table into the same constrained three-column data disks, they will be stored as follows.

Reading from a Columnar Database To get the total sum of ages, the computer only needs to go to one disk (Disk 3) and add up all the values there. There is no need to load extra memory, and the number of disks accessed is minimized. Although this is a bit oversimplified, it illustrates that by organizing data by columns, the number of disks that need to be accessed is reduced, and the amount of additional data that must be kept in memory is minimized. This greatly improves the overall speed of computation. There are other methods that can further enhance the performance of column-oriented databases.

The most expensive operation involving hard drives is the seek. To improve overall performance, the storage method of related data should minimize the number of seeks. This is known as locality of reference, a basic concept that appears in many different contexts. Hard drives are organized into a series of fixed-size blocks, usually large enough to store a few rows of a table. By organizing the table's data so that rows fit into these blocks and grouping related rows into contiguous blocks, the number of blocks that need to be read or sought, as well as the number of seeks, can be minimized in many cases.

Column-oriented databases serialize all the values of one column together, followed by the values of the next column, and so on. In this layout, any single column more closely matches the index structure in row-oriented systems. This can cause confusion, leading to the mistaken belief that columnar storage is "just" row storage with an index on each column. However, the data mapping is quite different. In row-oriented systems, the index maps column values to row IDs, whereas in column-oriented systems, the columns map row IDs to column values.

Whether a column-oriented system will run more efficiently largely depends on the workload being automated. Retrieving all data for a given object (an entire row) is slower. Row-oriented systems can retrieve the row in a single disk read, whereas a columnar database requires many disk operations to gather data from multiple columns. However, these full row operations are usually infrequent. In most cases, only a limited subset of data is retrieved. This is even more true when writing data to the database, especially when the data tends to be "sparse" and includes many optional columns. For this reason, partitioning, indexing, caching, views, OLAP cubes, and transaction systems (such as write-ahead logging or multi-version concurrency control) significantly influence the physical organization of either system. That is to say, RDBMS focused on OLTP are more row-oriented, while systems focused on OLAP strike a balance between row-oriented and column-oriented approaches.

Traditional OLTP databases typically employ row-based storage. In this approach, all columns are arranged sequentially to form a row, which is stored as a unit. This, coupled with indexing mechanisms such as B+ trees or SS-Tables (Sorted String Tables, although not a universally recognized term for indexing in databases, perhaps you meant something like LSM trees, which are used in systems like LevelDB and RocksDB), enables fast retrieval of corresponding row data through the primary key.

Row-oriented storage is natural for OLTP scenarios: most operations are performed on entities, meaning they typically involve inserting, deleting, updating, or querying entire rows of records. Clearly, storing a row's data in physically adjacent locations is a good choice.

However, for OLAP scenarios, a typical query needs to scan the entire table, performing grouping, sorting, aggregation, and other operations. In this case, the advantages of row-oriented storage disappear. Even worse, analytical SQL queries often do not use all columns but are only interested in performing operations on certain columns. Consequently, unrelated columns in a row still have to be scanned.

The comparison between row-oriented and column-oriented databases usually relates to the efficiency of hard disk access for a given workload, as seek time is very long compared to other bottlenecks in a computer. For example, the average seek time for a typical Serial ATA (SATA) hard drive is between 16 and 22 milliseconds, whereas DRAM access on an Intel Core i7 processor takes an average of 60 nanoseconds, which is nearly 400,000 times faster. Clearly, disk access is the main bottleneck in handling big data. Columnar databases improve performance by reducing the amount of data that needs to be read from the disk, effectively compressing similar columnar data and only reading the data necessary to answer queries.

In practice, columnar databases are very well-suited for OLAP workloads (e.g., data warehousing), which typically involve highly complex queries over all data (potentially petabytes). However, some effort is required to write data into columnar databases. Transactions (INSERTs) must be split into columns and compressed during storage, making them less suitable for OLTP workloads. Row-oriented databases are ideal for OLTP workloads, which involve heavier interactive transactions. For example, it is more efficient to retrieve all data from a single row when the data is located in one place (minimizing disk seeks), as is the case in row-oriented architectures. However, column-oriented systems have been developed to function as hybrids capable of handling both OLTP and OLAP operations. Some of the OLTP constraints faced by columnar systems are mitigated by using in-memory data stores (and other qualities).

Clearly, columnar storage is not friendly for OLTP, as writing a row of data requires modifying multiple columns simultaneously. However, it has significant advantages for OLAP scenarios: when a query involves only a subset of columns, only the relevant columns need to be scanned. Data in each column is of the same type and highly correlated, which allows for more efficient compression of column data.

Column data is of a uniform type; therefore, there are storage size optimization opportunities in columnar data that are not available in row-oriented data. For example, many popular modern compression schemes, such as LZW or run-length encoding, exploit the similarity of adjacent data for compression. Missing and duplicate values, common in clinical data, can be represented with two-bit markers.

Sorting rows can also help improve compression. For example, using bitmap indexing, sorting can increase compression by an order of magnitude. Column compression reduces disk space at the cost of retrieval efficiency. The more compression is applied to adjacent data, the harder random access can become, as the data may need to be decompressed for reading. To address this, columnar architectures are often designed to minimize the need to access compressed data.

Traditional relational databases are primarily designed for OLTP applications, which need to maintain good performance during frequent small-scale updates. Therefore, row-oriented storage has become the mainstream record organization format. However, for OLAP and data mining, which are query-intensive applications, the access patterns are completely different. These applications usually only access a small number of attributes in records and obtain summarized result sets by computing large amounts of data. If data is stored in a row-oriented format, accessing just a few columns still requires reading all the data in each record, leading to high I/O for a small amount of useful information. Therefore, for OLAP applications, two methods have emerged: precomputation and vertical partitioning. The precomputation method stores precomputed results in cubes or materialized views, allowing operations such as slicing, drilling down, and rolling up on the summarized results. However, this method has the disadvantage of only accelerating queries with known patterns and is not suitable for flexible ad hoc queries. Additionally, synchronization between the fact table and the precomputation result set must be considered. Vertical partitioning organizes data by columns. Since the values of queried attributes are stored in contiguous data blocks, operations like aggregation, grouping, sorting, and mapping can obtain more useful data with fewer accesses, greatly improving the efficiency of data analysis applications. Columnar storage enhances I/O efficiency, and by fully loading columnar storage into memory, it further avoids disk I/O access. This approach also leverages the advantages of parallel processing with multiprocessor, multicore systems, providing good performance for ad hoc queries.

Overall, the advantages of columnar storage are evident in two main areas: saving space and reducing I/O and optimizing computations through the columnar data structure.

1.5 Data Replication and Redundancy

High reliability and availability have always been fundamental requirements for database systems. In distributed systems, it is essential to provide reliable and efficient support at the data layer and data storage layer. Beyond the hardware and operating system layers, the database storage layer needs to implement multi-replica technology, a common strategy for enhancing data access reliability and system fault tolerance in cloud storage systems. Consequently, there are many technologies related to replicas. However, existing research often assumes that replica layout transformations happen automatically, concentrating on designing replica layout schemes, such as determining the number and location of data replicas. These studies rarely address the task scheduling challenges associated with these transformations. In multi-data center environments, task scheduling for complex operations—such as data replica splitting, merging, migration, deletion, backup, and recovery—can differ significantly in terms of space and time depending on the task scheduling strategies used, resulting in substantial variations in cost and efficiency. Therefore, studying a task scheduling model for data replica layout transformations in multi-data centers—focusing on optimizing scheduling problems under constraints of cost, space, and time—and exploring task scheduling strategies for replica layouts in multi-data centers is crucial for enhancing the performance of cloud storage systems.

Replication means keeping copies of the same data on multiple machines connected through a network. By replicating data, it brings the data geographically closer to users (thereby reducing latency); even if part of the system fails, the system can continue to operate (thereby improving availability); and it allows for scaling the number of machines that can accept read requests (thereby improving read throughput).

There are two common methods for distributing data across multiple nodes: replication and partitioning. A file system running on a large number of nodes is called a distributed file system, which is a widely used data storage method in parallel systems.

Data partitioning splits a large database into smaller subsets (called partitions), allowing different partitions to be assigned to different nodes (also known as shards). The advantage of partitioning is that when the database is partitioned across multiple nodes, we can use all the nodes to retrieve it. Similarly, during write operations, data can be written in parallel to multiple nodes. Therefore, I/O-parallel read and write operations on the entire database are significantly faster than non-I/O-parallel methods.

Data replication involves storing identical copies of data on several different nodes, possibly in different locations. Replication provides redundancy: if some nodes become unavailable, the remaining nodes can still provide data services. Similar to partitioning, replication also helps improve performance.

Since each partition is replicated, updates to the partition must be applied to all replicas, which can cause consistency issues. For data that has remained unchanged since its creation, read operations can be performed on any node, as all replicas are

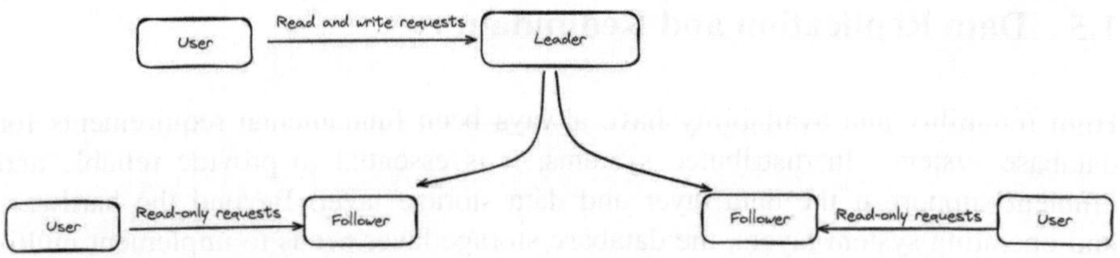

Fig. 1.11 Data redundancy and replication: single-leader

consistent at this point. However, the different replicas may hold divergent values if a replica is updated. To ensure all read operations get the latest value, all read operations should be sent to the primary replica.

If the replicated data remains unchanged over time, the replication process is straightforward: the data only needs to be copied to each node once, and the replication is complete. The challenge of replication lies in handling changes to the replicated data, which is the focus of this chapter. We will discuss three popular change replication algorithms: single-leader, multi-leader, and leaderless. Nearly all distributed databases use one of these three methods.

Each node that stores a copy of the database is known as a replica. When multiple replicas exist, an inevitable challenge arises: ensuring that all data is consistently present across all replicas. Every write operation to the database must be propagated to all replicas; otherwise, the replicas will diverge and contain inconsistent data. The most common solution to this problem is leader-based replication (also known as active/passive replication or master/slave replication), as illustrated in Fig. 1.11.

One of the replicas is designated as the leader, also referred to as the master. When a client needs to write to the database, it must send the write request to the leader, which then writes the new data to its local storage. The other replicas are called followers, also known as read replicas, slaves, standby replicas, or hot standbys. Whenever the leader writes new data to its local storage, it also sends the data changes to all followers, which is known as the replication log or change stream. Each follower retrieves the replication log from the leader and updates its local database replica by processing all writes in the same order as the leader. When a client wants to read data from the database, it can query either the leader or any of the followers. However, only the leader can accept write operations (from the client's perspective, the followers are read-only).

A critical factor in a replication system is whether replication is performed synchronously or asynchronously. In relational databases, this setting is typically configurable, whereas in other systems, the replication mode is often fixed to either synchronous or asynchronous.

The advantage of synchronous replication is that the replicas can ensure they have the latest data consistent with the leader. If the leader suddenly fails, we can be confident that this data is still available on the replicas. The drawback of synchronous replication is that the leader cannot process write operations if a synchronous replica becomes unresponsive due to a crash, network failure, or other issues. It must block all writes and wait until the synchronous replica becomes available again.

Therefore, it is impractical to set all replicas to synchronous: any interruption at one node would cause the entire system to stall. In practice, enabling synchronous replication in a database typically means that one replica is configured as synchronous while the remaining replicas operate asynchronously. If the synchronous replica becomes unavailable or slow, an asynchronous replica is promoted to run synchronously. This ensures you have the latest data copy on at least two nodes: the leader and the synchronous replica. This configuration is sometimes referred to as semi-synchronous.

Typically, single-leader replication is configured to be fully asynchronous. In this case, if the leader fails and cannot be recovered, any writes that have not yet been replicated to the replicas will be lost. This means that writes are not guaranteed to be durable, even if the client has received confirmation of success. However, a fully asynchronous configuration also has an advantage: the leader can continue to process writes even if all the replicas are lagging behind.

Single-leader replication has a significant drawback: there is only one leader, and all write operations must pass through it. If the leader becomes unreachable for any reason, such as a network interruption, no writes can be made to the database.

A natural extension of the single-leader replication model is to allow multiple nodes to accept writes. Replication still occurs in the same way: each node that processes a write must forward the data change to all other nodes. This is called a multi-leader configuration (also known as multi-master or multi-active replication). In this case, each leader is also a follower of the other leaders.

In a multi-leader configuration, there can be a leader in each data center. Figure 1.12 illustrates this architecture. Within each data center, standard leader-follower replication is employed. Between data centers, each leader replicates its changes to the leaders in other data centers.

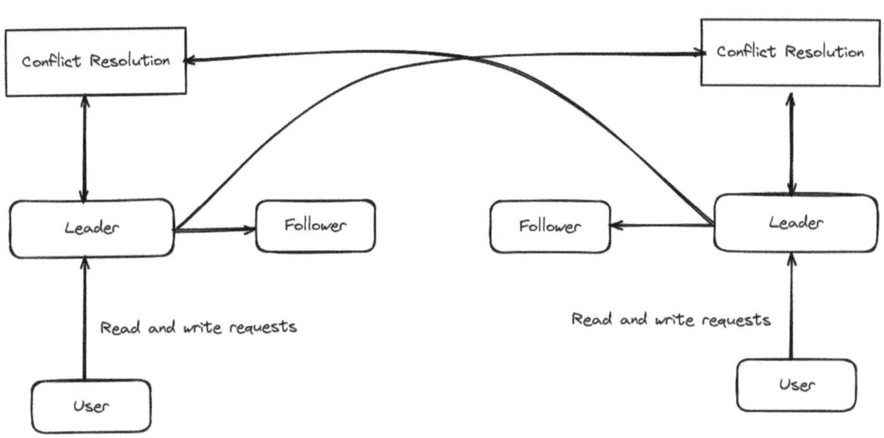

Fig. 1.12 Multi-leader replication

Despite the advantages of multi-leader replication, it has a significant drawback: two different data centers may simultaneously modify the same data, requiring a mechanism to resolve write conflicts. Since multi-leader replication is often an added feature in many databases, it frequently contains subtle configuration flaws and can exhibit unexpected interactions with other database functions. For instance, there may be complications with auto-increment primary keys, triggers, integrity constraints, and so on. As a result, multi-leader replication is generally considered a risky approach and should be avoided whenever possible.

Leaderless replication is a data replication strategy in distributed systems where all nodes can independently handle write requests without relying on a single leader. It uses a quorum mechanism to ensure data consistency, enhancing system availability and fault tolerance, making it well-suited for applications requiring high availability and high concurrency. However, this approach also has drawbacks. Since multiple nodes can write simultaneously, there is a potential for data conflicts, necessitating complex conflict resolution mechanisms. Maintaining consistency can be challenging, particularly during high-concurrency writes, which may introduce delays and lead to inconsistent read results. Furthermore, the coordination required among nodes introduces overhead, potentially impacting performance due to the need for communication with multiple nodes for write operations.

1.6 Summary

This chapter introduces the fundamentals of distributed systems, providing an overview of their architecture. In Sect. 1.1, we provide a brief overview of the requirements and application scenarios for distributed systems, highlighting their advantages: increased flexibility, enhanced reliability, improved speed, and support for geographical distribution. Furthermore, we summarize the various distributed system models, including client-server, mobile networks, and peer-to-peer networks. It's important to emphasize that compared to centralized systems, distributed systems exhibit several new characteristics: scalability, concurrency, availability/fault tolerance, transparency, heterogeneity, and replication. In Sect. 1.2, we delve into the design goals of distributed systems, categorizing them into four primary objectives: consistency, availability, partition tolerance, and scalability. We discuss each of these goals in detail. Section 1.3 focuses on data models in distributed systems, comparing and contrasting the three commonly used models: relational, document, and graph data models. We analyze their strengths and weaknesses in different usage scenarios, acknowledging that each model excels in its respective domain and possesses unique advantages. Recognizing the diversity of data representation, we refrain from concluding that one model is inherently better than another. Instead, our concern lies in correctly selecting or combining these models based on various business scenarios. In Sect. 1.4, we explore data storage mechanisms in distributed systems, starting with the most basic storage units: disks, hard drives, flash memory, main memory, and cache. Each has its unique application scenarios, necessitating selection based on actual needs and conditions. Notably, flash memory and the below are nonvolatile storage, ensuring data safety by requiring data to be written into nonvolatile storage. Additionally, we introduce mainstream RAID solutions, explaining their principles, applications, and scenario selection. We also briefly discuss the internal data structures of databases, highlighting two methods of organizing relational databases: row-oriented and column-oriented. Row-oriented databases create traditional database management systems for data storage, enabling fast retrieval of individual or group rows. However, when performing aggregations, they bring additional data (columns) into memory, making the process slower than selecting only the relevant columns for aggregation. In contrast, column-oriented database management systems (or column stores) store data tables by columns rather than rows, offering benefits such as more efficient data access when querying subsets of columns (by eliminating the need to read irrelevant columns) and additional data compression options. Finally, Sect. 1.5 examines data redundancy and replication, which are fundamental to achieving high availability, performance, and consistency in distributed systems. Distributing data across multiple nodes generally involves two common approaches: replication and partitioning. Data replication involves maintaining identical copies of data on several nodes, potentially in different locations. One of these copies is designated as the leader or primary database. When a client writes to the database, it must send the request to the leader, which then writes the new data into its local storage.

Chapter 2
Distributed Data Processing Languages

2.1 SQL

The earliest version of SQL, originally named Sequel, was developed by IBM. Since the 1970s, it underwent significant evolution and eventually became known as SQL. SQL stands for Structured Query Language, which is a specialized language designed to communicate with databases. It utilizes simple sentences to form basic syntax for accessing the content of databases.

SQL became an ANSI standard in 1986. The standard is updated every few years, resulting in terms like ANSI SQL-89 and ANSI SQL-92, which are different sets of SQL standards added in 1989 and 1992, respectively. Despite these standards, no two relational database management systems are exactly the same. They only partially conform to ANSI standards. Therefore, readers should be aware of the differences between various SQL implementations. Generally speaking, industry products build upon ANSI SQL and then extend their functionality to showcase the unique features of their products.

In the context of working within a single RDBMS, adherence to the ANSI standard is not obligatory. However, problems can arise if your code works in one RDBMS and you want to use it in another. Code that does not adhere to the ANSI standard may not run in the new RDBMS.

Although SQL is referred to as a query language, its functions include data querying, data definition, data manipulation, and data control. SQL is concise, convenient, and fully functional, and it has become the most widely used language in relational database systems. SQL is not a proprietary language of any specific database vendor, and almost all major DBMSs support SQL. And SQL is widely used in various large- and medium-sized databases, such as Sybase, SQL Server, Oracle, DB2, MySQL, and PostgreSQL. It is also used in various small databases, including Fox Pro, Access, and SQLite.

Z. Xu, *Distributed Data Services*, https://doi.org/10.1007/978-981-96-9392-4_2

The main classifications of SQL statements include:

- Data Definition Language (DDL). DDL is primarily used for creating, modifying, and deleting database objects (such as tables, views, indexes, etc.), including statements like CREATE, ALTER, and DROP.
- Data Query Language (DQL). DQL is primarily used for querying data in a database. Its main statement is SELECT, which is the most important part of SQL language. The SELECT statement typically includes clauses such as FROM, WHERE, GROUP BY, HAVING, and WITH.
- Data Manipulation Language (DML). DML is primarily used for updating data in tables within a database, including statements like INSERT for adding new records, UPDATE for modifying existing records, and DELETE for removing records.
- Data Control Language (DCL). DCL is primarily used for granting and revoking certain types of access permissions to a database. It includes statements like GRANT and REVOKE. Specifically, the GRANT statement is used to grant permissions to users, while the REVOKE statement is used to take back permissions from users.
- Transaction Control Language (TCL). TCL is primarily used for controlling transactions in a database to ensure data consistency. It includes statements like COMMIT and ROLLBACK. Specifically, COMMIT is used to commit a transaction, while ROLLBACK is used to roll back a transaction.

This chapter mainly introduces the usage and main functions of SQL, as well as various expressions of connections, views, transactions, integrity constraints, authorization, and more. Through the study of this chapter, readers should understand the characteristics of SQL and master its four major functions. Special emphasis is placed on learning how to use the data querying function effectively. It also deepens the understanding of the implementation principles of database management systems in terms of data querying, data definition, data manipulation, and data control functions.

2.1.1 SQL Basics

The basic concepts of SQL are as follows.

Basic Table A basic table exists independently, with one relation corresponding to one basic table, and one or more basic tables corresponding to one storage file. A row in the table represents a kind of connection. In the terminology of the relational model, **relation** is used to represent a table, **tuple** is used to represent a row in the table, and **attribute** is used to represent a column in the table.

View A view is a table derived from one or more basic tables, and it is a virtual table. Only the definition of the view is stored in the database, not the data corresponding to the view, which is still stored in the basic tables from which the

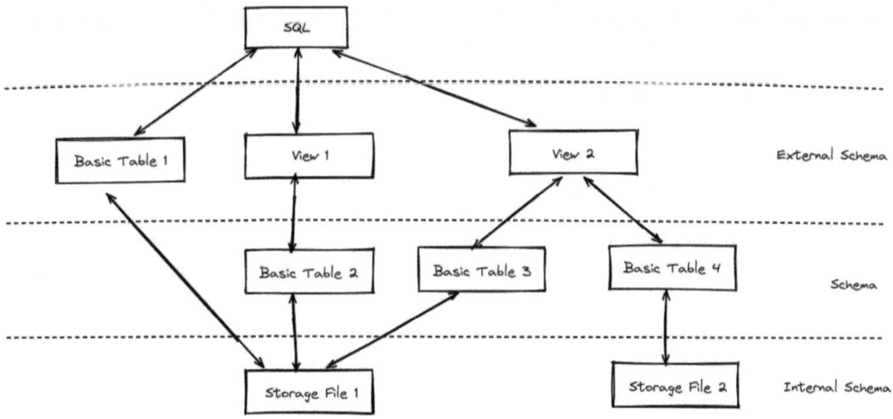

Fig. 2.1 Three-level schema architecture of relational databases supported by SQL

view is derived. When the data in the basic table changes, the data queried from the view also changes accordingly. From the user's perspective, a view provides different paths to view an actual table, allowing users to see only the content of the database that interests them. SQL supports the three-level schema structure of databases, as shown in Fig. 2.1. Among them, the external schema corresponds to views and some basic tables, the conceptual schema corresponds to basic tables, and the internal schema corresponds to storage files.

Transaction According to the SQL standard, when an SQL statement is executed, it implicitly initiates a transaction. A transaction is committed after completing all its steps, or all its actions are rolled back if it fails to complete all steps successfully. In this way, the database provides the property of **atomicity** for transactions, which signifies indivisibility. Either the transaction is fully committed to the database or nothing is committed at all.

Integrity Constraints Integrity constraints ensure that modifications made to the database by users do not violate data consistency. Therefore, integrity constraints prevent accidental damage to data. Generally speaking, an integrity constraint can be any predicate belonging to the database. Integrity constraints are typically considered part of the database schema design process and are declared as part of the CREATE table command used to create relations. However, integrity constraints can also be applied to existing relations using the "ALTER table table-name ADD constraint" command, where constraint can be any constraint on the relation. When executing the above command, the system first ensures that the relation satisfies the specified constraint. If it does, the constraint is applied to the relation; if not, the command is rejected.

Authorization We may grant a user several forms of permissions on certain parts of the database. Authorization of data includes:

- Authorization to read data.
- Authorization to insert new data.
- Authorization to update data.
- Authorization to delete data.

Each type of authorization is referred to as a **privilege**. We can grant these types of privileges to users on specific parts of the database, such as a relation or a view. When a user executes an SQL statement, the SQL execution first checks whether the query or update is authorized based on the privileges previously granted to that user. If the query or update is not authorized, it will be rejected. In addition to authorization on data, users can also be granted privileges on the database schema. For example, users can be allowed to create, modify, or delete relations. Users with certain privileges can also grant such privileges to other users or revoke granted privileges.

The SQL standard supports a variety of inherent data types, including:

- **char**(n). A fixed-length string where the length n is specified by the user.
- **varchar**(n). A variable-length string where the user specifies the maximum length n.
- **int**. Integer types.
- **smallint**. Small integer types.
- **numeric**(p, d). Fixed-point numbers, where the precision is specified by the user. This number has p digits (plus a sign bit), of which d digits are to the right of the decimal point.
- **real, double precision**. Floating-point numbers and double-precision floating-point numbers, whose precision is machine-dependent.
- **float**(n). Floating-point numbers with a precision of at least n digits.

The char data type stores strings of fixed length. For example, if the type of attribute A is char(5), When the string "ABC" is stored, two spaces will be appended to the string to make it a five-character string. Conversely, if the type of attribute B is varchar(5), and we store the string "ABC" in attribute B, no spaces will be added. When comparing the values of two char types, if their lengths are different, extra spaces will automatically be appended to the shorter string before the comparison to make their lengths equal, but this mainly depends on the database system, as some database systems do not perform alignment operations. So, even if attributes A and B store the same value "ABC," the comparison A=B may still return false. We can avoid such issues by using the varchar type instead of the char type.

SQL is composed of simple clauses, and all SQL statements have their own format. The structure of the query statement is shown in Fig. 2.2. The image displays a segment of SQL code that queries the sales quantity completed by each employee in 2021.

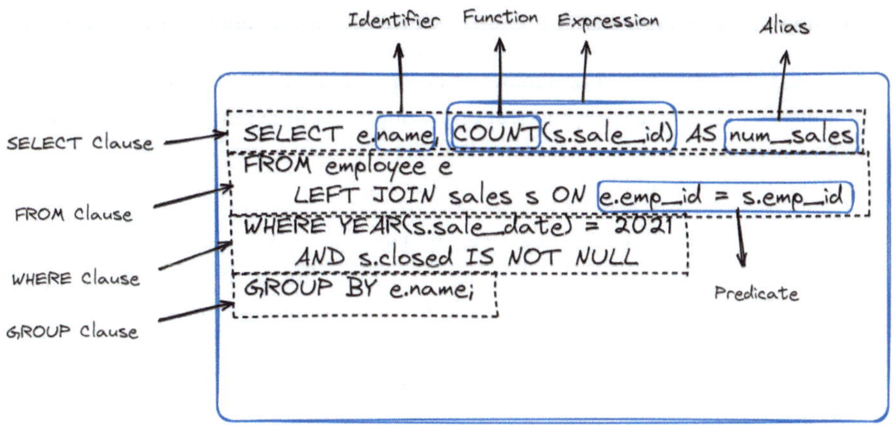

Fig. 2.2 SQL statement structure

Table 2.1 Common functions

Aggregate functions	Mathematical functions	String functions	Date and time functions
COUNT()	ABS()	LENGTH()	CURDATE()
AVG()	SQRT()	LTRIM()	CURTIME()
SUM()	LOG()	CONCAT()	DATE_ADD()
MAX()	ROUND()	SUBSTR()	DATE_FORMAT()
MIN()	CAST()	REGEXP()	DATE_SUB()

- **Statement**. The statement starts with a keyword and ends with a semicolon. If the statement begins with the keyword SELECT, the entire statement is referred to as a SELECT statement.
- **Clause**. A clause is a constituent part of a statement or query. Commonly used clauses typically begin with SELECT, FROM, WHERE, GROUP BY, HAVING, and ORDER BY.
- **Identifier**. An identifier is the name of a database object, such as a table or a column field. While SQL is case-insensitive, identifiers are typically written in lowercase and keywords in uppercase for better readability.
- **Function**. A function is a special type of construct in SQL that operates as a keyword. It takes zero or more inputs, processes them, and returns an outputf. SQL supports numerous categories of functions, including aggregate functions, mathematical functions, string functions, date and time functions, and others. Common functions are outlined in Table 2.1.
- **Expression**. An expression can be seen as a formula that produces a value. In the code block in Fig. 2.2, the expression is COUNT(s.sale_id). This expression includes a function COUNT and an identifier s.sale_id. Together, they form an expression that represents the calculation of the sales amount.
- **Alias**. An alias is a temporary renaming of a column or table during a query. In other words, the new alias will be displayed in the query results, but the original

column name or table name remains unchanged in the table being queried. The standard for renaming columns is to use AS, such as COUNT(s.sale_id) AS num_sales in Fig. 2.2. For tables, renaming is done by adding a space, e.g., employee e. However, in practical applications, both syntaxes are applicable to both columns and tables, meaning that the above two methods can be used for both columns and tables.

- **Predicate**. A predicate is a logical comparison that produces one of three values: TRUE, FALSE, or UNKNOWN. They are sometimes called conditional statements. Predicates are used in the search conditions of the WHERE clause and the HAVING clause, as well as in the join conditions of the FROM clause and other constructs that require a Boolean value.
- **Comments**. Comments are text that is ignored when the code runs. Writing comments in the code is useful so that other code reviewers can quickly understand the intent without having to read all the code.

  Single-line comment: - - These are my comments

  Multi-line comment: /* These are
  my comments */
- **Quotes.** In SQL, two types of quotes can be used: single quotes and double quotes. Single quotes are used when quoting string values. Double quotes are used when quoting identifiers. In practical use, single quotes are used much more frequently than double quotes.

2.1.2 SQL Query Statements

Writing SQL statements requires a good understanding of basic database design. To write efficient SQL, you must know which table the information is stored in and how the tables are related to each other. To help you better understand the concepts covered in this section, Fig. 2.3 shows a simple database containing three tables: the Student table, the Course table, and the Enrollment table. As seen in Fig. 2.3, it shows how the three tables are defined and how they are interconnected. All examples in this section are based on these three tables.

The SELECT statement is known as a query statement. It consists of six main clauses: the SELECT clause, FROM clause, WHERE clause, GROUP BY clause, HAVING clause, and ORDER BY clause. We will next detail each of these clauses and finally introduce the LIMIT clause, which is supported by MySQL, PostgreSQL, and SQLite. We will explain the syntax of SQL in detail through examples.

SELECT queries are divided into single-relation queries and multi-relation queries. A single-relation query refers to querying the needed information from a single table. When the information we need is stored in multiple tables and we need to combine these tables to obtain the desired information, we perform a multi-relation query.

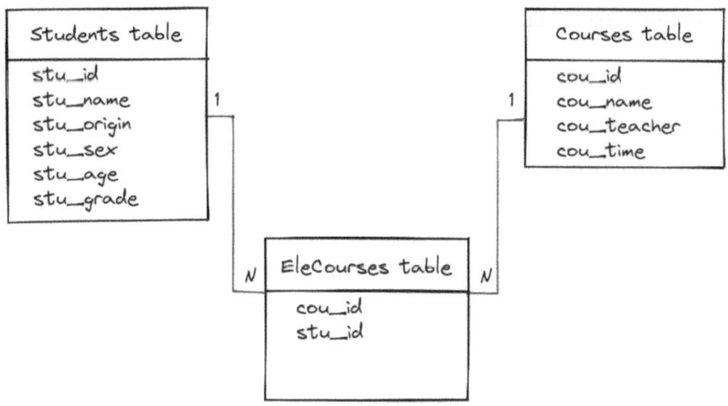

Fig. 2.3 Data model

Next, we will introduce single-relation queries. Querying a database means retrieving data from it, typically from one or more tables. When you query information from only one table, you are performing a single-relation query. In the SELECT clause, the SELECT keyword is followed by column names and expressions, separated by commas.

Let's consider a simple query using the Students table: "Query all students' IDs and names." The following query retrieves the columns named stu_id and stu_name from the Students table using the SELECT statement. The student IDs and names can be found in the Students relation, so we include this relation in the FROM clause. Since the student IDs and names are in the attributes stu_id and stu_name, we list these attributes in the SELECT clause, as shown in Listing 2.1. In general, the basic syntax of a SELECT statement is as follows: the desired column names are written after the SELECT keyword, and the FROM keyword indicates from which table to retrieve the data.

In addition to specifying the desired columns (one or more columns as described above), the SELECT statement can also retrieve all columns without listing them individually. This can be done by using the asterisk (*) wildcard in place of the actual column names, as shown in Listing 2.2.

Listing 2.1 SELECT retrieving multiple columns

```
Input:
        SELECT stu_id, stu_name
        FROM Students;
Output:
        stu_id        stu_name
        ----------------------
        10000         Xiao Chen
        10001         Ming Zhi
```

Listing 2.2 SELECT retrieving all columns

```
Input:
        SELECT *
        FROM Students;
Output:
        stu_id      stu_name        stu_origin      stu_sex     stu_age     stu_grade
        ---------------------------------------------------------------------------
        10000       Xiao Chen       Beijing         1           17          1
        10001       Ming Zhi        Shanghai        0           23          4
```

Listing 2.3 SELECT returning distinct rows

```
Input:
        SELECT DISTINCT stu_origin
        FROM Students;
Output:
        stu_origin
        ---------
        Beijing
        Shanghai
        Jiangxi
```

As mentioned earlier, the SELECT statement returns all matching rows. But what if you don't want the same values to appear multiple times in the results? The solution is to use the DISTINCT keyword, which instructs the database to return only unique values. For example, if you want to retrieve the origins of all students in the Students table, you would use the query shown in Listing 2.3.

SQL allows the use of logical connectors AND, OR, and NOT in the WHERE clause. The operands of logical connectors can be expressions that include comparison operators such as <, <=, >, >=, =, and <>. SQL allows us to use comparison operators to compare strings, arithmetic expressions, and special types like date types. Later in this section, we will also explore other features of the WHERE clause predicates.

Previously, our queries were based on a single relation, but how do we write queries when we need to combine information from multiple tables? The answer is to use multi-relation queries.

To consider the query "For all students, find their student ID, name, and the course IDs they have selected," we need to list the accessed relations in the FROM clause and specify the matching conditions in the WHERE clause, as shown in Listing 2.4.

As we have seen before, an SQL query consists of three types of clauses: the SELECT clause, the FROM clause, and the WHERE clause. A typical SQL query has the following form:

SELECT $A_1,A_2 \ldots A_n$
FROM $r_1, r_2 \ldots r_m$
WHERE P ;

Listing 2.4 Multi-relation query

```
Input:
       SELECT s.stu_id, s.stu_name, s.stu_grade, e.cou_id
       FROM Students AS s, Elecourses AS e
       WHERE s.stu_id = e.stu_id;
Output:
       stu_id         stu_name        stu_grade         cou_id
       -----------------------------------------------------
       10000          Xiao Chen           1               1
       10000          Xiao Chen           1               2
       10000          Xiao Chen           1               3
       10000          Xiao Chen           1               4
       10005          Yang Yan            2               2
       10005          Yang Yan            2               3
```

Listing 2.5 ORDER BY ascending order

```
Input:
       SELECT stu_id, stu_name
       FROM Students
       ORDER BY stu_name;
Output:
       stu_id         stu_name
       --------------------
       10006          Gan Yi
       10002          Hong Li
       10003          Li Zhang
       10001          Ming Zhi
       10000          Xiao Chen
       10005          Yang Yan
       10004          Zi Xiao
```

A_i represents an attribute, r_i represents a relation, and P represents a predicate.

Then, we will continue by teaching how to use the ORDER BY clause of the SELECT statement to sort the retrieved data as needed.

As previously mentioned, SQL statements return individual columns from a database table. However, the output does not have a specific order. If not sorted, the data will typically appear in the order it was initially added to the table or in the order it appears in the underlying table. To explicitly sort the data retrieved with a SELECT statement, you can use the ORDER BY clause. The ORDER BY clause takes one or more column names and sorts the output accordingly, as shown in Listing 2.5.

Sorting data is not limited to ascending order (from A to Z), which is the default sort order. You can also use the ORDER BY clause to sort in descending order (from Z to A). To sort in descending order, you must specify the DESC keyword, for example, ORDER BY stu_name DESC.

Table 2.2 WHERE clause operators

Operator	Description	Operator	Description
=	Equal to	>	Greater than
<>	Not equal to	>=	Greater than or equal to
!=	Not equal to	!>	Not greater than
<	Less than	BETWEEN	Between two specified values
<=	Less than or equal to	IS NULL	Is NULL value
!<	Not less than		

The DESC keyword applies only to the column name that immediately precedes it. In the ORDER BY clause, specifying DESC for the stu_name column sorts it in descending order. If you want to sort multiple columns in descending order, you must specify the DESC keyword for each column. Note that the opposite of DESC is ASC, which can be specified for ascending order. If neither ASC nor DESC is specified, the default is ASC, meaning the ORDER BY clause sorts in ascending order by default.

We have learned how to sort retrieved data using the ORDER BY clause in a SELECT statement. It is important to note that the ORDER BY clause must be the last clause in a SELECT statement. We can use ORDER BY to sort data on one or more columns as needed.

Next, we will introduce how to specify search conditions using the WHERE clause in a SELECT statement. In a SELECT statement, data is filtered according to the conditions specified in the WHERE clause, returning the rows that the user needs.

SQL supports all the conditional operators listed in Table 2.2. Some of the operators listed in Table 2.2 are redundant (e.g., <> is the same as !=, !< is equivalent to >=). Not all DBMSs support these operators. To determine which operators your DBMS supports, please refer to the corresponding documentation.

All WHERE clauses use single conditions to filter data. For more powerful filtering control, SQL allows for multiple WHERE clauses. These clauses can be used in two ways: with the AND clause or the OR clause. To filter by more than one column, you can add conditions to the WHERE clause using the AND operator. AND means that both expressions connected by AND must be satisfied for a result to be returned. For example, WHERE stu_age >= 17 AND stu_age <= 20 requires the attribute to satisfy both conditions stu_age >= 17 and stu_age <= 20 for a result to be returned.

The OR operator means that either one of the two expressions connected by OR must be satisfied for a result to be returned. It instructs the DBMS to retrieve rows that match either condition. In fact, many DBMSs do not evaluate the second condition if the first condition in the WHERE clause is already satisfied. The WHERE clause to query information for students who are either 17 or 20 years old would be WHERE stu_age = 17 OR stu_age = 20.

Listing 2.6 Using the BETWEEN keyword in the WHERE clause

```
Input:
        SELECT stu_id, stu_name, stu_age
        FROM Students
        WHERE stu_age BETWEEN 18 AND 20;
Output:
        stu_id          stu_name            stu_age
        -------------------------------------
        10002           Hong Li             20
        10003           Li Zhang            18
```

Listing 2.7 Using the IN keyword in the WHERE clause

```
Input:
        SELECT stu_id, stu_name, stu_origin
        FROM Students
        WHERE stu_origin IN ( 'Jiangxi', 'Shanghai' );
Output:
        stu_id          stu_name            stu_origin
        ----------------------------------------
        10001           Ming Zhi            Shanghai
        10002           Hong Li             Jiangxi
        10004           Zi Xiao             Shanghai
```

The WHERE clause can contain any number of AND and OR operators. Both can be combined to perform complex, advanced filtering. Whenever you use a WHERE clause with AND and OR operators, you should use parentheses to group the operators clearly. Do not overly rely on the default order of evaluation, as it can lead to errors.

To check for a range of values, the BETWEEN operator can be used. Its syntax differs slightly from other WHERE clause operators because it requires two values: the start value and the end value of the range. For example, the BETWEEN operator can be used to retrieve all students whose ages are between 18 and 20. Listing 2.6 illustrates how to use the BETWEEN operator.

The IN operator is used to specify a range of conditions, any of which can be matched. IN takes a set of valid values separated by commas and enclosed in parentheses. Listing 2.7 illustrates how to use this operator.

The SELECT statement in Listing 2.7 retrieves all students whose origin is Jiangxi or Shanghai. The IN operator is followed by valid values separated by commas, and these values must be enclosed in parentheses.

The NOT operator in the WHERE clause has only one function: to negate any condition that follows it. Because NOT is never used alone (it is always used with other operators), its syntax differs from other operators. The NOT keyword can be used in front of the column to be filtered. The SELECT statement in Listing 2.8 retrieves information for all students except those in the first grade.

Listing 2.8 Using the NOT keyword in the WHERE clause

```
Input:
        SELECT *
        FROM Students
        WHERE NOT stu_grade = 1;
Output:
        stu_id        stu_name        stu_origin        stu_sex     stu_age
      stu_grade
        ----------------------------------------------------------------------
        10001         Ming Zhi        Shanghai          0           23
      4
        10002         Hong Li         Jiangxi           1           20
      3
        10005         Yang Yan        Beijing           0           22
      2
        10006         Gan Yi          Beijing           1           17
      2
```

All the operators introduced previously filter based on known values. The commonality among them is that the values used for filtering are known, whether they involve matching a single value or multiple values, checking if a value is greater than or less than a known value, or checking a range of values. However, these methods of filtering are not always effective. For example, how do you find information for all students whose names contain "Li"? Simple comparison operators won't work—wildcards are needed. By using wildcards, you can create search patterns that match specific data.

The LIKE operator can be used on strings to perform pattern matching. We use two special characters to describe patterns:

- Percent sign (%). Matches any substring.
- Underscore (_). Matches any single character.

Pattern matching is case-sensitive by default in Postgres, SQLite, DB2, and Oracle databases, meaning uppercase characters do not match lowercase characters. However, in MySQL, pattern matching is case-insensitive by default. To illustrate pattern matching, consider the following examples:

- "SQL%" matches any string that starts with "SQL."
- "%SQL%" matches any string that contains the substring "SQL," such as "MYSQL5.0" and "SQL Server."
- "_ _ _" matches any string that contains exactly three characters.

In SQL, the comparison operator LIKE is used for pattern matching. For example, you can use the SELECT statement in Listing 2.9 to find information for all students whose names start with X.

We now know how to use the WHERE clause in a SELECT statement to filter returned data. We have also learned how to check for equality, inequality,

Listing 2.9 Using the LIKE keyword in the WHERE clause

```
Input:
        SELECT *
        FROM Students
        WHERE stu_name LIKE 'X%'
Output:
        stu_id        stu_name        stu_origin      stu_sex     stu_age
      stu_grade
        ----------------------------------------------------------------------
        10000         Xiao Chen       Beijing         1           17
     1
        10007         Xuan Li         Jiangxi         1           22
     4
```

Listing 2.10 Using the GROUP BY clause

```
Input:
        SELECT CASE WHEN stu_sex=1 THEN 'man' ELSE 'woman' END AS stu_sex,
                COUNT(stu_sex) AS nums
        FROM Students
        GROUP BY stu_sex;
Output:
        stu_sex         nums
        --------------------
        woman           3
        man             5
```

greater than, less than, and value ranges. Additionally, we learned how to combine conditions in a WHERE clause using AND and OR operators, explicitly manage the order of evaluation, use IN and NOT operators, and apply SQL wildcards.

Next, we introduce the GROUP BY clause. The GROUP BY clause groups rows and summarizes them. One or more attributes specified in the GROUP BY clause are used to form the groups. Tuples with the same values for all attributes are grouped together, and each group returns a single row in the final result.

Grouping is done using the GROUP BY clause in a SELECT statement. See Listing 2.10 for an example of grouping.

The SELECT statement in Listing 2.10 specifies two columns: stu_sex, which contains the gender of the students, and nums, a computed field created using the COUNT(stu_sex) function that represents the count of male and female students. The GROUP BY clause tells the DBMS to sort and group the data by stu_sex. With the GROUP BY clause, there is no need to manually specify each group; the system handles it automatically. The GROUP BY clause tells the DBMS to group the data and aggregate each group rather than the entire result set.

In addition to grouping data with GROUP BY, SQL also allows for filtering groups to specify which groups to include or exclude. For example, you might want to list all courses that have at least 100 students. To achieve this, you need to filter

Listing 2.11 HAVING clause to filter groups

```
Input:
        SELECT cou_id, COUNT(cou_id) AS stu_nums
        FROM EleCourses
        GROUP BY cou_id
        HAVING COUNT(cou_id) >= 2;
Output:
        cou_id          stu_nums
        ------------------------
        2               3
        3               3
        4               2
```

based on the complete groups rather than individual rows. We have already seen the function of the WHERE clause. However, WHERE cannot accomplish the task in this example because WHERE filters specify rows, not groups. In fact, WHERE does not have the concept of grouping.

What should be used instead of WHERE? SQL provides the HAVING clause for this purpose. HAVING is similar to WHERE, and all types of WHERE clauses can be replaced with HAVING. The key difference is that WHERE filters rows, and HAVING filters groups. Any non-aggregated attribute in the HAVING clause must also appear in the GROUP BY clause, or the query will be invalid.

So, how do we filter groups? See Listing 2.11 for a query that retrieves course IDs for courses with at least two students.

We have already discussed how to use the GROUP BY clause to summarize multiple groups of data and return results for each group. We also learned to use the HAVING clause to filter groups.

When performing database queries, sometimes we want to return only a limited number of rows. How do we limit the output? Different DBMSs use different syntax for this. MySQL, PostgreSQL, and SQLite support the LIMIT clause, while Oracle and SQL Server use different syntax. Listing 2.12 shows how to return the first three rows of the Students table on different DBMSs.

Let's review the order of clauses in a SELECT statement. Table 2.3 shows the order that must be followed when using clauses in a SELECT statement.

All the SELECT statements we've seen so far are simple queries—single statements that retrieve data from one table. SQL also allows subqueries, which are queries nested within other queries. Why use this? The best way to understand is by looking at examples.

The Elecourses table contains course IDs and student IDs. Course details are in the Courses table. Student information is in the Students table.

A query to find the names of all students enrolled in the course with ID 2 is shown in Listing 2.13.

Table 2.3 SELECT clauses and their order

Clause	Description
SELECT	Columns or expressions to return
FROM	Tables from which to retrieve data
WHERE	Row-level filtering
GROUP BY	Grouping specification
HAVING	Group-level filtering
ORDER BY	Sorting order of the output

Listing 2.12 Returning a limited number of rows in different databases

```
Input:
        -- MySQL, PostgreSQL, and SQLite
            SELECT *
            FROM Students
            LIMIT 3;
        -- Oracle
            SELECT *
            FROM Students
            WHERE ROWNUM <= 3;
        -- SQL Server
            SELECT TOP 3 *
            FROM Students;
Output:
        stu_id      stu_name      stu_origin     stu_sex    stu_age    stu_grade
        ------------------------------------------------------------------
        10000       Xiao Chen     Beijing        1          17         1
        10001       Ming Zhi      Shanghai       0          23         4
        10002       Hong Li       Jiangxi        1          20         3
```

Listing 2.13 SQL subquery example 1

```
Input:
        SELECT stu_id, stu_name
        FROM Students
        WHERE stu_id IN (SELECT stu_id
                         FROM Elecourses
                         WHERE cou_id = 2);
```

In a SELECT statement, subqueries are processed inside out. When processing the SELECT statement, the DBMS performs two operations. Firstly, it executes the query shown in Listing 2.14.

Listing 2.14 SQL subquery example 2

```
Input:
        SELECT stu_id
        FROM Elecourses
        WHERE cou_id = 2;
Output:
        stu_id
        ------
        10000
        10005
        10006
```

Listing 2.15 SQL subquery example 3

```
Input:
        SELECT stu_id, stu_name
        FROM Students
        WHERE stu_id IN (10000, 10005, 10006);
Output:
        stu_id          stu_name
        --------------------------
        10000           Xiao Chen
        10005           Yang Yan
        10006           Gan Yi
```

This query returns the IDs of all students enrolled in the course with ID 2 (10000, 10005, 10006). These values are then passed to the WHERE clause of the outer query in the comma-separated format required by the IN operator. The outer query is shown in Listing 2.15.

2.1.3 SQL Joins

In this section, we continue learning SQL. We will now explore more complex joins, their meanings, and usage. We will cover three types: SELF-JOIN, NATURAL JOIN, and OUTER JOIN.

1. SELF-JOIN

A self-join joins a table with itself, where the structure and data are the same. We usually rename the tables to distinguish between them before performing the join.

Suppose we want to find the student IDs and names of all students who are the same age as Xiao Chen. This query first finds Xiao Chen's age and then identifies all students of the same age. Listing 2.16 shows a way to solve this problem.

Listing 2.16 Subquery example

```
Input:
        SELECT stu_id, stu_name, stu_age
        FROM Students
        WHERE stu_age = (SELECT stu_age
                         FROM Students
                         WHERE stu_name = 'Xiao Chen');
Output:
        stu_id          stu_name          stu_age
        ----------------------------------------
        10000           Xiao Chen         17
        10004           Zi Xiao           17
        10006           Gan Yi            17
```

Listing 2.17 Self-join example

```
Input:
        SELECT s1.stu_id, s1.stu_name, s1.stu_age
        FROM Students AS s1, Students AS s2
        WHERE s1.stu_age = s2.stu_age AND s2.stu_name = 'Xiao Chen';
Output:
        stu_id          stu_name          stu_age
        ----------------------------------------
        10000           Xiao Chen         17
        10004           Zi Xiao           17
        10006           Gan Yi            17
```

This is the first solution that uses a subquery. Now let's see the same query using a self-join, as shown in Listing 2.17.

In this query, the two tables are the same; as you can see, the Students table appears twice in the FROM clause. Since the DBMS doesn't know which instance of the Students table you are referring to, you must use table aliases. For example, the SELECT statement uses the s1 prefix to clearly specify the desired columns. Without this, the DBMS would return an error because there are two columns named stu_id, stu_name, and stu_age. The DBMS wouldn't know which one to use (even though they are the same). The WHERE clause joins the tables and filters the data based on the stu_name in the second table, returning the desired results.

Although the final results are the same, many DBMSs handle joins faster than subqueries.

2. NATURAL JOIN

In our query examples, it is necessary to combine information from several tables that have at least one attribute with the same name in each table.

To simplify the work of SQL programmers in such general cases, SQL supports an operation called natural join, which we will discuss next. In fact, SQL also supports several other ways to join information from two or more relations.

Listing 2.18 Natural join

```
Input:
        SELECT s.stu_id, s.stu_name, s.stu_grade, e.cou_id
        FROM students AS s, Elecourses AS e
        WHERE s.stu_id = e.stu_id;
      -- This query can be rewritten using SQL's natural join operation:
        SELECT s.stu_id, s.stu_name, s.stu_grade, e.cou_id
        FROM students AS s NATURAL JOIN Elecourses AS e;
Output:
        stu_id        stu_name        stu_grade        cou_id
        ---------------------------------------------------------
        10000         Xiao Chen       1                1
        10000         Xiao Chen       1                2
        10000         Xiao Chen       1                3
        10000         Xiao Chen       1                4
        10005         Yang Yan        2                2
        10005         Yang Yan        2                3
```

Listing 2.19 JOIN...USING clause

```
        SELECT stu_id,stu_name,cou_name
        FROM (students NATURAL JOIN Elecourses) JOIN Courses USING (cou_id);
```

The natural join operation works on two relations and produces a single relation as a result. Unlike the Cartesian product of two relations, which joins every tuple of the first relation with all tuples of the second relation, a natural join only considers tuple pairs with the same values for attributes that appear in both relation schemas. Thus, in the example of the Students and Elecourses relations, the natural join of Students and Elecourses only considers tuple pairs with the same values for the common attribute stu_id. The natural join automatically matches records based on columns with the same name in the tables. There is no need to specify any equi-join conditions for a natural join.

Consider the query: "Find the student ID, name, and selected course IDs for all students." This query is shown in Listing 2.18.

In the FROM clause of an SQL query, multiple relations can be combined using a natural join, as shown below:

SELECT $A_1, A_2, \ldots, A_n$
FROM r_1 NATURAL JOIN r_2 NATURAL JOIN ... NATURAL JOIN r_m
WHERE P;

More generally, the FROM clause can be structured in the following form:

FROM $E_1, E_2, \ldots, E_n$

Each E_i can represent a relation or an expression. To leverage natural joins and avoid unintended equality attributes, SQL provides a construct to specify which columns should be equal for the join. The query in Listing 2.19 demonstrates this feature.

Listing 2.20 ON condition keyword

```
SELECT *
FROM Students AS s JOIN Elecourses AS e ON s.stu_id=e.stu_id;
```

Listing 2.21 WHERE clause

```
SELECT *
FROM Students AS s ,Elecourses AS e
WHERE s.stu_id=e.stu_id;
```

The JOIN ... USING operation requires specifying a list of attribute names, and both of its inputs must have attributes with the specified names.

We introduced natural joins and the JOIN ... USING clause, a type of natural join that matches values on specified attributes. SQL also supports another type of join that allows any join condition. The ON condition allows setting general predicates on the relations in the join. Its syntax is similar to the WHERE clause but uses ON instead of WHERE. Like the USING condition, the ON condition appears at the end of the join expression, as shown in Listing 2.20.

The ON condition in Listing 2.20 indicates that a tuple from Students and a tuple from Elecourses match if they have the same stu_id. The join expression in this example is almost the same as Students NATURAL JOIN Elecourses, since the natural join also requires matching tuples from Students and Elecourses.

The queries in Listings 2.20 and 2.21 are actually equivalent; they produce the same results.

The ON condition can represent any SQL predicate, allowing richer join conditions than natural joins. However, as shown in the previous example, a query with an ON condition can be replaced by moving the predicate to the WHERE clause. This might make the ON condition seem redundant. However, the ON condition has two advantages. First, in outer joins (which we will introduce next), the ON condition behaves differently from the WHERE condition. Second, when the join condition is in the ON clause and the other conditions are in the WHERE clause, these SQL queries are usually easier to read.

3. OUTER JOIN

Assuming we want to display a list of all students, including their student ID, name, grade, and enrolled course numbers. The query "SELECT * FROM Students AS s NATURAL JOIN Elecourses AS e" might seem to retrieve the desired information. However, this query does not achieve our goal. If some students have not enrolled in any courses, their tuples in the Students relation will not match any in the Elecourses relation, so their data will not appear in the result. As a result, we will not see information for students without course enrollments.

For example, in the Students and Elecourses relations, student Ming Zhi (stu_id 10001) has not enrolled in any courses. Ming Zhi appears in the Students relation, but their stu_id does not appear in the Elecourses relation. As a result, Ming Zhi will not be included in the natural join result.

More generally, some tuples in one or both of the relations might "lose" in the join. The outer join is similar to the joins we've learned, but it retains lost tuples by creating tuples with null values.

For example, to ensure that Ming Zhi from the previous example appears in the result, a tuple can be added with all attribute values from the Students relation set to Ming Zhi's values and the remaining attributes from the Elecourses relation, like cou_id, set to NULL.

There are actually three forms of outer joins:

- LEFT OUTER JOIN only retains the tuples that appear in the relation on the left side before the outer join operation.
- RIGHT OUTER JOIN only retains the tuples that appear in the relation on the right side after the right outer join operation.
- FULL OUTER JOIN retains the tuples that appear in both relations.

In comparison, the join operation that doesn't retain unmatched tuples is called an INNER JOIN, distinguishing it from the outer join.

We will now explain in detail how each form of outer join works. A left outer join can be computed as follows: first, compute the inner join result; then, for any tuple t in the left relation that doesn't match a tuple in the right relation, add a tuple r to the result. The tuple r is constructed as follows:

- The attributes of tuple r obtained from the left relation are assigned the values from t.
- The other attributes of r are assigned null values.

We can write the query "Student course information, including the course information of students who have not selected any courses," as shown in Listing 2.22.

As another example of an outer join, we can write a query to find all students who have not enrolled in any courses, as shown in Listing 2.23.

Listing 2.22 Left outer join example 1

```
Input:
        SELECT *
        FROM Students NATURAL LEFT OUTER JOIN Elecourses;

Output:
        stu_id    stu_name    stu_origin    stu_sex    stu_age    stu_grade    cou_id
-- ------------------------------------------------------------------------------
        10000     Xiao Chen   Beijing       1          17         1            4
        10001     Ming Zhi    Shanghai      0          23         4            NULL
        10002     Hong Li     Jiangxi       1          20         3            NULL
```

Listing 2.23 Left outer join example 2

```
Input:
        SELECT *
        FROM Students NATURAL LEFT OUTER JOIN Elecourses
        WHERE cou_id IS NULL;

Output:
        stu_id    stu_name    stu_origin    stu_sex    stu_age    stu_grade    cou_id
-- ----------------------------------------------------------------------------
        10001     Ming Zhi    Shanghai      0          23         4            NULL
        10002     Hong Li     Jiangxi       1          20         3            NULL
        10003     Li Zhang    Beijing       0          18         1            NULL
```

Listing 2.24 Right outer join example

```
-- Input:
        SELECT *
        FROM Students NATURAL RIGHT OUTER JOIN Elecourses;
```

The right outer join is symmetrical to the left join. Tuples from the right relation that don't match any tuple in the left relation are padded with null values and included in the result. If we rewrite the previous query using a right join and switch the order of the relations, as shown in Listing 2.24.

We get the same result, but the order of the attributes in the result is different.

A full outer join combines left and right joins. After the inner join, unmatched tuples from the left relation are padded with null values and added to the result. Similarly, unmatched tuples from the right relation are also padded with null values and added to the result.

To distinguish between regular joins and outer joins, SQL refers to regular joins as inner joins. Therefore, the JOIN clause can use INNER JOIN to indicate a regular join. However, the keyword INNER is optional; if the JOIN clause does not use the OUTER prefix, the default join type is INNER JOIN. Consequently, SELECT * FROM Students JOIN Elecourses USING (cou_id) is equivalent to SELECT * FROM Students INNER JOIN Elecourses USING (cou_id). Similarly, NATURAL JOIN is equivalent to NATURAL INNER JOIN. Any form of join (INNER JOIN, LEFT OUTER JOIN, RIGHT OUTER JOIN, or FULL OUTER JOIN) can be combined with any join condition (NATURAL JOIN, USING clause, or ON clause).

The summary of various table joins is shown in Table 2.4.

2.1.4 Other SQL Statements

This section covers SQL's INSERT, UPDATE, and DELETE statements, along with join expressions, views, transactions, integrity constraints, and authorization.

Table 2.4 Comparison of various join types

Join Type	Definition	Diagram	Example
Inner Join	Only matches rows	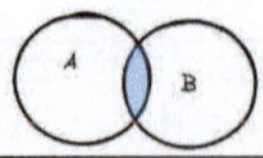	SELECT A.c1, B.c2 FROM A JOIN B ON A.c3 = B.c3
Left Outer Join	Includes all rows from the left table (regardless of whether there is a match in the right table) and all matching rows from the right table	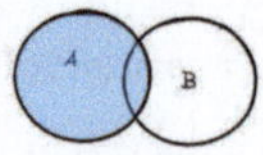	SELECT A.c1, B.c2 FROM A LEFT JOIN B ON A.c3 = B.c3
Right Outer Join	Includes all rows from the right table (regardless of whether there is a match in the left table) and all matching rows from the left table	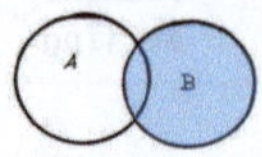	SELECT A.c1, B.c2 FROM A RIGHT JOIN B ON A.c3 = B.c3
Full Outer Join	Includes all rows from both the left and right tables, regardless of whether there is a match in the other table	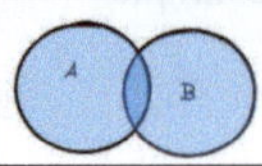	SELECT A.c1, B.c2 FROM A FULL JOIN B ON A.c3 = B.c3

INSERT is used to insert (or add) rows into a database table. The insertion can insert either a complete row or a partial row.

Insert a new student into the Students table. The data for each column is specified in the VALUES clause, and a value must be provided for each column. If a column has no value, a NULL value should be used (assuming the table allows NULL values for that column). Columns must be populated in the order they appear in the table definition, as shown in Listing 2.25.

Specifying the column names explicitly when using INSERT is recommended. This approach also allows omitting columns, meaning values can be provided for only certain columns. An example of this is shown in Listing 2.26.

INSERT is typically used to insert rows with specific column values into a table. Another form, INSERT SELECT, allows inserting the results of a SELECT statement into a table. For example, if you want to merge new students from another table into the Students table, you don't need to insert each row individually, as shown in Listing 2.27.

Listing 2.25 INSERT statement to insert a row

```
Input:
    INSERT INTO Students
    VALUES(10006, 'Tom', 'Beijing', 'man', 20, '3');
```

Listing 2.26 INSERT statement to insert some attributes of a row

```
Input:
        INSERT INTO Students(stu_id, stu_name, stu_origin)
        VALUES(10006, 'Tom', 'Beijing');
```

Listing 2.27 INSERT statement to insert the results of a SELECT statement

```
Input:
        INSERT INTO Students(stu_id, stu_name, stu_origin, stu_sex, stu_age, stu_grade)
        SELECT stu_id, stu_name, stu_origin, stu_sex, stu_age, stu_grade
        FROM StudentsNew;
```

Listing 2.28 UPDATE statement to modify attributes

```
Input:
        UPDATE Students
        SET stu_name = 'jerry'
        WHERE stu_id = 10006;
```

Listing 2.29 DELETE statement to remove a row

```
Input:
        DELETE FROM Students
        WHERE stu_id = 10006;
```

To update data in a table, use the UPDATE statement to modify specific rows. For example, the student with ID 10006 needs their name changed, as shown in Listing 2.28.

A delete request is similar to a query. We can only delete entire tuples, not specific attribute values. In SQL, deletion is represented by the statement: DELETE FROM r WHERE P. The P represents a predicate, and the r represents a relation. The DELETE statement identifies all tuples t in r that satisfy $P(t)$ and removes them. If the WHERE clause is omitted, all tuples in r are deleted. The DELETE command only operates on one relation. To delete tuples from multiple relations, use a DELETE command for each relation. The predicate in the WHERE clause can be as complex as in a SELECT command. Alternatively, the WHERE clause can be empty; DELETE FROM Students' will delete all tuples in the Students relation. The Students relation remains but becomes empty. Below are examples of SQL delete requests. To delete data from a table, use the DELETE statement to remove specific rows, as shown in Listing 2.29.

This statement is easy to understand. "DELETE FROM" requires specifying the table from which data will be deleted, while the "WHERE" clause filters the rows to delete. In this example, only the student with a student ID of 10006 is deleted.

View

A view is a virtual table defined by a query. Like a base table, a view contains named columns and row data. Views are not stored as data in the database. The data in a view comes from base tables referenced by the query and is dynamically generated when the view is accessed. Views act like filters on the underlying tables they reference. These filters can come from one or more tables in the current or other databases or from other views. Distributed queries can also define views using

Listing 2.30 Syntax format for creating a view

```
CREATE VIEW v AS <query expression>;
```

multiple heterogeneous data sources. For example, if different servers store data for a unit in different regions and need to combine similar data, this approach is useful. Views are often used to centralize, simplify, and customize each user's database view. They can be used as a security mechanism, allowing users to access data through views without direct access to the underlying tables. Views can also provide backward-compatible interfaces by simulating base tables whose schemas have changed.

You can use the SQL statement CREATE VIEW to create a view, and its syntax format is shown in Listing 2.30.

The <query expression> can be any valid SQL query statement, and v represents the name of the view.

After a view is created, only its definition is stored in the data dictionary, and the SELECT statement in the subquery is not executed until the view is queried. The data is retrieved from the base tables according to the view's definition only when the user operates on the view.

Transaction

A transaction consists of a sequence of query and/or update statements. The SQL standard specifies that executing an SQL statement implicitly starts a transaction:

- **Commit work**. Committing the current transaction permanently saves the updates made by that transaction in the database. After the transaction is committed, a new transaction automatically begins.
- **Rollback work**. Rolling back the current transaction undoes all the updates made to the database by the SQL statements within that transaction. This way, the database is restored to the state it was in before the first statement of the transaction was executed.

Rollback is useful when an error occurs during a transaction. Once a transaction executes **Commit work**, its effects cannot be undone with **Rollback work**. Database systems ensure that in case of failures like SQL errors, power failures, or system crashes, if a transaction hasn't completed **Commit work**, its effects will be rolled back. In the event of power failure or system crash, rollback is executed after the system restarts.

Integrity Constraints

The CREATE TABLE command is used to define a relational table. The CREATE TABLE command can also include integrity constraint statements. Besides the primary key constraint, many other constraints can be included in the CREATE TABLE command. Allowed integrity constraints include:

- **NOT NULL**: for some attributes, null values may not be appropriate. Consider a tuple in the Students relation where stu_name is NULL. Such a tuple provides student information for an unknown student, which lacks useful information. Similarly, we wouldn't want its value to be NULL. In such cases, we can prohibit null values by restricting the domain of the stu_name attribute using the following declaration: name varchar(20) NOT NULL. The NOT NULL declaration prevents inserting null values into this attribute. Any modification that attempts to insert a null value into a NOT NULL attribute will generate an error.
- **UNIQUE**: the UNIQUE($A_{m1}, A_{m2}, \ldots, A_{mn}$) which indicates that the attributes A_{m1},A_{m2},...,A_{mn} form a candidate key; that is, there are no two tuples in the relation that can have the same values on all listed attributes.
- **CHECK Clause**: which are commonly used to ensure attribute values meet specified conditions. The CHECK(P) clause specifies a predicate P, which every tuple in the relation must satisfy. This creates a powerful constraint system. For example, the CHECK(stu_age > 0) clause in the CREATE TABLE command for creating the Student relation ensures that stu_age values are positive.

Authorization

The SQL standard includes SELECT, INSERT, UPDATE, and DELETE permissions. A user who creates a new relation is automatically granted all permissions on that relation. The SQL Data Definition Language (DDL) includes commands for granting and revoking permissions. The GRANT statement is used to grant permissions. The basic form of the statement is shown in Listing 2.31 as follows.

The SELECT privilege on a relation is used to read tuples from the relation. The GRANT statement in Listing 2.32 grants database users A and B the SELECT privilege on the Students relation.

The UPDATE privilege allows users to modify tuples in a relation. It can be granted on all attributes or specific attributes of the relation. If the GRANT statement includes the UPDATE privilege, a list of attributes can follow the UPDATE keyword in parentheses. The attribute list is optional; the privilege applies to all attributes if omitted.

The GRANT statement in Listing 2.33 grants users A and B the UPDATE privilege on the stu_name attribute of the Students relation.

Listing 2.31 GRANT permission syntax format

```
GRANT <permission_list>
ON <table_or_view_name>
TO <user_or_role_list>;
```

Listing 2.32 Example of GRANT to grant permissions

```
GRANT SELECT ON Students TO A, B;
```

Listing 2.33 Example of GRANT to modify privileges

```
GRANT UPDATE (stu_name) ON Students TO A, B;
```

Listing 2.34 Syntax for REVOKE to withdraw privileges

```
REVOKE <privilege list>
ON <table or view name>
FROM <user/role list>;
```

Listing 2.35 Example of REVOKE to withdraw privileges

```
REVOKE SELECT ON Students FROM A, B;
REVOKE UPDATE (stu_name) ON Students FROM A, B;
```

The INSERT privilege on a relation allows users to insert tuples into the relation. The INSERT privilege can also specify a list of attributes.

The DELETE privilege on a relation allows users to delete tuples from the relation.

We use the REVOKE statement to withdraw privileges. As shown in Listing 2.34, the form of this statement is almost the same as that of GRANT.

To withdraw the privileges we granted earlier, as shown in Listing 2.35.

A view relation can be defined as a relation that contains the results of a query. Views are useful as they can hide unnecessary information and gather information from multiple relations into a single view.

A transaction is a sequence of queries and updates that are executed together to perform a specific task. A transaction can be committed or rolled back. When a transaction is rolled back, all the updates performed by that transaction are undone, reversing the effects of those updates.

Integrity constraints ensure that changes made to the database by authorized users do not result in the violation of data consistency. Referential integrity constraints guarantee that the values appearing in a given set of attributes of one relation also appear in a specific set of attributes of another relation. Domain constraints specify the set of possible values that can be taken on an attribute. Such constraints can also prohibit the use of null values on specific attributes.

Through the SQL authorization mechanism, database users can be differentiated based on the types of access allowed to different data values in the database. Users who have been granted a certain form of authorization may be permitted to pass this authorization on to other users. However, we must be cautious about how permissions are transferred between users to ensure that such permissions can be revoked at some point in the future.

2.2 NoSQL

As previously introduced, SQL is so irreplaceable that when we choose a database for a new project, it is merely a matter of selecting which vendor's implementation of a relational database to use, rather than questioning whether to use a relational database at all. So, given the existence of powerful relational databases and the corresponding SQL, why do we still need NoSQL databases and NoSQL?

Although relational databases are powerful, they still have many unsatisfactory aspects. For example, consider the data model shown in Fig. 2.4. When writing code for a shopping cart functionality, we need to write a complex SQL statement that uses JOIN operations to connect four tables (and it might be even more complex in an actual system). When this operation is performed in a clustered environment, the problem becomes even more severe: the tables we need to join might not be on the same server, which involves complex distributed join issues.

If relational databases are already slightly fatigued when faced with the problems described above, they are also quite clumsy when faced with data with complex relationships, as shown in Fig. 2.5, both in terms of creating data tables and writing query SQL for complex relationships. This is where a new database and database query language is needed.

Thus, NoSQL came into being. NoSQL is not "No SQL," but an abbreviation for "Not Only SQL." This means that it does not aim to completely replace SQL, but rather to complement it, playing its role in scenarios where SQL is hard to handle. Strictly speaking, NoSQL is not a specific language, but a concept, implying that data is stored without using a relational structure. The term NoSQL is used here to distinguish it from SQL. Next, we will introduce the processing languages used by different types of databases within NoSQL. For each type of database, we will introduce a common database management system. Table 2.5 shows each type of database and its corresponding database management system.

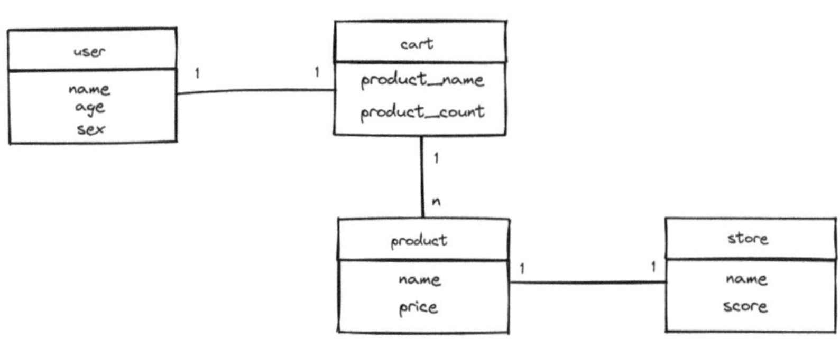

Fig. 2.4 Shopping cart relationship model

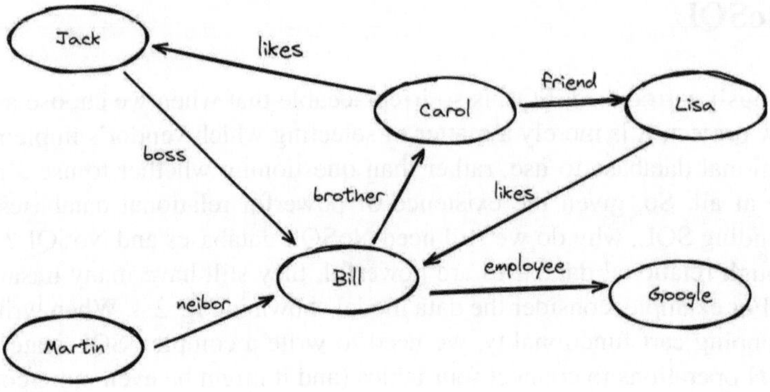

Fig. 2.5 Data with complex relationships

Table 2.5 Types of databases and their corresponding database management systems

Database type	Database management system	Version
Key-value database	Redis	3.0.504
Document database	MongoDB	4.2.21
Columnar database	Cassandra	4.0.6
Graph database	Neo4j	5.1.0

2.2.1 Key-Value Database Processing Languages

A key-value store is a simple hash table, mainly used in scenarios where all database access is performed using primary key operations. If this were a relational database, it would be a table with two columns: one column for keys to store keywords, and the other for values to store the corresponding values. Users can provide key and value pairs to persist the key-value pairs. If the key exists, the original value will be overwritten; otherwise, a new key-value pair is created.

Popular key-value databases include Riak, Redis, Memcached DB, BerkeleyDB, HamsterDB (especially suitable for embedded development), and Amazon DynamoDB (not open source). In key-value databases like Redis, the stored aggregates can be various data structures, such as lists, sets, and hashes. They also support operations such as "finding the intersection of sets," "finding the union of sets," and "retrieving data within a certain range." These operations make Redis more powerful than ordinary key-value databases and broaden its applicability.

This section mainly focuses on the processing language of key-value databases. There are many kinds of key-value databases, and new ones are constantly emerging. To simplify our discussion, we use Redis 3.0.504 as an example to introduce key-value database languages. The structure of the Redis database is shown in Fig. 2.6. Each Redis instance contains multiple databases, and each database includes a hash table to store key-value pairs.

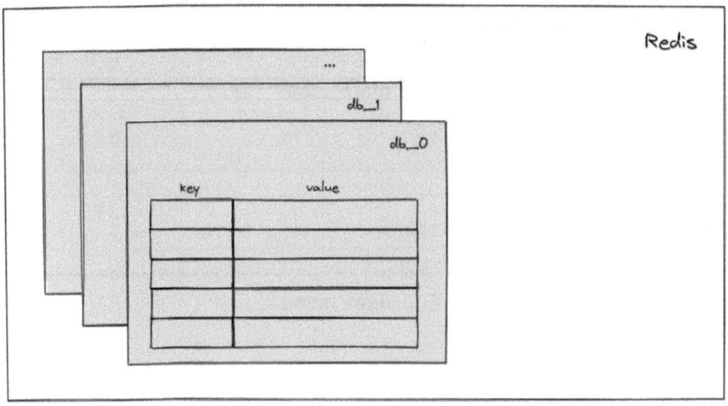

Fig. 2.6 Redis database structure

Listing 2.36 Redis syntax example

```
redis 127.0.0.1:6379> COMMAND key_name
```

Redis Syntax The syntax of Redis is very simple. COMMAND indicates an operation on a key, and key_name indicates the name of the key to be operated on. The syntax is as shown in Listing 2.36.

Redis String Command Introduction Redis databases support various data structures. We use the string data structure as an example to introduce Redis string commands. Below are some common Redis string commands:

- SET key value. Setting the value of the specified key to value.
- GET key. Getting the value of the specified key.
- DEL key. Deleting the key-value pair corresponding to the specified key.
- EXISTS key. Checking whether the specified key exists.
- GETRANGE key start end. Returning the substring composed of characters in the value of the key from position [start, end].
- GETSET key value. Setting the value of the given key to value and returns the old value of the key.
- STRLEN key. Returning the length of the string value stored by the key.

Redis String Command Examples

- Insertion Operation. Use the SET command to insert the key-value pair user_name, jerry, as shown in Listing 2.37.

Listing 2.37 Inserting a key-value pair

```
Input:
        redis 127.0.0.1:6379> SET user_name jerry
Output:
        OK
```

Listing 2.38 Querying the value of user_name

```
Input:
        redis 127.0.0.1:6379> GET user_name
Output:
        "jerry"
```

Listing 2.39 Modifying the value of user_name

```
Input:
        redis 127.0.0.1:6379> SET user_name tom
Output:
        OK
```

Listing 2.40 Querying the substring of the value of user_name

```
Input:
        redis 127.0.0.1:6379> GETRANGE user_name 1 2
Output:
        "om"
```

- Query Operation. Use the GET command to query the value corresponding to the key user_name, as shown in Listing 2.38.
- Modification Operation. Use the SET command to change the value corresponding to the key user_name from jerry to tom, as shown in Listing 2.39.
- Query Operation. Use the GETRANGE command to query the substring located at positions [1, 2] in the value corresponding to the key user_name, as shown in Listing 2.40.
- Modification Operation. Use the GETSET command to change the value corresponding to the key user_name to green and return the original value, as shown in Listing 2.41.
- Query Operation. Use the STRLEN command to query the length of the value corresponding to the key user_name, as shown in Listing 2.42.
- Deletion Operation. Use the DEL command to delete the key-value pair corresponding to the key user_name, as shown in Listing 2.43.

Listing 2.41 Modifying the value of user_name and returning the original value

```
Input:
        redis 127.0.0.1:6379> GETSET user_name green
Output:
        "tom"
```

Listing 2.42 Querying the length of the value of user_name

```
Input:
        redis 127.0.0.1:6379> STRLEN user_name
Output:
        (integer) 5
```

Listing 2.43 Deleting the key-value pair corresponding to user_name

```
Input:
        redis 127.0.0.1:6379> DEL user_name
Output:
        (integer) 1
```

Listing 2.44 Checking whether the key-value pair corresponding to user_name Exists

```
Input:
        redis 127.0.0.1:6379> EXISTS user_name
Output:
        (integer) 0
```

- Query Operation. Use the EXISTS command to check whether the key-value pair corresponding to the key user_name exists, as shown in Listing 2.44.

For more complex data types and related operations provided in Redis, readers can refer to the examples and usage in the official Redis documentation [110]; this book will not cover them in detail.

2.2.2 Document Database Processing Languages

A document store is a type of database similar to traditional relational databases, with "documents" being the main concept. Documents are presented in a hierarchical tree structure, typically represented in formats such as XML, JSON, and BSON. Documents in the database are similar to each other but do not need to be identical. There are no empty attributes in document database documents; if an attribute does not exist, it can be assumed that the attribute value is not relevant to the document. Adding attributes to a document does not require pre-definition or modification of existing content in the document. There are many popular document

Table 2.6 Comparison of MySQL and MongoDB terminology

MySQL	MongoDB
database	database
table	collection
row	document

Listing 2.45 MongoDB vocabulary

```
>>> db.collection.function(<parameter 1>, <parament 2>, ...)
```

databases, such as MongoDB, CouchDB, and RavenDB. This book introduces the document database processing language using MongoDB 4.2.21 as an example. Table 2.6 compares the terminology between MySQL and MongoDB.

MongoDB Syntax The MongoDB syntax is similar to that of JavaScript, Python, and other scripting languages, as shown in the program Listing 2.45. Users can easily use the various interfaces provided by MongoDB. In the statement in the program Listing 2.45, db is used to denote the name of the database, collection is used to denote the name of the collection, function is used to denote the name of the function to be used, and parameter in parentheses is the parameter required by the function.

An Introduction to MongoDB Add, Delete, Query, and Modify Operations MongoDB supports a lot of complex operations; we take the most basic add, delete, check, and modify operations as an example to demonstrate the syntax of MongoDB. The following is an example of MongoDB's add, delete, check, and modify operations:

- db.collection.insertOne(<document>). This operation can be used to insert a document into a collection, where <document> is the document to be inserted. It is important to note that there is a _id field for each document in MongoDB, which is automatically generated when a document is inserted if the _id field is not specified.
- db.collection.insertMany([<document 1>, <document 2>, ...]). This operation can be used to insert one or more documents into a collection, similar to the insertOne operation, where [<document 1>, <document 2>, ...] is the collection of documents to be inserted.
- db.collection.deleteOne(<query document>). This action can be used to delete the first document in the collection that matches the condition, where <query document> is the condition that needs to be met for the document to be deleted.
- db.collection.deleteMany(<query document>). This action can be used to delete all documents in the collection that match the condition, where <query document> is the condition that needs to be met for the document to be deleted.
- db.collection.find(<query document>, <projection document>). This action can be used to query a document that matches a condition and return, where <query document> is the condition to be satisfied by the query document and <projection

document> is the field to be displayed in the query result. When <projection document> <field>: 1 means include the field in the returned document, <field>: 0 means exclude the field in the returned document. The _id field is returned by default, so you can set _id: 0 in the <projection document> to not return this field.

- db.collection.updateOne(<query document>, {$set: <update document>}). This operation can be used to modify the first document in the set that matches the conditions, where <query document> is the condition to be met by the document to be modified, $set is the update operator that sets the value of the field in the document, and <update document> is the field and value to be modified, or created if the field does not exist.
- db.collection.updateMany(<query document>, {$set: <update document>}). This operation can be used to modify all matching documents in a collection, similar to the updateOne operation.
- db.collection.replaceOne(<query document>, <replacement document>). This operation can be used to replace the first document in the collection that matches the condition, where <query document> is the document to be modified to match the condition and <replacement document> is the replacement document.

Examples of Adding, Deleting, Querying, and Modifying MongoDB Operations

The following will introduce the basic operations of a document database using MongoDB with JSON document format as an example. The initial data of the database is shown in Table 2.7: the initial database contains one collection named "inventory," which has five document entries. The CRUD operations (Create, Read, Update, Delete) of MongoDB will be introduced below.

- Query operation. Use the find function, set the query condition to {}, and query all the documents in the collection inventory, as shown in the program Listing 2.46.

Table 2.7 Inventory set initial data

File
{ item: "journal", qty: 25, status: "A", tags: ["blank", "red"] }
{ item: "notebook", qty: 50, size: { h: 8.5, w: 11, uom: "in" } }
{ item: "paper", qty: 10, status: "D", size: { h: 8.5, w: 11, uom: "in" }, tags: ["red", "blank", "plain"] }
{ item: "planner", qty: 0, size: { h: 22.85, w: 30, uom: "cm" }, tags: ["blank", "red"] }
{ item: "postcard", qty: 45, status: "A", tags: ["blue"] }

Listing 2.46 Query all documents

```
Input:
        >>> db.inventory.find({})
Output:
        {"_id": ObjectId("633f89d3bef160b823f81064"), item: "journal",
    qty: 25, status:"A",          tags: [ "blank", "red" ] }
        {"_id": ObjectId("633f89d3bef160b823f81065"), item: "notebook",
    qty: 50, size:{ h:          8.5, w: 11, uom: "in" } }
        {"_id": ObjectId("633f89d3bef160b823f81066"), item: "paper", qty
    : 10, status:"D",           size: { h: 8.5, w: 11, uom: "in" }, tags
    : [ "red", "blank", "plain" ] }
        {"_id": ObjectId("633f89d3bef160b823f81067"), item: "planner",
    qty: 0, size: { h:         22.85, w: 30, uom: "cm" }, tags: [ "blank
    ", "red" ] }
        {"_id": ObjectId("633f89d3bef160b823f81068"), item: "postcard",
     qty: 45,status: "A",         tags: [ "blue" ] }
```

Listing 2.47 Query documents with status "D"

```
Input:
        >>> db.inventory.find({status:"D"})
Output:
        {"_id": ObjectId("633f89d3bef160b823f81066"), item: "paper", qty
    : 10, status: "D", size: { h: 8.5, w: 11, uom: "in" },
        tags: [ "red", "blank", "plain" ] }
```

Listing 2.48 Query documents with uom "in" in size

```
Input:
        >>> db.inventory.find( { "size.uom": "in" } )
Output:
        {"_id": ObjectId("633f89d3bef160b823f81065"),
        item: "notebook", qty: 50, size: { h: 8.5, w: 11, uom: "in" } }
        {"_id": ObjectId("633f89d3bef160b823f81066"),
        item: "paper", qty: 10,status: "D", size: { h: 8.5, w: 11, uom: "in" },
        tags: [ "red", "blank", "plain" ] }
```

- Query operation. Use the find function and set the query condition to {status: "D"} to query the documents in the collection inventory that satisfy status="D," as shown in Listing 2.47.
- Query operation. Use the find function to set the query condition to {"size.uom": "in"}, querying the documents in the inventory collection where the uom in size is "in," as shown in Listing 2.48.
- Query operation. Use the find function to set the query condition to {tags: ["blank," "red"]}, querying the documents in the inventory collection where the tags field contains "blank" and "red" in the specified order (i.e., "blank" before "red"), as shown in Listing 2.49.

Listing 2.49 Query documents with "blank," "red" in tags that match sequentially

```
Input:
        >>> db.inventory.find( { tags: [ "blank", "red" ] } )
Output:
        {"_id": ObjectId("633f89d3bef160b823f81064"),
        item: "journal", qty: 25, status: "A", tags: [ "blank", "red" ] }
        {"_id": ObjectId("633f89d3bef160b823f81067"),
        item: "planner", qty: 0, size: { h: 22.85, w: 30, uom: "cm" },
        tags: [ "blank", "red" ] }
```

Listing 2.50 Queries a document with status "A" and displays the itme and status fields in the document

```
Input:
        >>> db.inventory.find( {status:"A"}, { _id: 0, item: 1, status: 1 } )
Output:
        {"item":"journal", "status":"A"}
        {"item":"postcard", "status":"A"}
```

Listing 2.51 Insert a document data

```
Input:
        >>> db.inventory.insertOne({item: "flower", qty: 20,
            status: "A", tags: [ "black", "red" ]});
Output:
            {
              "acknowledged" : true,
              "insertedIds" : [
                ObjectId("633f89d3bef160b823f81069")
              ]
            }
```

- Query operation. Use the find function to set the query condition to {status:"A"} and the projection to { _id: 0, item: 1, status: 1 }, querying the documents in the inventory collection where the status is "A" and returning only the item and status fields while excluding the _id field, as shown in Listing 2.50.
- Insertion operation. Use the insertOne function to set a document as an argument and insert a document into the inventory collection, as shown in Listing 2.51.
- Insertion operation. Use the insertMany function to set two documents as arguments and insert two documents into the inventory collection, as shown in Listing 2.52.
- Deletion operation. Use the deleteOne function to set the deletion condition as {item:"flower"}, deleting the first document in the inventory collection that satisfies item being "flower," as shown in Listing 2.53.

Listing 2.52 Insert two document data

```
Input:
        >>> db.inventory.insertMany([
        ...     {item: "book", qty: 30, status: "C", tags: [ "blank", "blue" ] },
        ...     {item: "leaf", qty: 17, size: { h: 5.1, w: 1, uom: "in" } }
        ... ]);
Output:
        {
          "acknowledged" : true,
          "insertedIds" : [
            ObjectId("633f89d3bef160b823f81070"),
            ObjectId("633f89d3bef160b823f81071")
          ]
        }
```

Listing 2.53 Delete the document whose first item in the collection is "flower"

```
Input:
        >>> db.inventory.deleteOne({item:"flower"})
Output:
        { "acknowledged" : true, "deletedCount" : 1 }
```

Listing 2.54 Delete all documents in the collection that have a status of "A"

```
Input:
        >>> db.inventory.deleteMany({status:"A"})
Output:
        { "acknowledged" : true, "deletedCount" : 2 }
```

Listing 2.55 Change the qty of the first document in the collection with a qty of 50 to 1

```
Input:
        >>> db.inventory.updateOne({qty:50},{$set:{qty:1}})
Output:
        { "acknowledged" : true, "matchedCount" : 1, "modifiedCount" : 1
    }
```

- Deletion operation. Use the deleteMany function to set the deletion condition as {status:"A"}, deleting all documents in the inventory collection that have a status of "A," as shown in Listing 2.54.
- Update operation. Use the updateOne function to set the update condition as {qty:50}, and set the attribute and corresponding value to be updated as {qty:1}, updating the qty of the first document in the inventory collection that satisfies qty being 50 to 1, as shown in Listing 2.55.
- Update operation. Use the updateMany function to set the update condition as {tags: "blank"} and define the field and value to be updated as {qty: 1}. This will update the qty field to 1 for all documents in the inventory collection where the tags field contains "blank," as shown in Listing 2.56.

Listing 2.56 Change qty to 1 for documents with “blank" in tags

```
Input:
        >>> db.inventory.updateMany({tags:"blank"},{$set:{qty:1}})
Output:
        { "acknowledged" : true, "matchedCount" : 3, "modifiedCount" : 3
    }
```

Listing 2.57 Replace the first document in the collection that contains “blank" in its tags

```
Input:
        >>> db.inventory.replaceOne({tags:"blank"},
            {item: "book", qty: 11, status: "D", tags: [ "green" ] })
Output:
        { "acknowledged" : true, "matchedCount" : 1, "modifiedCount" : 1
    }
```

- Update operation. Use the replaceOne function to set the update condition as {tags:“blank”}, replacing the first document in the inventory collection that has “blank” in its tags with {item, “book”; qty, 11; status, “D”; tags, [“green”]}, as shown in Listing 2.57.

In addition, MongoDB also provides a more powerful and richer API; the reader can refer to the MongoDB official documents [92] in the examples and usage; this book will not be introduced in detail.

2.2.3 Column-Family Database Processing Languages

Column-Family database store data in a column family, which divides related data that usually need to be accessed together into groups, and then uses row keys to associate these groups. Column-family databases use a columnar storage format, where data within the same column family is stored together. This approach allows the database to avoid reading entire rows when only specific columns are needed, thereby reducing I/O overhead.

Popular column-family databases include Cassandra, HBase, and DynamoDB. Cassandra is designed to perform write operations quickly across a distributed cluster and is easily scalable to handle large volumes of data. In a Cassandra cluster, there is no master node; each node is capable of independently handling read and write operations. The following introduces the query language for column-family databases, using Cassandra version 4.0.6 as an example. Table 2.8 compares the terminology used in MySQL and Cassandra.

Cassandra Syntax Cassandra uses CQL (Cassandra Query Language) as its query language, and CQL is very similar to SQL. In fact, many of the basic concepts are

Table 2.8 Comparison of terminology between MySQL and Cassandra

MySQL	Cassandra
database	keyspace
table	column family
row	row
column(Each row corresponds to the same column)	column(Columns can be different for different rows)

Listing 2.58 Differences between CQL and SQL in database creation

```
/* Creating a new keyspace in CQL */
CREATE KEYSPACE myDatabase WITH replication =
    {'class': 'SimpleStrategy', 'replication_factor': 1};

/* Creating a new database in SQL*/
CREATE DATABASE myDatabase;
```

Listing 2.59 Setting data expiration time in CQL

```
/* Data expired after 24 hours */
INSERT INTO myTable (id, myField) VALUES (2, 9) USING TTL 86400;
```

even identical. We've already covered the SQL language, so we won't go into detail about the syntax of CQL, but only the differences between CQL and SQL:

- Creating a database. In Cassandra, keyspaces require more detailed specifications than databases in standard relational database systems. The following example shows the simplest form of keyspace creation. In a distributed production environment, the keyspace would typically be configured with different strategies and replication factors. The program Listing 2.58 demonstrates the difference between CQL and SQL when creating a database.
- CQL does not support features like JOIN, GROUP BY, and FOREIGN KEY constraints. By not implementing these relational features, Cassandra optimizes for faster read and write operations, enhancing performance and scalability in a distributed environment.
- Time to live. CQL allows you to set a Time to Live (TTL) on rows, meaning that rows can be configured to expire after a specified period, such as 24 hours from their creation time. This is achieved using the USING TTL clause, where the TTL value is specified in seconds. For example, to set a TTL of 24 hours, you would use USING TTL 86400 (since 24 hours equals 86,400 seconds), as demonstrated in Listing 2.59.
- Filtering. You may receive an error message when running a SELECT statement on a large range of values, as shown in the program Listing 2.60.

Listing 2.60 Querying large ranges of data in CQL

```
/* Selection of data in the range */
SELECT * FROM myTable WHERE myField > 5000 AND myField < 100000;

InvalidRequest: Error from server: code=2200 [Invalid query] message="
    Cannot execute this query as it might involve data filtering and
    thus may have unpredictable performance. If you want to execute this
     query despite the performance unpredictability, use ALLOW FILTERING
    "

Invalid request: error from server: code=2200 [invalid query] message="
    Unable to execute this query as it may involve data filtering and
    therefore may have unpredictable performance. If you want to execute
     this query without considering the unpredictability of performance,
     use ALLOW filtering"
```

Listing 2.61 Query all data

```
Input:
        cqlsh:test> SELECT * FROM emp;
Output:
         emp_id | emp_city       | emp_name |  emp_phone      | emp_sal
        --------+-----------+----------+------------+---------
              1 | Hyderabad  |      ram   | 9848022338 |   50000
              2 | Hyderabad  |    robin   | 9848022339 |   40000
              3 |    Chennai |   rahman   | 9848022330 |   45000

        (3 rows)
```

Examples of Cassandra Add, Delete, Query, and Modify Operations The initial database is shown in Table 2.9. The database contains one table, emp, with three rows of data. The following sections demonstrate how to perform add, delete, query, and modify operations using Cassandra:

- To query all the data in the emp table, use the SELECT * FROM emp statement. The FROM emp clause specifies that the query retrieves data from the emp table, while SELECT * indicates that all columns of the data will be returned. An example of this query is shown in Listing 2.61.
- Query operation. To retrieve the first two records from the emp table, use the LIMIT 2 clause in the query. This clause restricts the result set to the first two records that meet the query conditions, based on the example shown in Listing 2.61. The modified query is illustrated in Listing 2.62.

Table 2.9 Initial data for emp table

emp_id	emp_city	emp_name	emp_phone	emp_sal
1	Hyderabad	ram	9848022338	50000
2	Hyderabad	robin	9848022339	40000
3	Chennai	rahman	9848022330	45000

Listing 2.62 Query the first two records

```
Input:
        cqlsh:test> SELECT * FROM emp LIMIT 2;
Output:
         emp_id | emp_city      | emp_name |  emp_phone       | emp_sal
        --------+---------------+----------+------------------+---------
              1 | Hyderabad     |      ram | 9848022338 |   50000
              2 | Hyderabad     |    robin | 9848022339 |   40000

        (2 rows)
```

Listing 2.63 Query emp with emp_id = 1

```
Input:
        cqlsh:test> SELECT emp_name,emp_sal FROM emp WHERE emp_id = 1;
Output:
         emp_name | emp_sal
        ----------+---------
              ram |   50000

        (1 rows)
```

Listing 2.64 Insert new emp

```
Input:
        cqlsh:test> INSERT INTO emp (emp_id, emp_name, emp_city, emp_phone, emp_sal)
        VALUES(4,'lura', 'Google', 9848022371, 60000);
```

- Query operation. To query the emp_name and emp_sal columns from the emp table for records where emp_id = 1, use the WHERE clause to specify the condition emp_id = 1. The SELECT emp_name, emp_sal statement retrieves only the emp_name and emp_sal columns of the matching data. An example of this query is shown in Listing 2.63.
- Insert operation. To insert a new emp into the emp table, use the INSERT INTO statement. The INSERT INTO emp command specifies that data is being inserted into the emp table, while (emp_id, emp_name, emp_city, emp_phone, emp_sal) lists the columns for which data is being provided. The VALUES (4, "Lura," "Google," 9848022371, 60000) clause defines the corresponding values for each column. An example of this insertion is shown in Listing 2.64.
- Insert operation with specified survival time. To insert a new emp into the emp table with a specified survival time, use the USING TTL clause. The USING TTL 86400 clause sets the data's survival time to 86,400 seconds (24 hours). An example of this insertion is shown in Listing 2.65.
- Delete operation. To delete the emp_sal column for the record with emp_id equal to 3 in the emp table, use the DELETE statement. The DELETE emp_sal command specifies that the emp_sal column should be removed from the data where emp_id is 3. An example of this operation is shown in Listing 2.66.

Listing 2.65 Insert a new emp with a specified survival time

```
Input:
        cqlsh:test> INSERT INTO emp (emp_id, emp_name, emp_city, emp_phone, emp_sal)
        VALUES(9,'sally', 'Insgream', 9848022390, 41000) USING TTL 86400;
```

Listing 2.66 Delete emp_sal column for emp_id = 3

```
Input:
        cqlsh:test> DELETE emp_sal FROM emp WHERE emp_id=3;
```

Listing 2.67 Delete emp with emp_id = 2

```
Input:
        cqlsh:test> DELETE FROM emp WHERE emp_id=2;
```

Listing 2.68 Modify emp with emp_id = 1

```
Input:
        cqlsh:test> UPDATE emp SET emp_city='Delhi',emp_sal=50000
        WHERE emp_id=1;
```

- Delete operation. To delete the emp record from the emp table where emp_id is 2, use the DELETE statement without specifying a column name. Omitting the column name after DELETE means the entire data row will be removed. An example of this operation is shown in Listing 2.67.
- Modify operation. To modify the emp_city to "Delhi" and emp_sal to 50000 for the emp with emp_id equal to 1 in the emp table, use the UPDATE statement. The UPDATE emp command indicates that data in the emp table is being updated, while SET emp_city = "Delhi," emp_sal = 50000 specifies that emp_city should be set to "Delhi" and emp_sal to 50000. An example of this operation is shown in Listing 2.68.

Cassandra uses a query language called, which is very similar to SQL. Users who are already proficient in SQL will find it easy to get started with CQL. Additionally, Cassandra offers a wide range of data types and query functions. For detailed examples and usage, readers are encouraged to refer to Cassandra's official documentation [22], which will not be covered in this book.

2.2.4 Graph Database Processing Languages

The data model of a graph database is very simple, consisting of nodes and edges. A node represents an entity, while an edge describes the relationship between two entities. Once the graph structure is built with nodes and edges, specialized query operations designed for "graphs" can be used to search the network within the graph database. This is a significant difference between graph databases and relational databases. Although relational databases can also achieve such relationships through foreign keys, performing JOIN operations across various relationships is very time-consuming and usually inefficient. In contrast, traversing relationships in a graph database is very fast.

There are many kinds of graph databases, such as Neo4j, FlockDB, OrientDB, HugeGraph, JanusGraph, etc. This book introduces the graph database processing language with Neo4j 5.1.0 as an example.

Syntax for Neo4j Query Operations The following sections will introduce the graph database query language using the Neo4j database as an example. Neo4j uses Cypher as a query language, similar to SQL and a declarative query language. Cypher is designed to be human-readable and easy to understand, making it accessible for both beginners and experienced users.

When introducing the Cypher query syntax, we will focus solely on basic query operations, omitting complex join queries and subqueries. Figure 2.7 illustrates a basic Cypher query statement, which consists of three main parts: the MATCH clause, the WHERE clause, and the RETURN clause. The MATCH clause is used to match data in the graph, finding nodes and relationships that meet specified conditions. The WHERE clause functions similarly to the WHERE clause in SQL, filtering data by applying certain rules to the data found in the MATCH clause and retaining only the data that satisfies the conditions. The RETURN clause is similar to the SELECT clause in SQL; it is used to specify the data that should be returned as a query result. In the following sections, we will explore each component of the Cypher query statement in detail.

A basic graph is composed of nodes and relationships. To illustrate how Cypher represents nodes and relationships, we will use the data shown in Fig. 2.8 as an example.

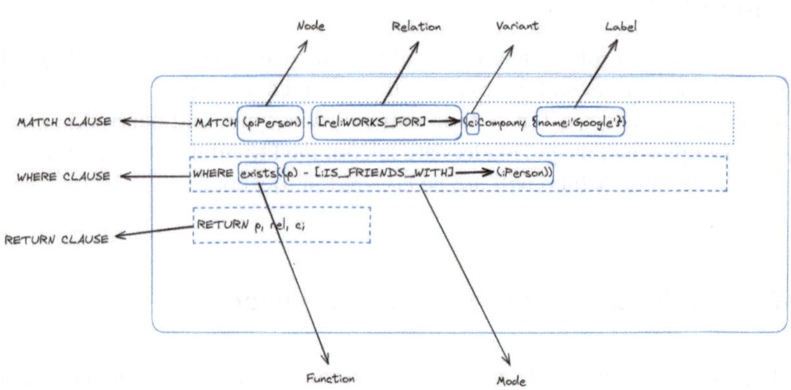

Fig. 2.7 Cypher query statement

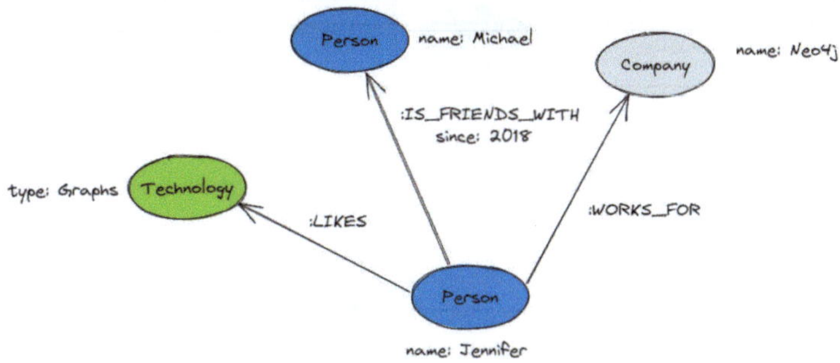

Fig. 2.8 Cypher graph data

Listing 2.69 Cypher variables and functions

```
MATCH (n)-->(b)
WHERE exists(n.name)
RETURN b
```

Variables and functions are important components in Cypher syntax. When referencing certain parts of a pattern or query, they can be named, and these names are called variables. Functions are essential tools in Cypher for querying and manipulating the graph, and there are two rules for functions. First, if a function's argument is null, the function returns null. Second, when the input to a function is a string, it is input as Unicode-encoded characters rather than char[]. As shown in Listing 2.69, *n* and *b* are variables, and "exists" is a function.

Nodes are data entities in a graph, and nodes can usually be identified by looking up nouns or objects in the data model. In our example, Jennifer, Michael, Graphs, and Neo4j are node data.

To describe a node in Cypher, we enclose the node in parentheses, for example (node). If you want to refer to the node later, we can give it a node variable, such as (*p*) for person or (*t*) for thing. Like in programming languages, you can assign names to variables in Cypher and refer to them by the same name later in the query. If the node is not relevant to your return result, you can specify an anonymous node () using empty parentheses. We can also group similar nodes together by assigning node labels so that certain types of entities can be looked up or created. In our example, Person, Technology, and Company are labels. Using labels helps Cypher differentiate between entities and optimize query execution. Whenever possible, it is recommended to use node labels in queries; otherwise, the query will need to scan all nodes in the database.

Taking the node in Fig. 2.9 as an example, nodes are represented in Cypher in the following ways, as demonstrated in Listing 2.70.

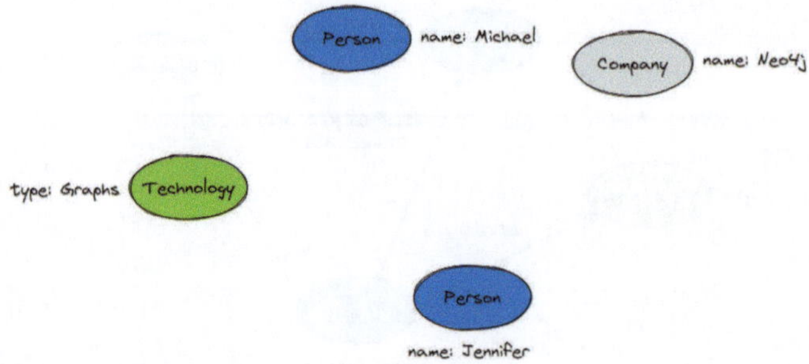

Fig. 2.9 Cypher graph node data

Listing 2.70 Cypher node

```
()                    //Anonymous nodes (no tags or variables) can
reference any node in the database
(p:Person)            //Using the variable p and the label Person
(:Technology)         //No Variables, Tagged Technology
(work:Company)        //Using the variable work and the label Company
```

Relationship refers to the data that represents connections between nodes. In our example, the connections between nodes labeled as LIKES, IS_FRIENDS_WITH, and WORKS_FOR are examples of relationships.

In Cypher, using arrows -> or <- indicates the existence of a directed relationship between two nodes, while an undirected relationship is represented without arrows, using -. Additional information about the relationship, such as the type of relationship between nodes and properties related to the relationship, can be inserted into square brackets [] in the middle of the arrow. This allows you to specify the relationship type and define any relevant properties directly within the brackets. Although the direction of the relationship must be specified when inserting it into the database, the inserted relationship can match undirected relationships; Cypher ignores any specific direction and retrieves relationships and connecting nodes regardless of the physical direction. This makes queries more flexible, as users do not need to know the physical direction of the relationships stored in the database. Like nodes, if we want to reference the relationship later in the query, we can give it a variable, such as [r] or [rel]. If the relationship does not need to be referenced, we can use <-, ->, or - to specify an anonymous relationship.

As an example, the relationship depicted in Fig. 2.10 can be represented in Cypher as follows, as demonstrated in Listing 2.71.

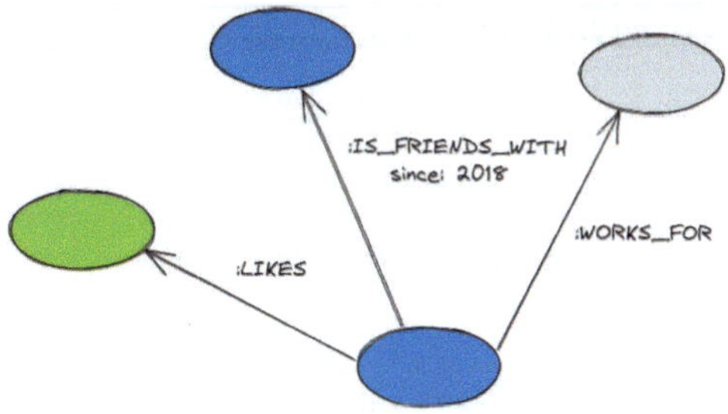

Fig. 2.10 Cypher relational data representation

Listing 2.71 Cypher relations

```
(p:Person)-[:LIKES]->(t:Technology)      //Relationship from node (p:
Person) to (t:Technology)
(p:Person)<-[:LIKES]-(t:Technology)      //Relationship from node (t:
Technology) to (p:Person)
(p:Person)-[:LIKES]-(t:Technology)       //irrelevant
(p:Person)-[like:LIKES]->(t:Technology) //Relationships using
variables
(p:Person)-->(t:Technology)              //Anonymous relationships
```

Listing 2.72 Cypher properties

```
(p:Person {name: 'Jennifer'})            //Node Properties
-[rel:IS_FRIENDS_WITH {since: 2018}]->  //Relationship Properties
```

Listing 2.73 Cypher patterns

```
(p:Person {name: "Jennifer"})-[rel:LIKES]->(g:Technology {type: "
Graphs"})
```

Attribute is the final part of the graph data model. Attributes are name-value pairs that provide more detailed information about nodes and relationships. To represent this in Cypher, we can use curly braces inside the parentheses of a node or relationship. Enter the name and value of the attribute inside the curly brackets. Our example graph has both a node attribute (name) and a relationship attribute (since), as demonstrated in Listing 2.72.

Patterns are made up of nodes and relations that can be combined to express simple or complex patterns. In Cypher, nodes and relations can be written as successive complex patterns, or complex patterns can be broken into smaller patterns and bound together with commas. The program Listing 2.73 demonstrates patterns in Cypher.

Listing 2.74 Add a node

```
CREATE (<node-name>:<label-name>
{
  <Property1-name>:<Property1-Value>
   ......
  <Propertyn-name>:<Propertyn-Value>
});
```

Listing 2.75 Add a relationship rel for node1 and node2

```
MATCH (node1:<label-name>)
MATCH (node2:<label-name>)
CREATE (node1)-[rel:<relationship-label>{
    <Property1-name>:<Property1-Value>
    ......
    <Propertyn-name>:<Propertyn-Value>
}]->(node2);
```

Listing 2.76 Delete nodes and relationships that satisfy the conditions

```
MATCH (node1:<label-name>)-[rel:<relationship-label>]->(node2)
DELETE node1, rel, node2;
```

Listing 2.77 Modify the attributes of the nodes that satisfy the conditions

```
MATCH (node1:<label-name>)-[rel:<relationship-label>]->(node2)
SET node1.<property1-name> = val1
    node1.<property2-name> = val2
    ...
    node1.<propertyn-name> = valn;
```

Syntax for Neo4j Add, Delete, and Modify Operations

The following introduces Neo4j's add, delete, and modify operations:

- Add operation. Add a node and set its appropriate attributes. An example of this operation is shown in Listing 2.74.
- Add operation. Add a relationship rel between node1 and node2 and set the corresponding attributes for rel. An example of this operation is shown in Listing 2.75.
- Delete operation. Delete nodes and relationships that satisfy specific conditions. An example of this operation is shown in Listing 2.76.
- Modify operation. Modify the attributes of a node that meets specific conditions. An example of this operation is shown in Listing 2.77.

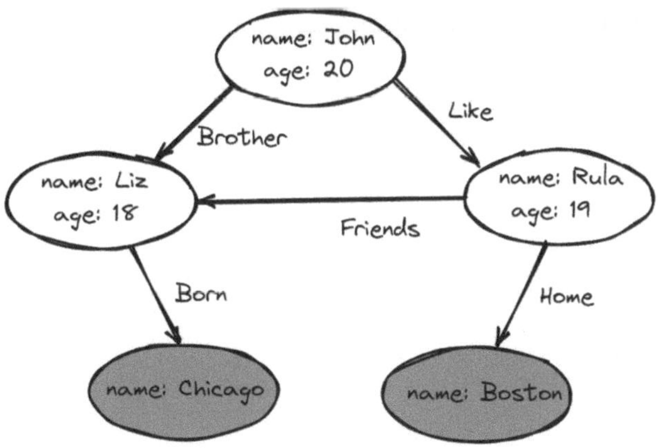

Fig. 2.11 Neo4j initial data

Listing 2.78 Query all data

```
Input:
        MATCH (entity)
        RETURN entity;
Output:
        "entity"
        {"name":"Liz","age":18}
        {"name":"Rula","age":19}
        {"name":"Chicago"}
        {"name":"Boston"}
        {"name":"John","age":20}
```

Examples of Neo4j Add, Delete, Query, and Modify Operations

The initial state of the Neo4j database is illustrated in Fig. 2.11. The database contains two types of nodes: City (e.g., Chicago), represented by gray ovals, and Person (e.g., John), represented by white ovals. The nodes are connected by relationships, depicted as lines with arrows. The following sections provide examples of Neo4j operations, including adding, deleting, querying, and modifying data.

- Query operation. To query all nodes in the database, use the "MATCH (entity)" clause, which places no restrictions on the variable entity and matches all nodes. The "RETURN entity" clause specifies that the information of each node is returned. An example of this query is shown in Listing 2.78.
- Query operation. To query Person nodes with an age greater than 18, use the "MATCH (person:Person)" clause to specify that the person variable represents nodes of the Person type. The "WHERE person.age > 18" clause adds a condition that the person node must have an age property greater than 18. An example of this query is shown in Listing 2.79.

Listing 2.79 Search for Person older than 18

```
Input:
        MATCH (person:Person)
        WHERE person.age > 18
        RETURN person;
Output:
        "person"
        {"name":"Rula","age":19}
        {"name":"John","age":20}
```

Listing 2.80 Insert Person Bob

```
Input:
        CREATE (:Person {name:"Bob",age:23});
Output:
        Added 1 label, created 1 node, set 2 properties, completed in
    less than 1 ms.
```

Listing 2.81 Insert Person Glory born in City Portland

```
Input:
        CREATE (:Person {name:"Glory",age:25}) -[:Born]-> (:City {name:"
    Portland"});
Output:
        Added 2 labels, created 2 nodes, set 3 properties,
        created 1 relationship, completed in less than 1 ms.
```

- Insert operation. To create a Person node with the attributes {name, "Bob"; age, 23}, use the CREATE keyword in Cypher. An example of this operation is shown in Listing 2.80.
- Insert operation. To insert a Person node with attributes {name, "Glory"; age, 25} and a City node with attributes {name: "Portland"}, and to specify a Born relationship from the Person node to the City node, use the CREATE keyword in Cypher. An example of this operation is shown in Listing 2.81.
- Delete operation. To delete a Person node with {name: "Bob"}, the MATCH (person:Person name: "Bob") clause means that the variable person is a node that satisfies the type of Person with the attribute {name: "Bob"}, and the DELETE person clause means to delete person's information, as shown in the programme Listing 2.82.
- Delete operation. To delete a Person node with {name: "Glory"}, the associated City node, and any relationship between them, use the MATCH and DELETE clauses in Cypher. The "MATCH (person:Person {name: "Glory"})" clause specifies that the variable person represents a Person node with the attribute {name: "Glory"}. The " (city:City)" clause defines city as a City node. The "-[relation]->" clause indicates that there is a relationship named relation from the person node to the city node. The "DELETE person, relation, city" statement

Listing 2.82 Delete Person Bob

```
Input:
        MATCH (person:Person {name:"Bob"})
        DELETE person;
Output:
        Deleted 1 node, completed after 1 ms.
```

Listing 2.83 Delete Person Glory and the City and relation associated with it

```
Input:
        MATCH (person:Person {name:"Glory"})-[relation]->(city:City)
        DELETE person, relation, city;
Output:
        Deleted 2 nodes, deleted 1 relationship, completed after 2 ms.
```

Listing 2.84 Modify Person John

```
Input:
        MATCH (person:Person {name:"John"})
        SET person.phone = "987489";
Output:
        Set 1 property, completed after 1 ms.
```

will delete the person node, the relation, and the city node. An example of this operation is shown in Listing 2.83.

- Modify operation. To modify the phone attribute of the Person node with {name:"John"} to "987489," use the SET clause in Cypher. The statement "SET person.phone = '987489'" assigns the new phone number "987489" to the phone attribute of the person node. An example of this operation is shown in Listing 2.84.

Graph database is very different from the previous relational database, Neo4j provides some of the operations on the database, and the reader needs to practice using more in order to understand more deeply. The above is only a list of common Neo4j add, delete, check, and change operations; for readers who want to learn more complex operations, please refer to the official Neo4j documentation [96]; this book will not be introduced in detail.

2.3 Chapter Summary

This chapter provides an overview of the syntax and usage examples of SQL and NoSQL. In Sect. 2.1, we introduced SQL, which is used for operating relational databases. SQL is divided into five categories, Data Definition Language, Data Query Language, Data Manipulation Language, Data Control Language, and

Transaction Control Language, each serving different database operations. We primarily focus on Data Query Language, the most important part of SQL. Using a student and course enrollment database, we demonstrate the functions of different parts of SQL query statements step by step, providing examples of how to use SQL for comprehensive database query operations. We believe that readers will find this very beneficial. After introducing basic query statements, we explored table join operations in SQL, which include three types: self-join, natural join, and outer join. These join operations address the issue of multi-table relational queries. Finally, we introduced other SQL statements, including those for operating on views, transaction-related statements, and authorization-related statements. For these different statements, it is necessary to analyze our requirements reasonably and use the corresponding statements to achieve our functions.

In Sect. 2.2, we introduced the NoSQL language, which is mainly used for operating non-relational databases. Since there are many types of non-relational databases, each with significant differences, for convenience, we divided non-relational databases into four categories: key-value databases, document databases, column-family databases, and graph databases. For each of these four types, we chose a database management system to provide examples and demonstrate how to use their respective NoSQL languages to operate the databases. When introducing the key-value database language, we used Redis to demonstrate the basic operations of key-value databases, primarily covering the add, delete, query, and update operations for string-type data. When introducing the document database language, we chose MongoDB to demonstrate the basic operations of document databases, highlighting that document database query languages are more complex than key-value databases and allow different filtering conditions to obtain the desired data. Then, we introduced the column-family database language, using Cassandra to demonstrate the basic operations of column-family databases. The CQL used by column-family databases is quite similar to SQL, with only a few differences to remember. Finally, we introduced the graph database language, using Neo4j to demonstrate the basic operations of graph databases. Unlike all other databases, graph databases use the Cypher language, which is close to natural language, and understanding this language is easier when combined with actual graph data.

Lastly, for all the language examples in this chapter, we recommend that readers with the necessary conditions perform these operations on real database systems. Doing so will help you gain a deeper understanding of the language details, and successfully executing these examples will provide a sense of accomplishment, which is essential for learning distributed languages.

Chapter 3
Distributed Query Processing

Distributed query refers to the process of retrieving large datasets in a distributed environment. Compared to centralized storage structures, the relational tables in distributed queries are fragmented or replicated and stored on different physical nodes. As shown in Fig. 3.1, during the implementation of a distributed query, the system needs to extract relevant data from various physical storage nodes and transmit the data over the network to the computation nodes for join processing, thus introducing additional network communication overhead and increasing the query processing response time [100].

During the query processing, the nodes involved in query processing are divided into two roles. The compute nodes are responsible for **data reception** and **query computation**, whereas the data storage nodes primarily handle **local data extraction** and **data filtering**. A simple distributed query process can be roughly divided into the following steps [132]:

- The compute nodes send data requests to the relevant data storage nodes based on the query statements entered by the user.
- The data storage nodes perform local data extraction and filtering in response to the requests. Afterward, the filtered data is transmitted to the compute nodes via the network.
- After obtaining the table data from the storage nodes, the compute nodes perform query computation uniformly on the data based on the semantic analysis of the input query statements.

According to the above process, the time overhead T of the distributed query can be expressed by Eq. 3.1:

$$T = T_{\textbf{data extraction and filtering}} + T_{\textbf{data network transmission}} + T_{\textbf{data reception}} + T_{\textbf{query computation}} \tag{3.1}$$

The time overhead for data reception by compute nodes and data extraction by data storage nodes is primarily influenced by the respective nodes' speed of

Z. Xu, *Distributed Data Services*, https://doi.org/10.1007/978-981-96-9392-4_3

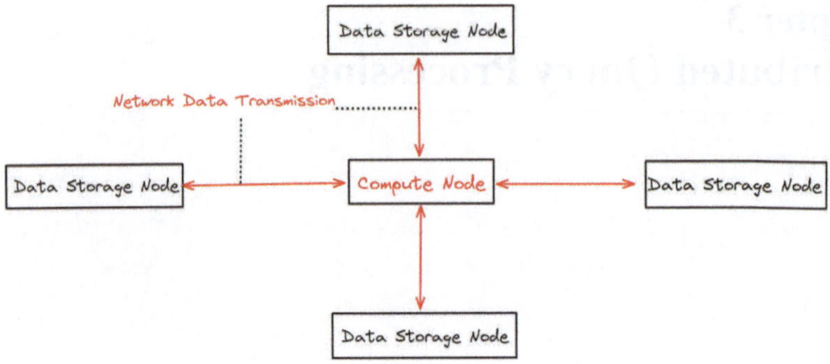

Fig. 3.1 Distributed join query

reading table data. Therefore, the reasons affecting the efficiency of distributed queries can be analyzed from the following three aspects: local data reading, data network transmission, and query computation. The following focuses on analyzing the specific factors affecting efficiency in these three areas and their corresponding optimization directions:

- **Local data readout**. Local data reading primarily consists of two steps: extracting and filtering the data requested by the compute nodes from the data storage nodes. Data extraction involves loading the data from local disks into memory and then conditionally filtering the extracted table data according to the compute nodes' requests. Finally, the filtered data needs to be serialized and packaged for network transmission to the compute nodes. The process of reading and filtering local data on storage nodes can be achieved through two data reading methods: index-based selection and range scanning. Index-based selection filters data using primary keys and indexes, allowing for relatively fast retrieval of the requested data. In contrast, range scanning depends on the size of the data and accelerates data reading by increasing parallelism.
- **Network transmission of data**. In distributed queries, network transmission overhead mainly depends on the size of the data being transmitted. Particularly when the tables involved in the queries contain large amounts of data, filtering data to reduce the volume of invalid data can enhance the efficiency of distributed queries. In a distributed storage environment, mainstream technologies include semi-join operations and Bloom filters. By filtering the data of the query relations, these technologies eliminate data that cannot be joined or matched, thereby reducing data transmission volume.
- **Query calculation**. After all data has been transmitted to the compute nodes, the compute nodes need to perform semantic analysis based on the input query statements and conduct corresponding relational algebra operations on the data. The most common query operation is relational table join queries, which involve joining multiple relational tables through join columns to obtain the required information. Depending on different scenario requirements and whether the

relational tables have index structures, the compute nodes will choose appropriate join algorithms to improve query processing efficiency.

The above roughly analyzes the steps required in the distributed query process and the various factors affecting query performance. Compared to centralized storage, the most significant characteristic of distributed storage is that relational table data is stored on different physical nodes according to certain rules, thereby introducing additional network transmission overhead during distributed queries involving cross-node operations. How to generate an appropriate query execution strategy based on the input query statements and the distribution of each storage node will greatly impact the efficiency of distributed query processing. The following chapters will further elaborate on the relevant algorithm strategies implemented in the distributed query process, focusing on data network transmission and query join processing, and analyze the corresponding performance overhead.

3.1 Distributed Join Problems

As previously mentioned, in a distributed environment, relational tables may undergo partitioning and replication, and be stored on different physical nodes. In distributed join queries involving multiple relational tables, it is necessary to access different network nodes and transfer the corresponding data to computation nodes for join processing. Due to large data volumes and network fluctuations, network transmission overhead contributes significantly to the communication costs in distributed query processing. To address this, the query processor can leverage the characteristics of distributed storage during query processing. It semantically analyzes the input query statements, splitting them into multiple relational algebra operations. Then, it selects the appropriate order of relational operations. Based on the analysis of data transmission methods and the required network transmission overhead, the processor chooses the optimal nodes for executing the operations. This generates an optimized distributed query execution plan. To further illustrate the importance of the order of relational algebra operations and execution nodes in distributed query processing, we will demonstrate and analyze the costs of processing the query described in Eq. 3.2 using different strategies:

$$\Pi_{SNAME,CNAME,SCORE}(STUDENT \bowtie_{SID} (\sigma_{CID=\text{“2”}}(CSCORE))) \tag{3.2}$$

The join query involves the relational tables STUDENT and CSCORE, which are joined based on the column SID. To facilitate analysis, we assume that the STUDENT and CSCORE tables contain 2000 and 1000 tuple records, respectively, with 100 records in the CSCORE table having a CID attribute value of 2. Both relational tables are horizontally partitioned and stored based on the size of SID in distributed storage. The partitioning results and storage location arrangements are as follows:

$$STUDENT_1 = \sigma_{SID \leq \text{“1000”}}(STUDENT) \quad \textbf{Stored at node 1}$$

$$STUDENT_2 = \sigma_{SID \geq \text{“1000”}}(STUDENT) \quad \textbf{Stored at node 2}$$

$$CSCORE_1 = \sigma_{SID \leq \text{“1000”}}(CSCORE) \quad \textbf{Stored at node 3}$$

$$CSCORE_2 = \sigma_{SID \geq \text{“1000”}}(CSCORE) \quad \textbf{Stored at node 4}$$

Based on the sequence of different relational algebra operations and the choice of execution node locations, we can accomplish the distributed query of Eq. 3.2 using the following two equivalent strategies. For convenience in our explanation, we will ignore the projection operator analysis in the query strategy analysis below. As shown in Fig. 3.2, query strategy A simply transfers the data stored in various physical nodes to the computation node for data localization. Afterward, the computation node uniformly performs the join query computation on the data. On the other hand, as shown in Fig. 3.3, query strategy B leverages the characteristic of the relational tables STUDENT and CSCORE being partitioned and stored in the same manner, dividing the query operations into multiple relational algebra operations for processing, and executes projection, selection, and join computations in parallel. The results are then transmitted to the computation node for assembly.

We analyze the query costs for strategies A and B, respectively. First, we establish a simple cost measurement method, assuming that the cost of reading each tuple is $cost_\alpha$, which is 1ms. The network transmission cost for one tuple record is set to $cost_\beta$, which is 10ms. Finally, we assume that the relational tables STUDENT and CSCORE have indexes established on the storage nodes based on SID and CID, respectively. This allows for efficient access to the corresponding tuple data based on the values of the attributes SID and CID.

Based on the above cost assumptions and the query processing steps of strategy A shown in Fig. 3.2, we can calculate the costs required for strategy A step by step:

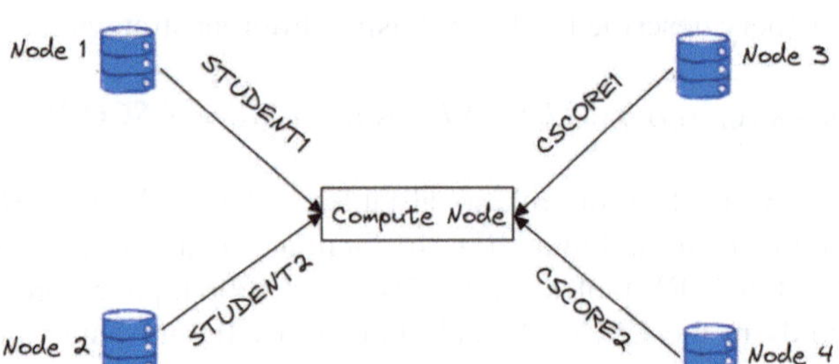

Fig. 3.2 Query strategy A

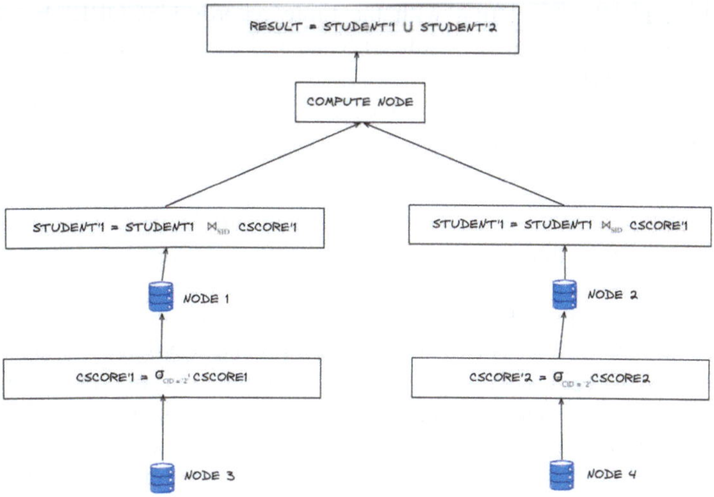

Fig. 3.3 Query strategy B

1. Transmitting the tuple records of the STUDENT relation from nodes 1 and 2 to the computation node requires $2000 \times cost_\beta = 20000$ ms.
2. Transmitting the tuple records of the CSCORE relation from nodes 3 and 4 to the computation node requires $1000 \times cost_\beta = 10000$ ms.
3. Since there is no index on the CID attribute of the CSCORE relation in the computation node, it is necessary to scan all records in the CSCORE relation to find tuples matching the condition CID = "2," which requires $1000 \times cost_\alpha = 1000$ ms.
4. Similarly, since there is no index on the SID attribute of the STUDENT relation in the computation node, it is necessary to loop through the STUDENT relation for each filtered tuple from the CSCORE relation to complete the join computation, which costs $2000 \times 100 \times cost_\alpha = 200000$ ms.
5. The total cost is 231000 ms.

On the other hand, based on the characteristics of distributed storage, query strategy B decomposes the query computation into multiple relational algebra operations. The costs required for strategy B can be calculated step by step as follows:

1. Perform the selection relational algebra on the CSCORE relation to filter tuples with the attribute CID = 2, denoted as CSCORE', which requires $(50 + 50) \times cost_\alpha = 100$ ms.
2. Transmit the filtered tuple records CSCORE' to nodes 1 and 2 that store the STUDENT relation, with a network transmission cost of $(50 + 50) \times cost_\beta = 1000$ ms.

3. Join the STUDENT relation with the tuple records CSCORE' based on the attribute SID to generate the result set STUDENT', which requires $(50 + 50) \times cost_{\alpha} \times 2 = 100\,\text{ms}$.
4. Return the generated result set STUDENT' to the computation node, requiring $(50 + 50) \times cost_{\beta} = 1000\,\text{ms}$.
5. The total cost is 2200 ms.

From the above cost analysis, it is evident that the implementation cost of strategy B is much lower than that of strategy A. The greatest discrepancy between the two stems from the costs associated with performing join computations and network transmission. In strategy B, the join computation between the tuple record set CSCORE' and the STUDENT relation utilizes the index based on SID established on nodes 1 and 2 to quickly complete the join computation. In contrast, in strategy A, all data is unified and transmitted to the computation node for processing. During network transmission, we assume the worst-case scenario where the index access methods for STUDENT and CSCORE based on attributes SID and CID are lost, causing the join computation in step four of strategy A to use a nested-loop algorithm for the join computation [100]. For 100 tuples of CSCORE that satisfy the join conditions, each requires looping through the STUDENT table to find matching tuples for the join computation. In terms of network transmission costs, strategy A directly transmits all data from the involved relations to the computation node, including a large amount of tuple data that cannot participate in the join computation, significantly increasing the data transmission volume. On the other hand, strategy B divides the query computation into multiple relational operations that are processed on the respective storage nodes. Layered filtering and computation reduce the data volume for network transmission, thereby optimizing network transmission costs.

Compared to strategy A, strategy B provides better workload distribution among nodes, thereby reducing the data volume for network transmission and the cost required for join processing, optimizing the cost by nearly 47 times compared to strategy A. If the network transmission speed during the query process is slower or the number of relation slices is greater, the cost difference between the two strategies will be even larger. This demonstrates that in distributed query processing, choosing the appropriate nodes to handle computations and the execution order of relational operators has a significant impact on the processing costs of distributed queries [65].

In the subsequent chapters, the focus will be on introducing several commonly used algorithms for implementing distributed queries. By analyzing the implementation costs of various algorithms, the optimization directions and strategies for distributed query processing will be further elaborated.

3.1.1 *Direct Join Algorithm*

Direct join algorithm is a simple implementation method for distributed queries. To emphasize the performance cost analysis of different algorithm strategies for implementing distributed queries, we assume that after localizing the data involved in the query, the compute nodes uniformly adopt the nested-loop algorithm for join processing. The direct join algorithm selects the side involving relatively fewer tuples in the joined tables as the join compute node. The compute node imports all data in the locally stored relation table into memory in the form of blocks. Thereafter, each data block is traversed, during which the data of the other join table stored remotely is transferred to the compute node by page through the network, and matched with each tuple in the blocks [1]. To facilitate the analysis of the algorithm execution costs, we assume the memory loading cost is $cost_\alpha$, the network transmission cost is $cost_\beta$, and the message passing cost during the network connection is ignored. Therefore, the cost of implementing join queries using the direct join algorithm can be divided into:

1. **Local Table 1 memory loading overhead**: $num\_block_1 * cost_\alpha$.
2. **Offsite Table 2 network delivery and memory loading overheads**: $num\_blocks_1 * num\_pages_2 * (cost_\alpha + cost_\beta)$.

It is evident that the cost of implementing distributed join queries with this algorithm is enormous. To perform join matches between two tables stored on different nodes, it is necessary to repeatedly transfer the data of relational tables stored at different locations to the local nodes during each traversal, leading to excessive network transmission and memory loading overhead. To address the issue of repeated data transmission, the direct join algorithm can be optimized. As shown in Fig. 3.4, node 1 sends all tuple data of the relational table R that needs to be joined to node 2. Subsequently, the join match processing is executed on this node. The memory modules of Table S are iteratively traversed and matched with the data of Table R, thereby reducing the network overhead of repeated data transmission. The cost of the optimized direct join algorithm can be divided into:

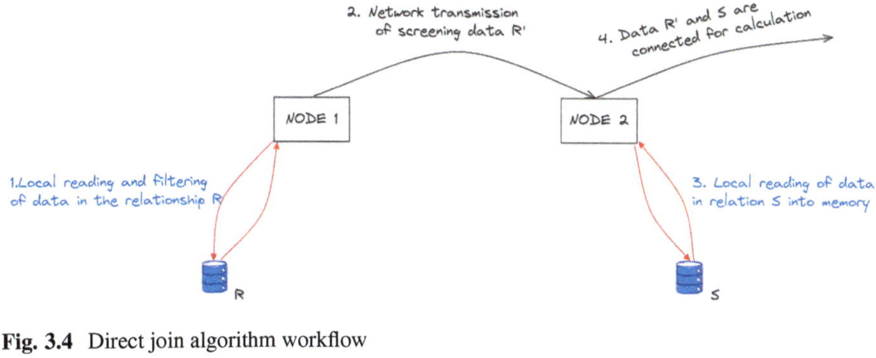

Fig. 3.4 Direct join algorithm workflow

1. **Local table R memory loading overhead**: $num\_block_1 * cost_\alpha$.
2. **Offsite Table S Network Delivery Overhead**: $num\_pages_2 * cost_\alpha$.
3. **Table S loaded into memory matching overhead**: $num\_blocks_1 * num\_pages_2 * cost_\alpha$.

Compared to the original version, the optimized direct join algorithm significantly reduces the network transmission overhead of the data in relational table R, avoiding the inefficiencies of repeatedly transmitting data to another node during the join process.

3.1.2 Semi-Join Algorithm

As introduced by the direct join algorithm in Sect. 3.1.1, the greatest challenge in implementing distributed join queries is the need to transfer relational table data between different storage nodes, thereby causing additional network transmission overhead. To reduce this overhead, the distributed query processor needs to minimize the amount of data transferred between different nodes as much as possible. For this purpose, researchers have proposed using the **semi-join algorithm** as a data compressor to filter the tuples that need to be transmitted, eliminating tuples that cannot be matched and thus reducing the volume of data transmitted over the network.

First, let's analyze from the perspective of relational algebra. Suppose relational tables R and S, stored respectively at nodes 1 and 2, are to be joined based on attribute A. The join operation on these two relational tables can be processed using the semi-join of one of the relational tables according to Eq. 3.3:

$$\begin{aligned} R \bowtie_A S &\Longleftrightarrow (R \ltimes_A S) \bowtie_A S \\ &\Longleftrightarrow R \bowtie_A (S \ltimes_A R) \\ &\Longleftrightarrow (R \ltimes_A S) \bowtie_A (S \ltimes_A R) \end{aligned} \tag{3.3}$$

The implementations of the three equivalent semi-join algorithms listed above can specifically choose the strategy with the smallest performance overhead for query execution based on the number of tuples in the relational tables to be processed and the location of data storage nodes. Since the idea of the semi-join algorithm is to filter out tuple data that cannot be joined by screening, thereby reducing network transmission latency, the semi-join algorithm can fully play its role and improve query processing efficiency only when the cost of sending the entire filtered table is greater than the overhead of performing the semi-join calculation and transmitting the filtered result. To further analyze the advantages of the semi-join algorithm, we compare the overhead of implementing distributed join queries using the $R \bowtie_A S$ and $(R \ltimes_A S) \bowtie_A S$ strategies, assuming that the number of tuples in the relational table R $size(R)$ is greater than the number of tuples in the relational table S $size(S)$.

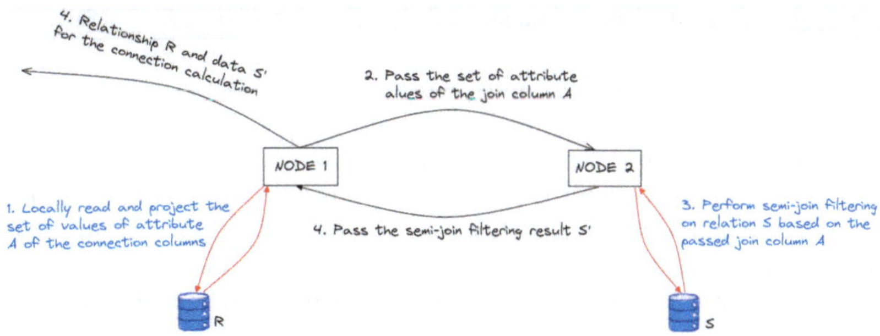

Fig. 3.5 Semi-join algorithm workflow

The $R \bowtie_A S$ strategy uses the direct join algorithm to localize the data processing to a single node, after which the join computation is performed centrally. The steps can be divided into:

1. Since the number of tuples in the relational table R is larger, in order to reduce network transmission overhead, we transfer the data of the relational table S to node 1.
2. Node 1 computes $R \bowtie S$ based on the relational tables R and S.

On the other hand, the implementation of a distributed join query based on the semi-join algorithm is shown in Fig. 3.5, where the steps can be specifically divided into:

1. Project the set of distinct attribute values Π_A on the join column A of the relational table R, and pass the set of attribute values to node 2.
2. Node 1 performs a semi-join computation $R' = R \ltimes_A S$ on relational table S based on the transmitted set of attribute values A to filter tuple records.
3. Send the subset of tuples from relational table S that meet the join conditions in the semi-join result, S', to node 1.
4. Perform the join computation $R \bowtie_A S'$ with the filtered S' and relational table R.

For simplicity of comparison, we assume that the network transmission overhead for data transfer is sufficiently large, thereby ignoring the overhead of establishing communication between the two nodes T_{MSG} and the overhead of node 2 performing the semi-join filtering T_{SEMI} during the network transmission. Therefore, we can intuitively compare the network transmission overhead based on the size of the data that needs to be transferred under the two strategies. The main overhead of the direct join algorithm is transferring the data from relation table S from node 2 to node 1, with the data size being $size(R)$. In contrast, the network transmission overhead for the semi-join-based strategy mainly comes from transferring the projection result $\Pi_A(R)$ to node 2 and transferring the semi-join result $R \ltimes_A S$ to node 1. Thus, this strategy requires transferring a data size of $size(\Pi_A(S)) + size(R \ltimes_A S)$:

$$size(\Pi_A(S)) + size(R \ltimes_A S) < size(R) \tag{3.4}$$

In summary, when Eq. 3.4 is satisfied, the semi-join algorithm effectively compresses data by filtering out unnecessary tuples, thus reducing network transmission delays in distributed join queries. However, if most of the tuples from relation R still participate in the join after semi-join processing, then the condition described by Eq. 3.4 does not hold, and a direct join would be more efficient. This is due to the additional overhead introduced by the semi-join, including extra data processing and network communication. The distributed query processor needs to combine these two algorithms based on the specific requirements of the actual problem to achieve efficient distributed query processing.

3.1.3 Bloom Filter Join Algorithm

In Sect. 3.1.2, the network transmission overhead in the implementation of distributed join queries using the semi-join algorithm mainly arises from transmitting the projection value set R' of relation table R on join column A to node 2 and transmitting the semi-join filtering result from node 2 back to node 1. During this process, the number of projection values generated from relation table R, $size(\Pi_A(R))$, can be very large, leading to excessive network transmission overhead. To address this issue, we can introduce a Bloom filter to replace the transmission of the projection value set, allowing node 2 to filter relation table S, thereby reducing the network transmission overhead.

A Bloom filter is a probabilistic data structure that can quickly determine whether an element is a member of a given set. The filter consists of an array of m bits and k independent hash functions. Initially, all bits in the array are set to 0. When an element is inserted, it is processed by the k hash functions, and the k hash values obtained are used to set the corresponding bit positions in the array to 1. Unlike standard hash tables, the advantage of a Bloom filter is its space efficiency, allowing it to represent a set with an arbitrary number of elements using a fixed size [76]. However, the drawback is that the filter's judgment results are probabilistic, meaning there is a chance of false positives. A false positive result occurs when the filter indicates that an element is in the set, but it is not actually present. The probability of false positives increases as more elements are inserted into the filter, especially when all bits in the array are set to 1, causing any query to return a positive result regardless of the actual presence of the queried element. Therefore, there is a trade-

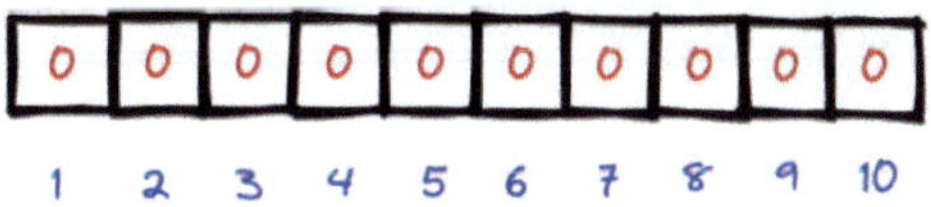

Fig. 3.6 Empty Bloom filter

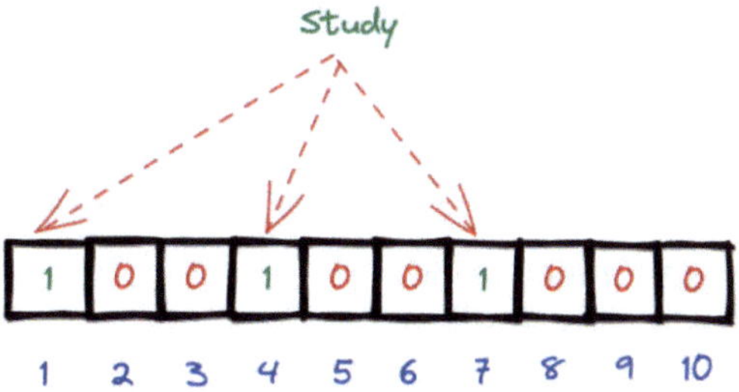

Fig. 3.7 Insert the element "Study"

off between the bit size m and the false positive probability p. Assuming n elements are inserted, the size of the false positive rate can be expressed by Eq. 3.5 [14]:

$$p = \left(1 - \left(1 - \frac{1}{m}\right)^{kn}\right)^{k} \approx \left(1 - e^{\frac{kn}{m}}\right)^{k} \tag{3.5}$$

For the Bloom filter, we will use an example to illustrate how to store and determine whether an element is in the set. First, as shown in Fig. 3.6, initialize an array of length 10 and set all bits to 0.

Next, we need to define k hash functions $h_1(x), h_2(x), \ldots, h_k(x)$ to hash the input element. The results of these hash functions will determine which bit positions in the array are set to 1. In this example, we use three hash functions to compute and map to the respective positions in the array. Assuming the input element is the string "Study," the hash calculations are as follows:

$$h_1(\text{"}Study\text{"})\%10 = 1$$

$$h_2(\text{"}Study\text{"})\%10 = 4$$

$$h_3(\text{"}Study\text{"})\%10 = 7$$

As shown in Fig. 3.7, based on the hash calculation results of the inserted data above, we need to set the number of bits in the array index in positions 1, 4, and 7 to 1.

Next, follow the same steps to insert the string "System," with the hash calculations as follows:

$$h_1(\text{"}System\text{"})\%10 = 2$$

$$h_2(\text{"}System\text{"})\%10 = 6$$

$$h_3(\text{"}System\text{"})\%10 = 9$$

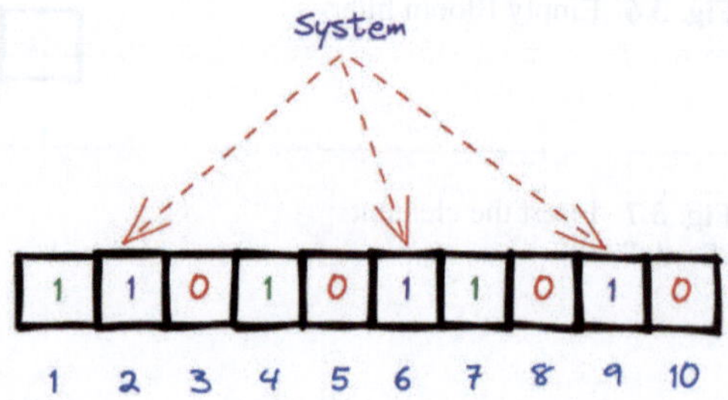

Fig. 3.8 Insert the element "System"

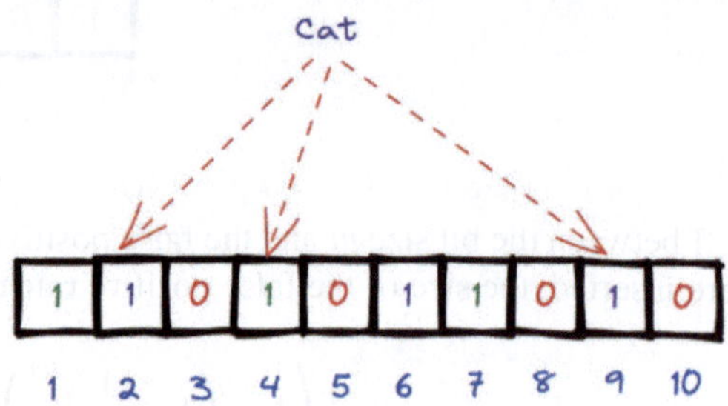

Fig. 3.9 False positive results misreported

As shown in Fig. 3.8, again we need to set the array index value of 3, 5, 4 position of the number of bits to 1.

After the insertion operation is completed, to check if the element "Study" is in the set, we need to calculate the hash values of the input element using the defined hash functions. Then, we check the corresponding positions in the Bloom filter array. If all these positions are set to 1, the filter indicates that the element might be in the set. Conversely, if any of these positions are not set to 1, the filter indicates that the element is not in the set.

Regarding the **cases of false positive results**, suppose we need to query the element string "Cat." After processing the input element with the hash functions, the resulting indices are 2, 4, and 9. As shown in Fig. 3.9, the values at these three index positions are all 1, so the filter will indicate that the element might be in the set. However, we have never inserted the element "Cat," thus resulting in a false positive.

To reduce the false positive rate, we can adjust the size of the Bloom filter, increasing the bit space to decrease the probability of false positives. However, using a larger bit array and more hash functions will also reduce space and time efficiency. Therefore, to achieve an optimal balance, we need to evaluate according to the formula mentioned above (Eq. 3.5) and the number of elements to be represented in the set, and then choose an appropriate combination of filter bit array size and the number of hash functions.

The **Bloom join algorithm** optimizes the semi-join algorithm by using the Bloom filter introduced above. First, it scans the join column A of relation R and inserts the distinct attribute values into the Bloom filter. Subsequently, node 1 only

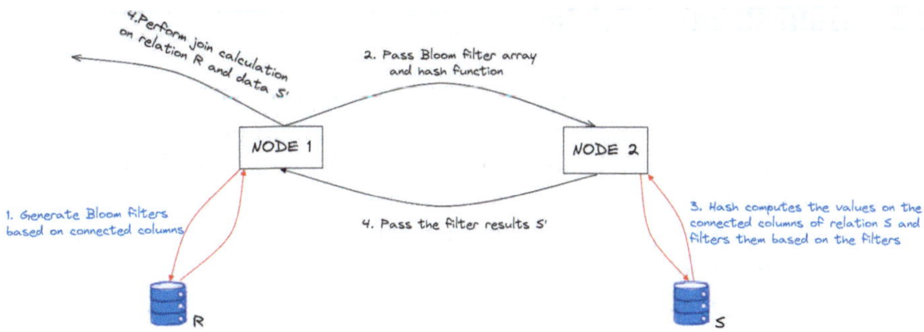

Fig. 3.10 Bloom join algorithm workflow diagram

needs to transmit the generated Bloom filter and the corresponding hash functions to node 2 for semi-join processing [66]. By replacing the actual projection of the join column A with a bit array of a certain size and hash functions, the data size transmitted over the network can be significantly reduced. For the specific implementation process of this algorithm, we assume that there are two physical nodes storing relation R and relation S, respectively, and that $size(R) > size(S)$. As shown in Fig. 3.10, the entire implementation process can be divided into the following steps:

1. Generate a Bloom filter based on relation table R by scanning each value in the join column A of relation table R and performing hash function calculations to fill the bit array of the filter appropriately.
2. Transmit the generated Bloom filter bit array and hash functions to node 2 over the network.
3. Scan the join column A of relation table S, perform hash calculations, and use the Bloom filter to check for join matches, storing the matching tuples in the data stream S'.
4. Transmit the filtered data stream S' back to node 1 over the network and perform the join operation with relation table R.

In summary, the essence of the semi-join and the Bloom join algorithms is to reduce the amount of data transmitted over the network during distributed join queries, thereby optimizing query response time. The semi-join algorithm focuses on filtering out tuples that cannot be join-matched, thus reducing the amount of data that needs to be transmitted to another node for join processing. On the other hand, the Bloom join algorithm uses a Bloom filter to replace the transmission of the projected key set, allowing the other node to match and filter the relation table, thereby reducing the key data size for filtering.

3.2 Multi-Relational Joins

In distributed relational databases, a query usually involves more than just one or two tables, and the distributed storage of data incurs transmission costs between them. Therefore, efficiently joining multiple tables located at different sites with minimal transmission cost to perform high-efficiency queries becomes a key aspect of distributed database query technology, known as the multi-join query problem. In simple terms, a multi-join query is a relational algebra expression composed of join operations of N base relations obtained after optimization under projection and selection.

Multi-relation joins can be divided into two types:

1. Multi-relation join on the same attribute $R_1 \infty A R_2 \infty A R_3$.
2. Multi-relation join on different attributes $R_1 \infty A R_2 \infty B R_3$.

For multi-relation join operations, there are generally two strategies: one is based on direct join strategy, and the other is based on semi-join strategy.

3.2.1 *Goals of Distributed Query Optimization*

Query optimization is an important task in distributed query processing. Since there exist multiple correct transformations for the same query operation, only the strategies that consume fewer resources are more likely to be retained. The issue lies in how to measure resource consumption. There are two common metrics for resource consumption.

Total cost indicates the sum of all time factors during the query processing. This total cost is the sum of the time spent on processing query operations at various sites and the time for transmission communication between sites, which is

$$Total\_time = T(CPU) * insts + T(I/O) * I/Os + T(MSG) * msgs + T(TR) * bytes \tag{3.6}$$

In Eq. 3.6, *insts* is the total number of instructions accessing memory, I/Os is the total number of instructions accessing the disk, *msgs* is the total number of messages, and *bytes* is the total number of bytes of all transferred data. CPU overhead comes from operations on memory data; I/O overhead comes from operations on accessing the disk; and communication overhead comes from the latter two parts of Eq. 3.6: T(MSG) is the time from sending a message to completing its reception, and T(TR) is the time it takes for a unit of data to be transmitted from one site to another. Here, the unit of data is measured in bytes, but other units (such as packets) can also be used. T(TR) is affected by the network and is generally assumed to be constant, but in a wide area network, due to significant variations in distances between sites, the value may fluctuate.

Response time indicates the time elapsed to complete a query. Considering that data query operations in distributed database systems can be executed in parallel at different sites, the query response time might be less than the total cost. A general formula for response time is

$$Response\_time = T(CPU) * seq\_insts + T(I/O) * seq\_I/Os + T(MSG) * seq\_msgs + T(TR) * seq\_bytes \quad (3.7)$$

In Eq. 3.7, seq_x indicates the maximum number of queries required to complete serially. x can be instructions (insts), messages (msgs), or bytes.

From the above formula, it can be seen that, overall, the memory access operations, I/O operations, and inter-site communications that affect the total cost also influence the response time to some extent. However, there is not a strictly monotonic relationship where both increase or decrease together. Increasing the degree of parallelism can minimize the response time, but it may lead to an increase in the total time. This is because increased parallelism generates more parallel local processing time and transmission.

To distinguish between total cost and response time, consider the following example: a query request is initiated by node 3, and the required query data comes from node 1 and node 2, as shown in Fig. 3.11. Thus, node 1 transmits x units of data to node 3, and node 2 transmits y units of data to Node 3. In this example, for simplicity, CPU cost and I/O cost are not considered.

The Total_time of the query is

$$Total\_time = 2T(MSG) + T(TR) * (x + y)$$

Since transfers can be done in parallel, the response time for queries *Response_time*:

$$Response\_time = max\{T(MSG) + T(TR) * x, \quad T(MSG) + T(TR) * y\}$$

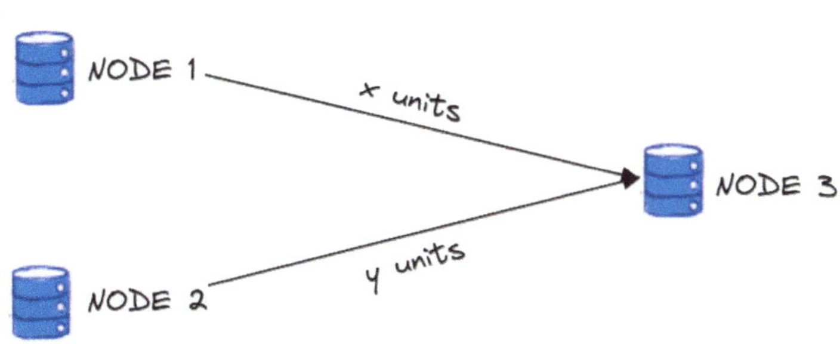

Fig. 3.11 Data transmission diagram

From the above, it can be seen that irrespective of the metric used to measure the query cost in a distributed system, CPU, I/O, and communication costs are always involved. However, the contribution of these three types of costs to the total query cost varies depending on the environment. In early distributed database systems, communication between sites was slow, and local query times were far less than communication times; hence, local overhead, namely, CPU and I/O costs, were often ignored. With the advancement of communication technology, we now have much faster communication networks, and communication overhead has started to approximate local overhead. Therefore, it is necessary to adjust the contribution proportions of these three cost components to the total cost to measure the system's query cost more accurately.

In summary, minimizing the total cost means optimizing the utilization of system resources, while minimizing response time means pursuing higher parallelism. In practical applications, we often aim to enhance both system resource utilization and parallelism simultaneously. Therefore, by using weighting to combine total cost and response time, we can achieve a balanced optimization strategy.

3.2.2 Basic Methods of Distributed Query Optimization

In a distributed database system, query optimization includes two aspects: local processing optimization and global strategy optimization.

Local optimization refers to optimizing query transformations, that is, executing relational operations in a more efficient sequence. Local queries in distributed database systems are similar to local queries in centralized databases. Therefore, we can use optimization strategies from centralized databases to optimize the local processing in distributed databases. In high-speed local area networks (LANs), the data transmission time in distributed databases is much shorter than the local processing time, making local processing optimization particularly critical.

Global strategy optimization refers to optimizing query mapping, using a series of efficient algorithms to access various devices and perform relational operations. This optimization involves characteristics like fragments, cardinality, and connection order.

3.2.3 Localized Processing Optimization

In most optimization mechanisms, a more traditional approach is query optimization based on relational algebra equivalence transformations. We know that relational query processors typically transform relational calculus expressions (high-level query statements, such as SQL) into equivalent relational algebra expressions. A relational calculus expression can have many equivalent and correct relational algebra transformations, each of which incurs a different amount of system overhead, sometimes significantly varying. Thus, we can consider selecting the equivalence transformation in relational algebra that minimizes resource consumption.

The issue lies in how we can minimize system overhead while ensuring relational algebra equivalency. The usual method is first to construct a relational expression tree and then optimize it according to the rules of relational algebra equivalence transformations. This is akin to query optimization in centralized databases. However, the difference with distributed databases is that they can utilize data partitioning mechanisms, enabling queries to operate on specific partitions rather than transmitting irrelevant data. This section first briefly introduces the basic methods of partitioning, then lists some commonly used relational algebra equivalence rules, and demonstrates how to combine both for query optimization through several examples.

There are three basic approaches to **data partitioning**, which are implemented through the basic operations of relational algebra:

1. Horizontal Partitioning: Execute a selection operation on the global relation, grouping tuples with the same properties into several mutually exclusive subsets.
2. Vertical Partitioning: Execute a projection operation on the global relation to obtain several subsets. Each attribute of the global relation maps to at least one vertical partition, and each vertical partition contains the key of the global relation. The global relation can be reconstructed by performing a join operation on these partitions.
3. Hybrid Partitioning: Alternately use horizontal and vertical partitioning.

There are four common rules for performing transformations on relational algebra:

Cautious Cartesian Product Operation In query processing, the most time-consuming and space-consuming operations are the Cartesian product and join operations, which should be performed as late as possible. Avoid performing Cartesian products directly; instead, merge a series of selection and projection operations before and after the Cartesian product.

Early Projection Early projection means performing the projection operation as early as possible. It can filter out unnecessary column data, reducing the transmission of intermediate information.

Condition Pushdown Condition pushdown means equivalently placing selection conditions into operators executed earlier. It transforms global queries into local queries by performing local selection operations first, filtering out unnecessary intermediate information before continuing with subsequent join operations.

Subquery Elimination A subquery refers to the subquery module following "from" in the query statement. In distributed databases, this often means additional execution threads or pipelines, extra data memory, and inter-execution data transmission overhead. Subquery elimination directly "embeds" the subquery into the parent query, avoiding unnecessary overhead and serving as a local processing optimization method.

Fig. 3.12 Horizontal partitioning

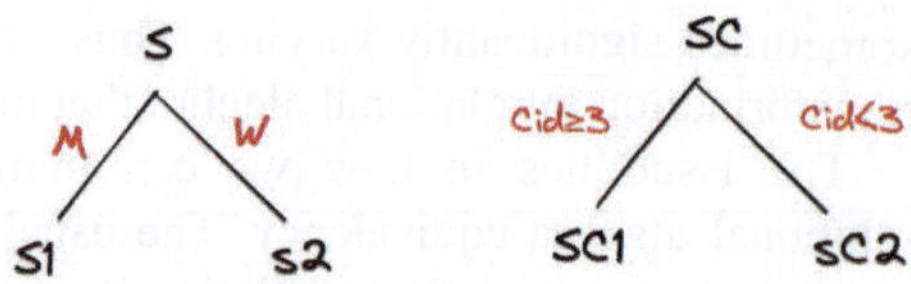

Listing 3.1 Horizontal partitioning

```
SELECT distinct Sname
FROM S,SC
WHERE S.Sid = SC.Sid and Sex = 'M' and Grade > 90;
```

Below, two examples illustrate how distributed database systems optimize queries by combining data partitioning with relational algebra transformation rules.

Example 3.1 There are two tables, S(Sid, Sname, Age, Sex) and SC(Sid, Cid, Grade), both of which use horizontal partitioning, as shown in Fig. 3.12. The user query statement is shown in Listing 3.1.

First, convert the user query statement into a relational expression:

$$\prod_{Sname}(\sigma_{\text{Sex}='M'\wedge\text{Grade}>90}(\sigma_{\text{S.Sid}=\text{SC.Sid}}(S \times SC)))$$

Then, transform the relational algebra expression into a query tree for optimization, as Fig. 3.13 shown.

Example 3.2 There is a global relation EMP(Eid, Ename, Sal, Dept, Dname), which takes vertical sharding, as shown in Fig. 3.14. The user query statement is shown in Listing 3.2.

First, convert the user query statement into the relational expression $\prod_{Ename,Sal}(EMP)$, and then transform the relational algebra expression into a query tree for optimization, as shown in Fig. 3.15.

Overall, the optimization algorithm first uses relational algebra transformation rules to push down selection and projection conditions in the query tree to the fragments as much as possible, and to pull up join and merge operations as much as possible, which means delaying their execution. Next, the partitioning method of the data is determined. Horizontal partitioning is used to eliminate fragment operations with contradictory conditions, and vertical partitioning is used to remove attribute data that are not relevant to projection operations. This reduces the generation of intermediate data, thereby optimizing the query.

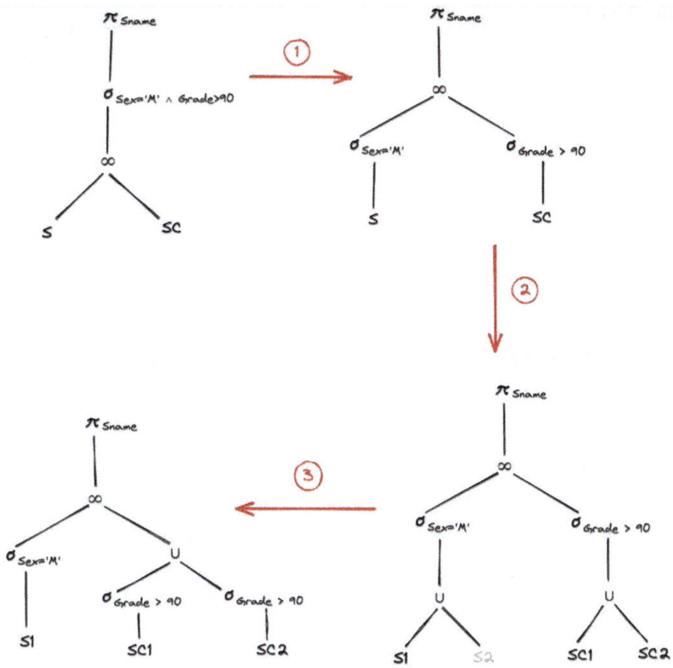

Fig. 3.13 Horizontal shard query optimization tree

Fig. 3.14 Vertical partitioning

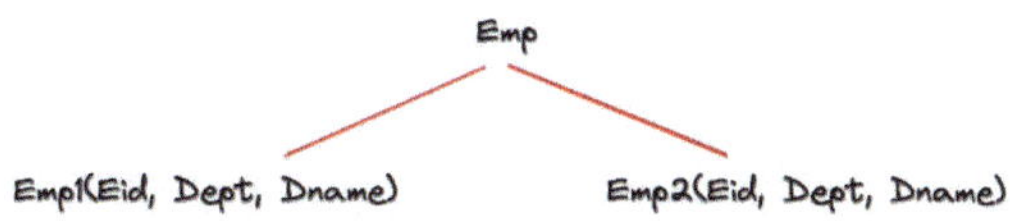

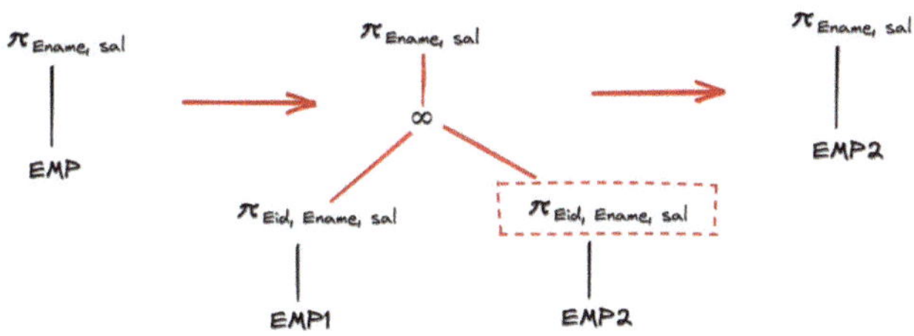

Fig. 3.15 Vertical shard query optimization tree

Listing 3.2 Vertical partitioning

```
SELECT Ename,Sal
FROM EMP;
```

Fig. 3.16 Site dependency

		Site	
		S1	S2
Relation	R1	F11	F12
	R2	F21	F22

3.2.4 *Multi-Join Query Optimization Based on Direct Joins*

The semi-join operation tends to improve data query speed by reducing the amount of data transfer over the network. However, it also increases the number of communications and local processing time. In high-speed local area networks, data transfer time between sites is typically shorter than local processing time, making the reduction of local processing time a key issue in query optimization. Under such circumstances, the use of direct join algorithms is more effective. The following introduces three representative algorithms based on direct join optimization.

The definition of the **site dependency algorithm** is as follows: let $R_i[A]$, $R_j[A]$, $S_i[A]$, and $S_j[A]$ represent the values of attribute A of relations R and S at sites i and j, respectively. If for $i \neq j$, we have $R_i[A] \cap S_j[A] = \emptyset$, then relations R and S are said to have site dependency on attribute A. Refer to Fig. 3.16.

Properties: if relations R and S are site-dependent on property A, then we have Eq. 3.8:

$$R \bowtie S = \bigcup (R_i \bowtie S_i) \tag{3.8}$$

Generally speaking, the site dependency algorithm divides the join operation of a large relation into join operations of fragments at different sites, and then takes the union of the results of these fragment join operations. The conditions for this algorithm to hold are as follows: the intersection of projections on the natural join attribute A must be an empty set, which satisfies the following four conditions:

- $\prod_A(F_{11}) \cap \prod_A(F_{12}) = \varnothing$
- $\prod_A(F_{21}) \cap \prod_A(F_{22}) = \varnothing$
- $\prod_A(F_{11}) \cap \prod_A(F_{22}) = \varnothing$
- $\prod_A(F_{12}) \cap \prod_A(F_{21}) = \varnothing$

Based on the above conclusions, there are the following inferences:

1. If R and S have site dependency on attribute A, and $B \supseteq A$, then R and S have site dependency on attribute B.
2. If R and S have site dependency on attribute A, and S and T have site dependency on attribute B, then:

$$R \bowtie S \bowtie T = \bigcup (R_i \bowtie S_i \bowtie T_i) \tag{3.9}$$

By Eq. 3.9, we can implement the join operation of multiple relations.

The advantages of using site-dependent algorithms to implement direct join operations are mainly the following three aspects:

1. No data transfer. Since the data of different relations that need to be joined exist in various sites, it is only necessary to directly issue execution commands to the data at each site. Even if there is data transfer, it is only the transfer of the computation results.
2. Parallel computation can be performed. There is no interference between computations at different sites; only the final collection and union of the results are required.
3. Local indexes can be used to improve efficiency.

The principle of the **shard replication optimization algorithm** is to select a set of sites, distribute all fragments of a certain relation in the query to these sites, and then replicate the remaining relations in the query to each of the selected sites.

Assuming there are two sites, S_1 and S_2, and two relations, R_1 and R_2 (both containing attribute A). R_1 is divided into two parts and stored on S_1 and S_2, respectively. Similarly, R_2 is also divided into two parts and stored on S_1 and S_2, as shown in Fig. 3.17.

If a join operation is performed on two relations R_1 and R_2 based on attribute A, the first approach keeps R_1 fragmented and R_2 complete. The processing method is as follows: Copy fragment F_{21} to site S_2 and merge it with F_{22} to get R_2; copy fragment F_{22} to site S_1 and merge it with F_{21} to get R_2, then perform the join operation at sites S_1 and S_2 respectively. Conversely, the second approach keeps R_2 fragmented and R_1 complete, and the calculation method is similar to the first approach.

Given the sizes of each fragment: $|F_{11}| = 50$, $|F_{12}| = 50$, $|F_{21}| = 100$, $|F_{22}| = 200$; the cost of data communication is $C(x) = x$; the cost of local union operation is $U(x_1, x_2) = 2(x_1 + x_2)$; the cost of local join operation is $J(x_1, x_2) = 5(x_1 + x_2)$.

Assuming That R_1 Remains Sliced The cost consumption at site S_1 includes the following parts: first, transmitting F_{F22} consumes 200; next, $F_{21} \cup F_{22}$ consumes

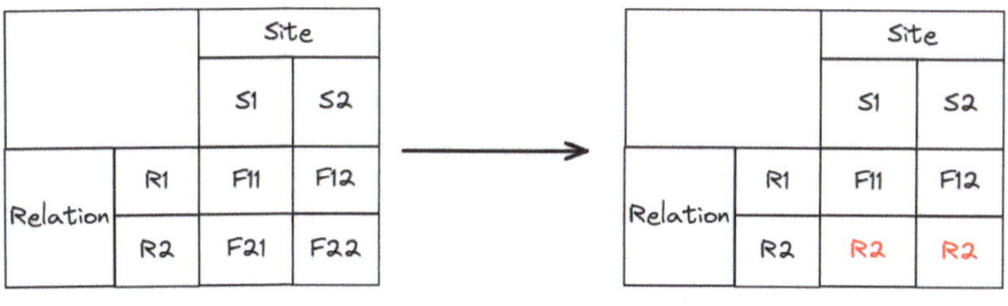

Fig. 3.17 Site dependencies to shard replication

$2 \times (100+200) = 600$; finally, $F_{22}\infty(F_{21} \cup F_{22})$ consumes $5 \times (50+300) = 1750$. Therefore, the final consumption is $FT(Q, S_1, R_1) = 2550$. The cost consumption at site S_2 includes the following parts: first, transmitting F_{21} consumes 100; next, $F_{21} \cup F_{22}$ consumes $2 \times (100 + 200) = 600$; finally, $F_{12}\infty(F_{21} \cup F_{21})$ consumes $5 \times (50+300) = 1750$. Therefore, the final consumption is $FT(Q, S_2, R_1) = 2450$. Since the operations at the two sites can be performed in parallel, the total cost is the maximum consumption of the two sites. Therefore, the final total cost is 2550.

Assuming That R_2 Remains Sliced The cost consumption at site S_1 includes the following parts: first, transmitting F_{12} consumes 50; next, $F_{11} \cup F_{12}$ consumes $2 \times (50 + 50) = 200$; finally, $F_{21}\infty(F_{11} \cup F_{12})$ consumes $5 \times (100 + 100) = 1000$. Therefore, the final consumption is $FT(Q, S_1, R_2) = 1250$. The cost consumption at site S_2 includes the following parts: first, transmitting F_{11} consumes 50; next, $F_{11} \cup F_{12}$ consumes $2 \times (50 + 50) = 200$; finally, $F_{22}\infty(F_{11} \cup F_{12})$ consumes $5 \times (200 + 100) = 1500$. Therefore, the final consumption is $FT(Q, S_2, R_2) = 1750$. Since the operations at the two sites can be performed in parallel, the total cost is the maximum consumption of the two sites. Therefore, the final total cost is 1750.

Between the two different fragmentation methods, fragmenting R_2 results in a lower total cost. **Therefore, it is concluded that R_2 should remain fragmented.**

In traditional distributed database systems, it is difficult for the data to meet the condition of being dependency-free and nonredundant across sites. Therefore, when a query cannot be processed without data transmission, a fragmentation replication algorithm can be used. Although this method introduces some data redundancy, it has a wider practical application coverage.

The main advantages of using fragmentation replication algorithm are as follows:

1. Parallel computation can be performed.
2. Local indexes can be utilized to a certain extent. For example, indexes can be established on fragments, or a local index can be created for each replica of a relation.

Hash Division Optimization Algorithm [35] The general idea is to choose a suitable Hash function and then apply the Hash function to the tuple values of an attribute or set of attributes of a relation, and place the tuples with the same Hash result in the same site, so that the horizontal partitioning of the corresponding relation can be easily obtained.

When performing multi-relation joins, rehashing can improve query speed (rehashing means performing hash partitioning multiple times). For multi-relation joins on the same attribute $R_1 \bowtie AR_2 \bowtie AR_3$, rehashing enables parallel execution compared to not rehashing. Instead of finding R_2 through R_1 and then R_3, R_1 and R_2, as well as R_1 and R_3, can perform joins in parallel and then merge results. Apart from the final result merging, data transmission can be basically ignored. For multi-relation joins on different attributes $R_1 \bowtie AR_2 \bowtie BR_3$, there are various approaches to consider. One method is to first hash-partition on attribute A, and then within the partitions obtained from A, further partition on attribute B. However, rehashing on B within the partitions obtained from A can result in different site locations for attribute B, necessitating constant data movement, which incurs high communication costs. Therefore, the key to solving this problem is to avoid the communication costs caused by rehashing as much as possible.

There are many approaches to solving the re-optimization problem, among which the most common is the Kruskal algorithm. The Kruskal algorithm is a commonly used algorithm in data structures for constructing a minimum spanning tree. Its basic idea is to continuously select the edge with the smallest weight to construct an acyclic graph until all the vertices are connected. Analogous to the optimization problem of rehashing, we can view the edges between two points as the join operations between two relations, replace the weight with the cost of rehashing, and use the Kruskal algorithm for optimization.

3.2.5 Multi-Join Query Optimization Based on Semi-Joins

SDD-1 Query Optimization Algorithm It originally stems from the "hill climbing" algorithm, which belongs to greedy algorithms. The algorithm starts with an initial feasible solution and always chooses the current least cost solution for iteration. Its biggest problem is that it may get stuck in a local optimum and fail to find the global optimum solution.

The algorithm is mainly divided into two parts: the basic algorithm and the post-optimization algorithm.

The general idea of **SDD-1 Basic Algorithm for Query Optimization** is to iteratively obtain and rank beneficial semi-connected operations whose costs are less than their benefits, based on the key factors in comparing reduction procedures—cost, efficiency, and benefit estimates.

The cost of a semi-join is a transmission to the semi-join attribute A, as shown in Eq. 3.10:

$$Cost(R \ltimes_A S) = T_{MSG} + T_{TR} * size(\Pi_A(S)) \tag{3.10}$$

The gain is then the irrelevant tuple transfer to R avoided by the semi-join, as shown in Eq. 3.11:

$$Benefit(R \ltimes_A S) = (1 - SF_{SJ}(S.A)) * size(R) * T_{TR} \tag{3.11}$$

where SF is the selection factor and SJ is the semi-join operation.

The algorithm ultimately derives a set of programs for semi-join reduction, from which the optimal cost-effective execution strategy is determined. Then, all the data from the sites are aggregated to the site with the largest amount of data for final assembly. The processing mainly includes the following three steps:

- Initialization. Generate a beneficial set of semi-joins from all possible semi-joins among the relations.
- Selection of beneficial semi-joins. From the beneficial semi-join set, identify the most beneficial semi-join, add it to the execution strategy, and accordingly modify the statistics (selectivity factor, relation size, etc.) of the affected relations.
- Selection of assembly site. Repeat the second step continuously until all beneficial semi-joins are added to the execution strategy. After reducing the relations through the above steps, choose the site with the largest amount of stored data for assembly.

The **SDD-1 post-optimization algorithm** is an algorithm that achieves a better solution by correcting the solution obtained from the SDD-1 basic algorithm. It makes the resulting optimal execution strategy more reasonable and time-saving. Finally, by integrating the data obtained from all sites, the final query result can be obtained.

The post-optimization for the algorithm is shown in Listing 3.3.

3.3 Relational Join Algorithms

In Sects. 3.1 and 3.2, we mainly introduced how various algorithm strategies can reduce the size of the data that needs to be transmitted over the network during the implementation of distributed join queries, thereby reducing communication overhead. When the queried data is transmitted through the network to the computing nodes, the computing nodes need to perform join computation on the data. In the previous examples, to analyze and explain the network transmission overhead, we uniformly used the nested-loop join algorithm as the join processing algorithm. However, there are actually three methods for implementing join computations, which are:

1. Nested-Loop Join.
2. Hash Join.
3. Sort-merge Join.

Each of the three algorithms has its specific application scenarios, and selecting the appropriate join algorithm based on the actual situation will improve the efficiency of join processing. The principles and usage scenarios of the three join algorithms are introduced below.

3.3.1 Nested-Loop Join Algorithm

The basic idea of a **nested-loop join** is to use one of the join inputs as the outer input table (shown as the top input in graphical execution plans), and the other join

Listing 3.3 SDD algorithm pseudocode,

```
Input: Query graph (QG), relationship location, statistics of the
     relationship
Output: Execution strategy(ES)
begin:
    // Initialisation
    // Generate an execution plan with only local processing ES
    ES<---local-operation(QG);
    // Initialise the set of beneficial semijoins (BS)
    BS<---NULL;
    // Finding beneficial half-connections from half-connections in the
     full relationship{SJ_1,SJ_2,...,SJ_k}
    for each semijoin SJ in QG do:
        if cost(SJ)<benefit(SJ) then
            BS<---BS+SJ;

    // Select the most beneficial half-connections to add to the
     execution plan
    while BS is not null do:
        SJ<---most_beneficial(BS);
        BS<---BS-SJ;
        ES<---ES+SJ;
        // Modify the statistical values of the affected relationships (
     selection factor, relationship size, etc.)
        modify statistics;
        BS<---BS-non_BS;
        BS<---BS+new_BS;

    //Selection of the site with the highest volume of data as the
     assembly site(assembly site,AS)
    AS<---site i having the largest amount of datas;
    ES<---ES+AS;

    //post-optimisation stage
    for each relation R_i at AS do
        for each semijoin SJ of R_i by R_j do
            if cost(ES)>cost(ES-SJ) then
                ES<---ES-SJ
end
```

input as the inner (bottom) input table. The outer loop consumes the outer input table row by row [89]. For each outer row, the inner loop is executed, searching for matching rows in the inner input table. In the simplest case, the entire table or index is scanned during the search; this is called a simple nested-loop join. If an index is used during the search, it is called an indexed nested-loop join. If the index is created as part of the query plan (and destroyed immediately after the query completes), it is called a temporary indexed nested-loop join. The pseudocode is presented as shown in Listing 3.4.

Listing 3.4 Nested-loop join

```
for each row R1 in the outer table
    for each row R2 in the inner table
    if R1 joins with R2
        return (R1, R2)
```

Nested-loop join suitable when the number of records in the inner input table is less than 1,000, or the inner input table has an effective access method (index) with good index selectivity.

Nested-loop join often performs inner join, left outer join, left semi-join, and left anti-semi join logical operations.

Nested-loop join typically uses an index to search for each row of the outer table in the inner table. Based on the estimated cost, Microsoft SQL Server decides whether to sort the outer input to change the search position of the inner input index.

Return all rows that satisfy the (optional) predicates within the Argument column based on the logical operations performed.

As shown by the pseudocode of the nested-loop join algorithm mentioned above, each tuple in the inner relation is compared with each tuple in the outer relation for matching. The time complexity of the nested-loop join algorithm is $O(n*m)$, where n and m represent the number of tuples in the outer and inner relations, respectively. As the size of these relations increases, the join operation's efficiency decreases significantly. To address this limitation, two optimized versions of the nested-loop join algorithm have been developed:

Index Nested-Loop Join An index on the join attribute is built for the inner table in the nested-loop join algorithm, allowing the inner loop to directly use the index lookup instead of a full scan in the inner loop [51]. As shown in Fig. 3.18, the join match condition of the outer relation is directly matched with the index of the inner relation, and then the corresponding tuple data is found through the inner table index for matching. This reduces the number of matches on the inner table and improves the efficiency of the nested-loop join algorithm. As shown in Eq. 3.12, compared to the nested-loop join algorithm, the index nested-loop join algorithm

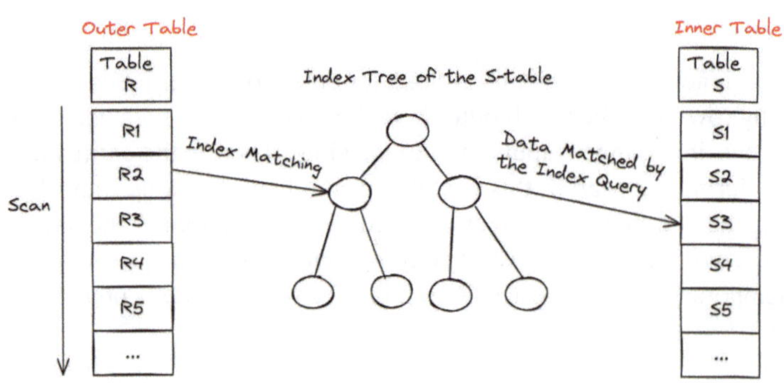

Fig. 3.18 Index nested-loop join

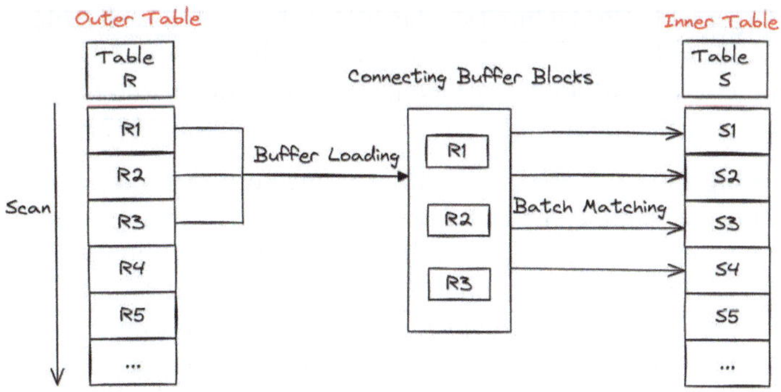

Fig. 3.19 Block nested-loop join

reduces the number of iterations in the inner loop through the index tree structure, thereby lowering the time overhead.

$$
\begin{aligned}
\text{Nested Loop Algorithm Join Match Count} &= \text{Number of outer table records} \\
&\quad \times \text{Number of inner table records} \\
\text{Index Nested-Loop Algorithm Join Match Count} &= \text{Number of outer table records} \\
&\quad \times \text{Height of the inner table index tree}
\end{aligned}
\tag{3.12}
$$

Block Nested-Loop Join Block nested-loop join loads multiple records from the outer table into a join buffer simultaneously and then uses this buffered data to batch-match against the inner table's data, reducing the number of full scans of the inner table [25]. As illustrated in Fig. 3.19, before executing the join operation, a fixed-size memory block is pre-allocated as the join buffer. A subset of records from the outer table is loaded into this buffer, with the number determined by available memory. During the subsequent scan of the inner table, multiple records from the outer table are matched concurrently, thereby decreasing the frequency of reloading the inner table into memory for matching. This approach significantly enhances the efficiency of the nested-loop join algorithm.

3.3.2 Hash Join Algorithm

The basic idea of the **hash join algorithm** is to create a hash table in memory based on the smaller table (Build Input) in the join computation. Then, scan each row of the relatively larger table (Probe Input) and map it to the created hash table according to the join condition. This can quickly obtain the matching rows of the two tables. The mainstream hash join algorithms generally consist of three phases: the partitioning phase, the hash table building phase, and the probing phase [132]:

1. **Partitioning Phase**. If there is insufficient memory to store the generated hash table in the hash area completely, the concept of partitioning is introduced. The build input and probe input are divided into multiple partition parts based on the join column (i.e., the key column). Each partition includes independent, paired build and probe inputs. This divides a large hash join task into multiple independent hash join tasks that do not interfere with each other. Each partition can build and explore the hash table in memory and complete the join computation through mapping.
2. **Hash Table Building Phase**. During the building phase, each tuple record is scanned from the build input, and the join field values are hashed using a hash function. The hash value is then mapped to corresponding hash buckets in the hash table. Once the entire build input has been scanned, a hash table representing all tuple records of the build input is generated.
3. **Probing Phase**. Similar to the build input, the probe input tuple records are scanned. The join column values are hashed using the same hash function used in the build phase to obtain the hash values. These values are mapped to the hash table to search for the corresponding hash buckets and perform the join matches.

The pseudocode for hash join algorithms [89] is presented in the Listing 3.5.

Assume there are two relational tables *R* and *S* being joined using partitioned hash join, where table *R* serves as the build input and table *S* serves as the probe input. As shown in Fig. 3.20, in the partitioning phase, table *R* is divided into four

Listing 3.5 Hash joining

```
    // Hash table construction phase
    for each tuple r in R do
        {
            hash on join attributes r(b)
            place tuples in hash table based on hash values
        };

    // probe stage
    for each tuple s in S do
        {
            hash on join attributes s(a)
            if r hashes to a nonempty bucket of hash table for R
            then {
                if r matches any s in bucket
                concatenate s and r
                place in output relation Q
            };
        };
```

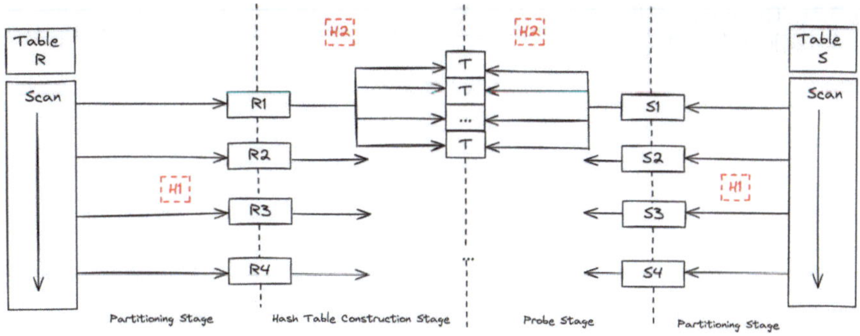

Fig. 3.20 Partitioned hash join

partitions, R_1, R_2, R_3, and R_4, using hash function H_1. Then, for each partition, hash tables are built in memory using hash function H_2. Similarly, table S needs to be partitioned, and using hash function H_1, its data is divided into four partitions, S_1, S_2, S_3, and S_4. Finally, each data partition of table S is probed using the same hash function H_2 as table R to complete the join match computation.

The efficiency of the hash join algorithm is primarily determined by the design of its hash function. An optimal hash function ensures that when a tuple from the probe input relation is hashed to a non-empty entry in the hash table, it can immediately join and match a set of tuple records. Under ideal conditions, the time complexity of this algorithm is $O(n + m)$, where n and m represent the number of tuples in each relation, as each relation table needs to be scanned only once for hashing and comparison. Conversely, an ill-suited hash function used to construct the hash table from the build input can result in numerous hash collisions. These collisions occur when multiple distinct keys are mapped to the same hash value. Consequently, during the probing phase, each tuple hashed to a non-empty entry in the hash table necessitates additional comparisons to verify if the join columns match. This extra step significantly diminishes the efficiency of the hash join algorithm, potentially approaching the time complexity of less efficient join methods.

The hash join algorithm is particularly effective when there is a substantial disparity in data volume between the two tables being joined. However, attention must be paid to the aforementioned issues. If the build input data volume is excessively large, the resulting hash table may exceed available memory capacity. In such cases, the table must be partitioned into multiple segments to facilitate the hash join. The partitioning operation introduces additional I/O overhead, thereby reducing the efficiency of the join computation. Therefore, when choosing whether to use the hash join algorithm, the hardware resources of the compute node, especially memory, are also important influencing factors.

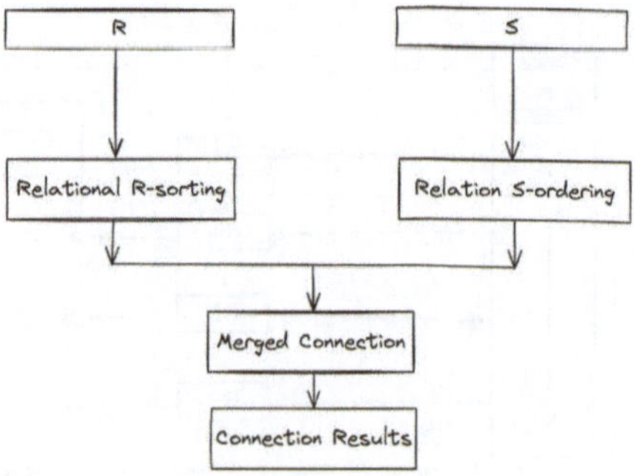

Fig. 3.21 The process of the sort-merge join algorithm

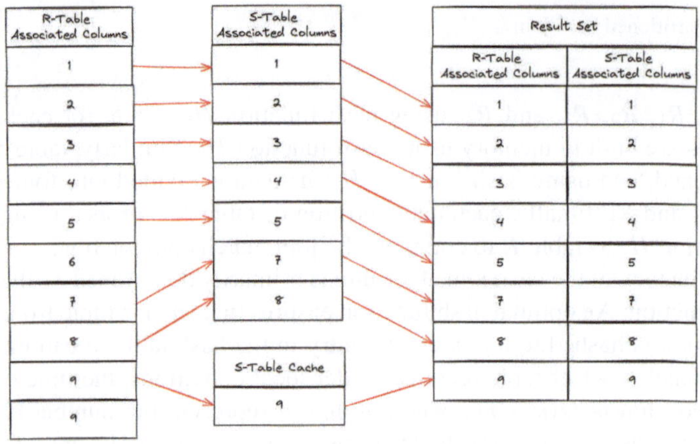

Fig. 3.22 Merge join operation

3.3.3 *Sort-Merge Join Algorithm*

Sort-merge join is an algorithm in database systems that performs relation joins using sorting. The algorithm first sorts the tuple data in two relational tables based on the join attributes and then performs a sequential scan to find and merge tuples that meet the join conditions.

Figure 3.21 shows the workflow of the sort-merge join algorithm. The compute node sorts the data in tables R and S. After that, it performs the merge join operation on the sorted data to obtain the result set. Figure 3.22 is a simple example of a merge join. After sorting the tuple data in tables R and S in ascending order, the join operation is performed, and new tuples that meet the join conditions are generated in the result set. The sort-merge join algorithm scans each table's sorted data only once; when the scan of one table is complete, the join operation stops, and there is no need to continue scanning the remaining data of the other table. The pseudocode for the sort-merge join algorithm [89] is shown in Listing 3.6.

The efficiency of the sort-merge join algorithm is primarily determined by its sorting and merging phases. Suppose the relational tables involved in the

Listing 3.6 Sort-merge join algorithm

```
    // Sorting stage
    sort R on r(a);
    sort S on s(b);

    // merger phase
    read first tuple from R;
    read first tuple from S;
    for each tuple r do
    {
        while s(b) < r(a)
        read next tuple from S;
        if r(a) = s(b) then
            join r and s
            place in output relation Q
    };
```

join are already sorted based on the join attributes. In that case, this algorithm significantly outperforms the nested-loop join algorithm's brute-force approach, as each relational table only requires a single scan during the join computation to perform the merge operation. However, when the relational tables are not presorted, the primary computational overhead stems from sorting the tables based on the join columns, typically with a time complexity of $O(nlogn)$, where n represents the number of tuples in the table. Despite this potential sorting overhead, the sort-merge join often remains more efficient than nested-loop join for large datasets, as it reduces the number of comparisons needed. Furthermore, the algorithm's performance scales well with increasing data size, making it particularly suitable for large-scale data processing tasks.

The sort-merge algorithm is usually used when the data tables do not have indexes but are already sorted. Generally, the efficiency of the hash join algorithm is better than that of the sort-merge join algorithm. However, if the table rows have already been sorted, then using the sort-merge join subsequently does not require additional sorting. Thus, the performance efficiency will be higher than that of the hash join algorithm.

3.4 Chapter Summary

This chapter mainly introduces the process of distributed query processing. In a distributed storage environment, since relational table data is stored on different physical nodes according to certain rules, the query processing process needs to gather the data involved to a specific node for computation, thereby introducing additional network transmission overhead. The distributed query process can be mainly divided into the following steps: extracting and filtering the relational table data involved in the query at the local node, serializing and packaging the data for network transmission, the corresponding node receiving the transmitted data, and the receiving node integrating the query data to complete the query computation.

Regarding data network transmission, Sect. 3.1 introduces three algorithm strategies for implementing data transmission processes and overhead on different nodes. The direct join algorithm transmits the relational data tables involved in the join processing to a computing node for join processing. This algorithm is very simple to implement, but the data in the relational tables may contain unmatchable tuple records, resulting in invalid network transmission overhead. To solve this issue, the semi-join algorithm has been proposed as a filter. It projects the join column of one table to generate a dataset, which is then transmitted to another node. This dataset of the join column is used to filter the other relational tables involved in the join operation. It selects only those tuples that can participate in the matching join, thereby eliminating invalid transmission overhead. However, it should be noted that the semi-join algorithm is only more efficient than the direct join algorithm when Eq. 3.4 is satisfied. To further optimize the semi-join algorithm, we introduce the use of Bloom filters to replace the projected value set of the join column as a filter, reducing the data volume in network transmission. However, due to false positives in Bloom filter judgments, we need to choose appropriate bit arrays and hash functions based on the data size to be represented.

Section 3.2 mainly introduces different strategies for distributed query optimization. We first analyze several essential factors that affect query costs in distributed systems: CPU overhead, I/O overhead, and communication overhead. We then introduce two common metrics used to measure query costs in distributed systems: total processing overhead and response time. Total processing overhead indicates the utilization of system resources, while response time indicates the parallelism of system operations. In practical applications, we often optimize by applying different weights to total processing overhead and response time to achieve both efficient resource utilization and parallelism. Second, we introduce the two parts of query optimization: local optimization, which explains some common relational algebra transformation rules used for query optimization, and global strategy optimization, which focuses on three representative algorithms based on direct join optimization, site-dependent, replicated fragment, hash partitioning optimization algorithms, and one optimization algorithm based on semi-join, the SDD-1 query optimization algorithm.

In terms of join computation, Sect. 3.3 introduces various join processing algorithms. Join computation performance can be enhanced by selecting the appropriate algorithm based on the characteristics of different relational data and index structures. The nested-loop join algorithm processes two relational tables as outer and inner input tables respectively, executing a two-layer loop to perform the matching and join operations. The hash join algorithm is suitable for cases where the data volumes of the two tables being processed are significantly different. It first builds a hash table in memory based on the relatively smaller table, and then scans the relatively larger table, mapping it to the hash table for fast join matching based on the join condition. The merge-sort join algorithm first sorts the two relational tables based on the join attributes, then sequentially scans, and merges tuples that meet the join conditions. When the data tables are already sorted, the merge-sort join algorithm can greatly improve join processing efficiency.

In a distributed environment, join processing involves multiple factors, resulting in a complex and challenging problem. However, we can decompose this problem into several comprehensible subproblems for analysis. By addressing the requirements and environmental conditions of each subproblem individually, and selecting appropriate algorithmic strategies, we can optimize the distributed query processing workflow and minimize associated overhead [100].

Chapter 4
Transaction Processing in Distributed Environments

Transactions and data services are intrinsically linked. A comprehensive understanding of data services necessitates a thorough grasp of transaction concepts. Transaction processing mechanisms in distributed environments exhibit distinctive characteristics. This chapter introduces various transaction types and their associated properties. By examining atomic commitment and concurrency control, readers will gain deep insights into the intricacies of transactions and appreciate the crucial role of distributed transactions in distributed systems.

Before understanding distributed database transactions, we need first to comprehend what a database transaction is. Let us consider a banking scenario as illustrated in Fig. 4.1: at time T_0, Account A has an initial balance of 200 units, and Account B has an initial balance of 100 units. At time T_1, A transfers 100 units to B's account. After T_1, B's balance should show 200 units, and A's balance should show 100 units. This is the logical outcome we expect in real life. However, how should we define the correctness of this transfer operation in a database system? Furthermore, in actual business systems, software or hardware may fail at any time, leading to system crashes or business vulnerabilities. If the business system is deployed in a distributed manner, how should we address network partition issues caused by disconnections between remote nodes? Database transactions are the common mechanism used to solve these problems.

How Database Transactions Address These Issues As illustrated in Fig. 4.1, when Client A performs a transfer operation by invoking $add(x, -100)$, the balance of Account A decreases by 100 units. Similarly, Client B's operation $add(y, 100)$ increases the balance of Account B by 100 units. Both operations are part of a single transaction. This ensures that all operations within the same transaction either succeed or fail as a whole. Database transaction mechanisms ensure that complex operations, such as fund transfers between accounts, maintain atomicity. This means there will never be a scenario where the debit operation on Account A succeeds while the credit operation on Account B fails, or vice versa. Such partial failures are inherently complex and resolving them can be extremely cumbersome. By leveraging database transaction mechanisms, developers can significantly reduce

Z. Xu, *Distributed Data Services*, https://doi.org/10.1007/978-981-96-9392-4_4

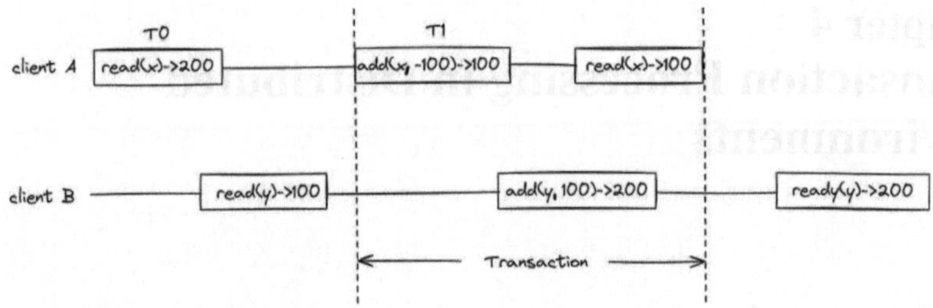

Fig. 4.1 Business scenario flowchart

the complexity of application development, as these mechanisms guarantee the atomicity of operations.

4.1 Deep Understanding of Transactions

All read and write operations within a transaction constitute a unified execution unit. Every operation in a transaction must either succeed entirely or fail as a whole. As the handling of partial failures is not required, the application design built upon transactions can be significantly streamlined. Before clearly articulating database transaction concepts, programmers working with databases faced numerous data anomalies. To address these challenges, the third revision of the SQL database query language standard (ANSI SQL-92) [4] introduced a unified definition of database transaction isolation levels for the first time: read uncommitted, read committed, repeatable read, and serializable. These isolation levels were proposed to resolve the aforementioned data anomalies. Each level addresses specific anomalies, with higher isolation levels mitigating a broader range of potential issues. Subsequent sections of this chapter will delve into more detailed discussions of these concepts.

4.1.1 Local Transactions

Local transactions, also known as "partial transactions," are characterized by using a single data source within a single service. For instance, when you utilize JDBC or a SQL Server connection to a database, you are essentially creating a database transaction, but these transactions only involve a single data source. Thus, they are referred to as local transactions or partial transactions.

Local transactions are the most fundamental transaction mechanism, embodying the core characteristics of database transactions. The global and distributed transactions that will be introduced subsequently are optimized realizations grounded in local transactions. Subsequently, we will delve into the pertinent properties of transactions from the viewpoint of local transactions, aiming to enhance the comprehension of transactions among scholars and practitioners alike.

Transaction States The database transaction primarily has the following states:

- Active. The state indicating that a transaction is currently being executed.
- Partially Committed. The state after the execution of the last statement in a transaction.
- Failed. The state where a transaction cannot continue to execute.
- Aborted. The state after a transaction rollback and the database has been restored to its state before the transaction began.
- Committed. The state after a transaction has partially committed and written data to disk, and upon completing the writing of the last piece of data.
- Terminated. If a transaction is rolled back into an "Aborted" or comes from a "Committed," the transaction will terminate, and the system will be ready to accept a new transaction.

What we refer to as transaction commit means the transaction enters the commit state, whereas transaction rollback means the transaction enters the abort state. Both of these outcomes will eventually lead to the terminated state. The process of database transaction state changes is illustrated in Fig. 4.2.

The first state a database transaction enters is the active state. After executing the last read or write operation, the transaction moves to the partially committed state. If the transaction can complete the data persistence operation, i.e., writing the transaction information to disk, it then enters the committed state and finally reaches the terminated state.

The state transition path described above is for a successfully committed transaction. For a failed transaction, the state transition also starts from the active state. If the transaction fails to complete all read and write operations or the persistence operation fails after completing the read and write operations, the transaction enters the failed state. At this point, the transaction cannot proceed and will roll back to the terminated state, meaning it returns to the state before the transaction begins. Finally, the transaction enters the terminated state.

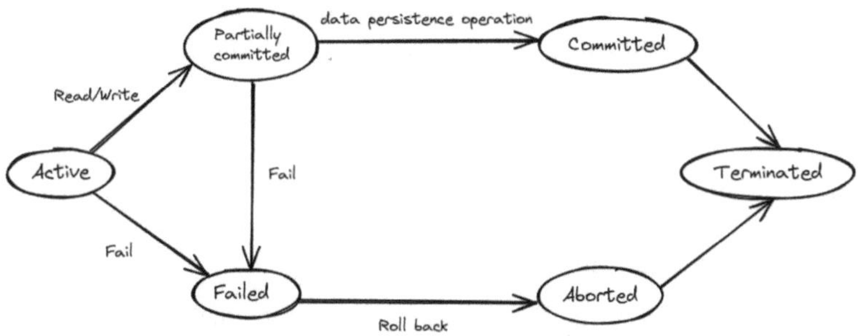

Fig. 4.2 The changes in the state of a database transaction

ACID Properties The core of database transactions lies in its ACID properties, around which almost all transaction-related research revolves. "ACID" is an acronym for four English words:

- Atomicity. In a transaction, all operations are executed as an indivisible unit, so we distinguish between transaction states but not between the states of operations within the transaction. During the execution of a database transaction, either all operations of the transaction succeed or all fail.
- Consistency. The data state remains consistent before and after transaction execution. Here, "consistent" in maintaining the data state before and after a transaction does not mean the same as consistency in data consistency. In the context of transactions, consistency refers to ensuring the data remains in a stable state. For instance, in the earlier example of a bank transfer transaction, the sum of balances in Account A and Account B should remain consistent before and after the transfer. If a transaction starts from a valid state, it should also end with the data in a valid state. The consistency of data states is typically specified by the application, and what constitutes a valid data state can vary between different applications.
- Isolation. When multiple transactions execute concurrently, they should be isolated from each other. Each transaction should not be able to see updates from other transactions until it is completed. When multiple users simultaneously read and write to the same table, isolating their transactions ensures that concurrent transactions do not interfere with or affect each other. Each request can occur as if it were happening one after another, even though they are actually happening simultaneously. Achieving isolation requires transactions to be executed serially. In essence, transactions should not see intermediate states of other transactions but only the final completed transaction results. Isolation, as one of the most critical properties of transactions, will be detailed further in subsequent discussions.
- Durability. Ensures that the changes made by successfully executed transactions to data are saved permanently. Even in the presence of hardware or software failures, this property guarantees that the changes made by transactions to data will not be lost.

ACID transactions ensure the highest possible data reliability and integrity. They prevent data from falling into an inconsistent state due to partially completed operations. For example, without ACID transactions, if a user is writing data to a database table and a hardware failure occurs, only part of the data might be saved. The database is now in an inconsistent state, making recovery very difficult and time-consuming. A more detailed description of the ACID properties is that Consistency (C) is the goal, while Atomicity (A), Isolation (I), and Durability (D) are the means to achieve consistency.

Transaction Isolation Levels We previously mentioned that the isolation property is the most crucial attribute in ACID properties. The ultimate manifestation of isolation is that concurrent transactions execute in a serializable manner. Failure

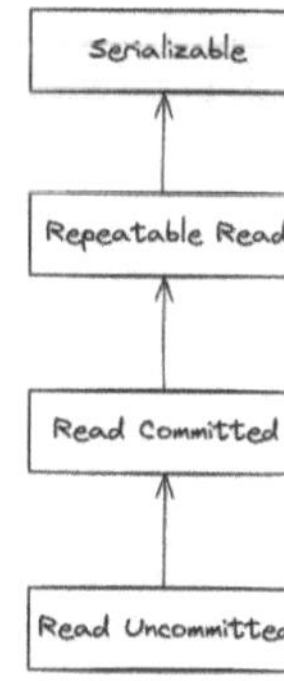

Fig. 4.3 Transaction isolation levels defined by the SQL standard

Listing 4.1 Transaction execution pseudocode

```
BEGIN Transaction
        Select * from Ticket WHERE ID=1;
        Update Ticket set Available_Seat=9 WHERE ID=1;
        Payment;
        Rollback;
END Transaction
```

to maintain isolation can lead to data anomalies. Each type of data anomaly is "captured" and defined at different levels. Generally, the higher the level of data anomaly, the higher the isolation level required to prevent it. Figure 4.3 illustrates the isolation levels defined by the SQL standard.

Before introducing these transaction isolation levels, it is necessary first to discuss *data anomalies*. This will help to understand the differences among various transaction isolation levels.

Dirty Read The term "dirty read" refers to a transaction reading data that is considered invalid because it has been modified by another transaction that has not yet been committed. A common scenario used to illustrate dirty reads is in a ticket booking system. For example, suppose there is a ticket booking system where one customer tries to reserve a ticket when there are ten seats available. Before completing the payment, another customer attempts to reserve a ticket, causing the system to show nine seats available to the second customer temporarily. If the first customer's transaction is then rolled back due to insufficient funds in their debit card or wallet, the second customer's transaction would have read the information of nine available seats, which is considered a dirty read. Pseudocode for the execution of the first customer's transaction is shown in Listing 4.1.

The timeline of two transaction executions is shown in Fig. 4.4. After the first user's transaction executes the Update statement, the second user's transaction begins execution. The pseudocode for the transaction is similar to the previous

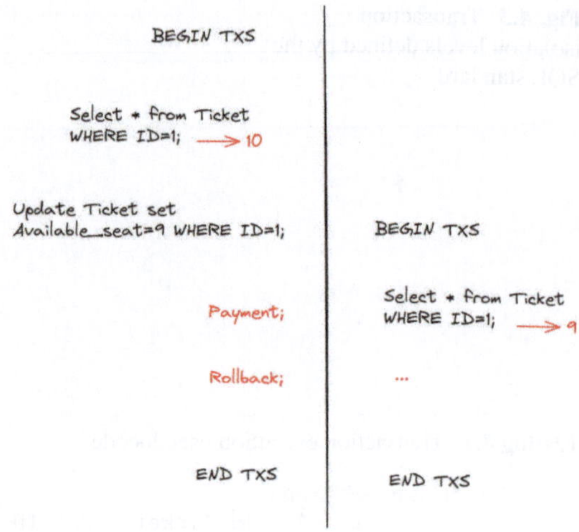

Fig. 4.4 Example of dirty read phenomenon

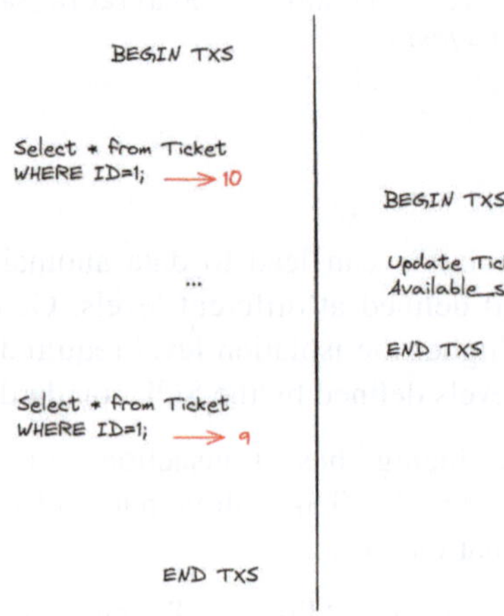

Fig. 4.5 Example of non-repeatable read phenomenon

example, but the Select statement retrieves the Available_Seat value for ID = 1 as 9, which originates from the first user's uncommitted transaction.

Non-repeatable Read In transaction *A*, there are two consecutive reads of the same data item "data." Between these two reads, another transaction *B* accesses data item "data" and performs an update that changes its value. Transaction *B* then commits. If transaction *A*'s second read operation retrieves an inconsistent value due to transaction *B*'s update, this phenomenon is known as a non-repeatable read, as illustrated in Fig. 4.5.

Phantom Read The phenomenon of phantom reads is very similar to non-repeatable reads, where one transaction's write operation changes the query results of another transaction. However, while non-repeatable reads primarily concern updates, phantom reads involve a more complex scenario. In phantom reads,

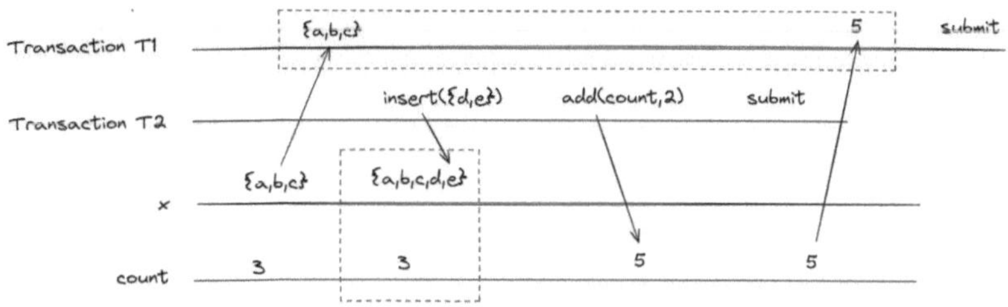

Fig. 4.6 Example of phantom read phenomenon

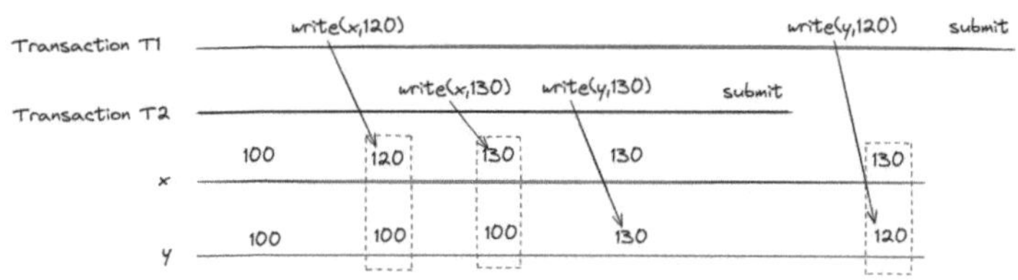

Fig. 4.7 Example of dirty write phenomenon

transaction *A* reads all data within a specific range and makes modifications, while transaction *B* inserts a new row within the same range. When transaction *A* reads the same range again, it discovers a new "phantom" row. Therefore, phantom reads primarily involve insert operations, whereas non-repeatable reads focus on update operations. The phenomenon of phantom reads is illustrated in Fig. 4.6.

The three abnormal phenomena mentioned above originate from the definition in ANSI SQL-92. However, in practical business systems, more abnormal phenomena are likely to occur. The paper [10] identifies some issues overlooked by ANSI SQL-92 and proposes higher requirements for ANSI SQL-92 isolation levels under the implementation of two-phase locking (2PL) in databases. This article provides two interpretations: using P (Phenomenon) to indicate *possible abnormal phenomena* and using A (Anomaly) to indicate *already occurred abnormalities*. Because P only represents potential abnormalities, it is also referred to as the broad interpretation, while A is referred to as the strict interpretation.

In the realm of abnormal phenomena, P1–P3 originate from ANSI SQL-92 and correspond to dirty reads, non-repeatable reads, and phantom reads, respectively. The paper also supplements other abnormal phenomena as follows.

Dirty Write A dirty write occurs when two transactions simultaneously write to the same key, causing a conflict, as illustrated in Fig. 4.7. If transaction T_2 commits successfully while transaction T_1 rolls back, the rollback target is ambiguous—it's unclear whether to revert to the initial value or to the value committed by transaction T_2. If both T_1 and T_2 commit successfully, it may violate the system's data consistency constraints. In Fig. 4.7, the constraint condition is $x = y$, but after both transactions T_1 and T_2 commit successfully, this constraint is violated,

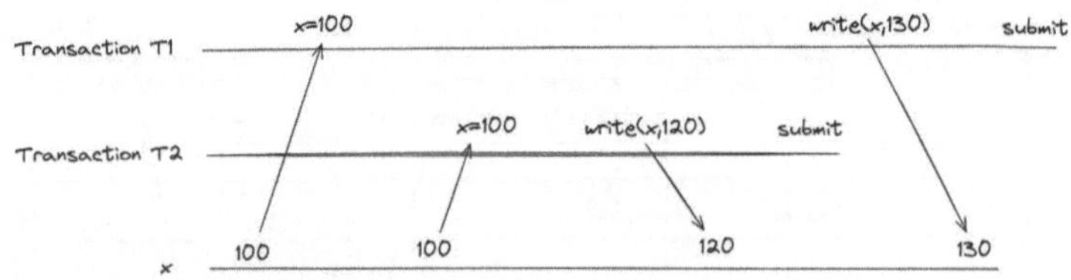

Fig. 4.8 Example of lost update phenomenon

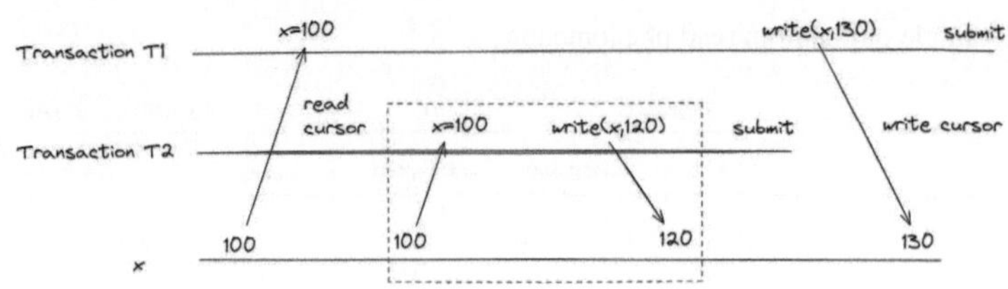

Fig. 4.9 Example of cursor lost update phenomenon

even though each transaction's operations individually appeared to comply with the constraint logic.

Lost Update Lost update refers to the scenario where transaction *A* attempts to write based on data it has read, but between the read and write operations of transaction *A*, transaction *B* performs another write operation. If transaction *B* successfully commits before transaction *A* completes its write operation, transaction *A*'s write operation will overwrite the write operation of transaction *B*, resulting in the loss of transaction *B*'s update. The lost update phenomenon is illustrated in Fig. 4.8, where the write operation of transaction T_2 is lost even though it has been successfully committed.

Cursor Lost Update A cursor is a database system's data buffer used to store the results of SQL statement execution. Database read operations can be implemented in two ways: with locking and without locking. After transaction *A* performs a read operation, the cursor records this action. If the cursor's current data is locked using a locking mechanism, transaction *B* cannot write to this data while transaction *A* is executing. This mechanism helps prevent the occurrence of cursor lost update phenomena. Cursor lost update is a variant of the lost update phenomenon involving cursors, illustrated in Fig. 4.9.

Read Skew Read skew refers to a transaction's read operation observing data that violates application rules. It is somewhat similar to non-repeatable reads, but unlike non-repeatable reads which involve reading different versions of the same key, read skew occurs when reading different keys where the observed data violates constraints. An example of read skew is illustrated in Fig. 4.10. The process is similar to the dirty write example shown in Fig. 4.7, but read skew occurs when a read operation observes a constraint violation (in Fig. 4.10, the data constraint is

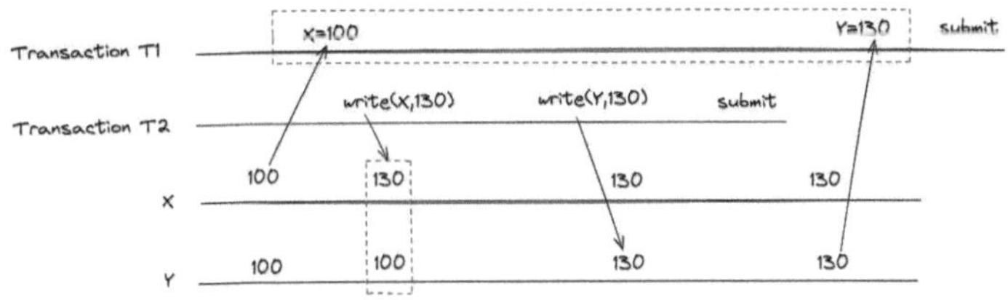

Fig. 4.10 Example of read skew anomaly

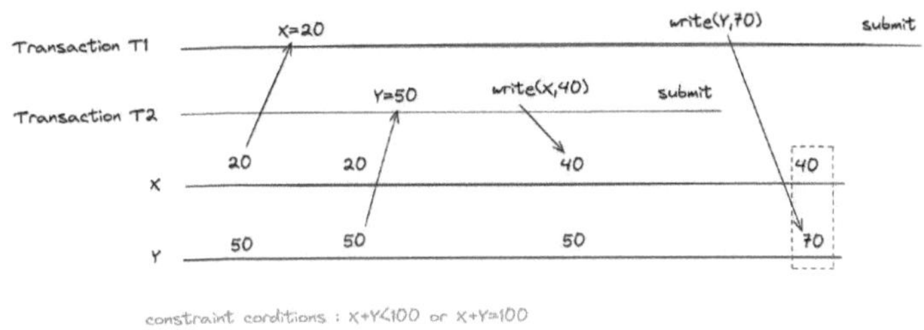

Fig. 4.11 Example of write skew anomaly

$X = Y$). Therefore, read skew is an anomaly that has already occurred, whereas dirty write is a potential anomaly.

Write Skew Write skew is an anomaly that occurs when multiple transactions write to the same object. As shown in Fig. 4.11, the data constraint rule is $X + Y \leq 100$. During the execution of transaction T_1, another transaction T_2 inserts, and from each transaction's perspective, their operations comply with the data constraint. However, after both transactions are committed, write skew causes a violation of transaction consistency.

The above describes the data anomalies defined in the paper [10]. It can be observed that each type of anomaly presents different challenges in resolution. Therefore, providing different isolation levels for transactions can prevent the occurrence of these data anomalies. ANSI SQL-92 addresses the data anomalies they identified (dirty reads, non-repeatable reads, and phantom reads) by proposing different transaction isolation levels to mitigate these anomalies:

- Read Uncommitted.
- Read Committed.
- Repeatable Read.
- Serializable.

Read Uncommitted Read uncommitted, also known as dirty read, allows a database transaction at this isolation level to see data that has been modified by other transactions but not yet committed. This exposes transactions to the risk of reading data that may ultimately be rolled back by the other transaction, resulting in the data

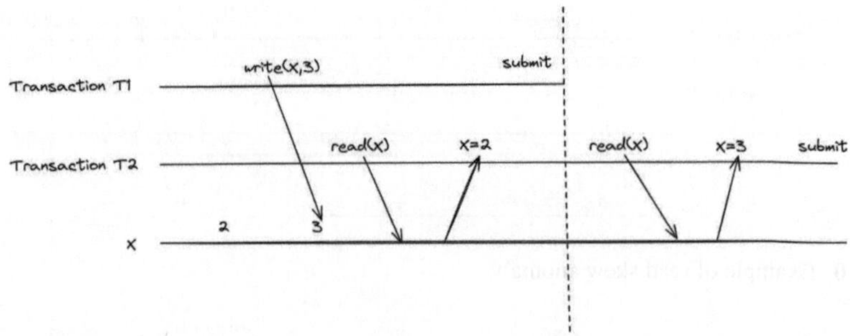

Fig. 4.12 Read committed prevents dirty reads

anomaly known as dirty read. As the lowest isolation level, read uncommitted does not prevent any of the three data anomalies specified in ANSI SQL-92.

Read Committed Read committed is often considered the most basic transaction isolation level because read uncommitted does not prevent data anomalies. It provides two guarantees:

- When a transaction performs a read operation, it can only access data that has been successfully committed.
- When a transaction performs a write operation, it can only modify data that has already been successfully committed.

The first guarantee prevents the occurrence of the P1 phenomenon (dirty read), while the second guarantee prevents the occurrence of the P0 phenomenon (dirty write). As shown in Fig. 4.12, transaction T_2 will not read the data that transaction T_1 has not yet committed, such as $x = 3$. Before transaction T_1 commits, the read operation by transaction T_2 will still return the old value: $x = 3$.

When a transaction needs to update multiple objects, it is essential to prevent dirty reads, which means at minimum, the system should use the Read-Committed isolation level. When Transaction A updates multiple objects, there will always be a period where only some updates are completed. This is very common in real-world applications, such as in e-commerce systems. After a user clicks purchase, the system may need to simultaneously add a record to the order table and update the merchant's inventory. Without ensuring the Read-Committed isolation level, users might observe partial updates, like seeing inventory decrease without a corresponding order, or an order being created without the inventory change. This could lead users to take unexpected actions.

In addition to updates involving multiple objects, transactions may also face aborts and rollbacks. Without preventing dirty reads, another transaction might read data that has already been rolled back, leading to unpredictable consequences due to these anomalies.

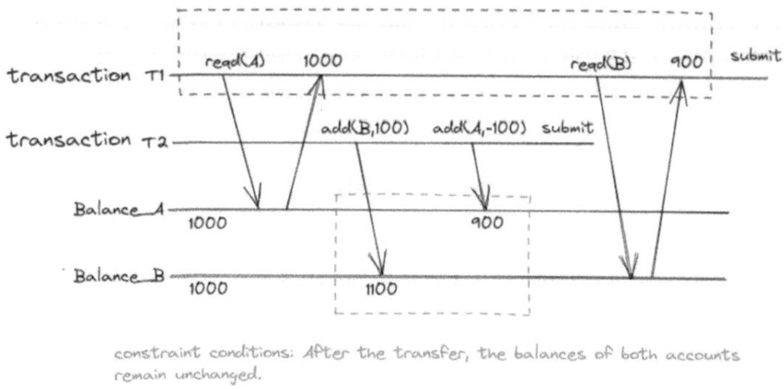

Fig. 4.13 The read committed isolation level cannot prevent read skew

Repeatable Read In Fig. 4.12, a significant issue with the read-committed isolation level is evident. While it prevents dirty reads, it cannot avoid the two anomalies mentioned earlier: non-repeatable reads and read skew. A non-repeatable read occurs when data read within the same transaction differs at different times. Read skew happens when the data read does not satisfy a constraint due to concurrent updates; these anomalies are similar. Therefore, some sources [64] equate these two anomalies. Figure 4.13 illustrates an example of read skew under the read committed isolation level.

Figure 4.13 illustrates a scenario of a bank transfer. Initially, both A and B have a balance of 1000 yuan. There are two concurrent transactions: Transaction T_2 initiated by A to transfer funds to B and simultaneously Transaction T_1 initiated by the banking database system to check the account balances of A and B. Transaction T_1 starts before T_2 and initially sees A's account balance as 1000 yuan. Transaction T_2 begins execution where the transfer process involves first validating business resources (details omitted here), followed by increasing B's account by 100 yuan and decreasing A's account by 100 yuan. This ensures that the total balance of A and B remains unchanged before and after the transfer, satisfying the business data constraints. After Transaction T_2 commits, Transaction T_1 continues and checks B's account balance as 900 yuan. Ultimately, from Transaction T_1's perspective, the account balances of A and B do not satisfy the business data constraints. *Using the read committed isolation level implies that non-repeatable reads and read skew are acceptable.*

Although non-repeatable reads and read skew are not permanent and may not appear during subsequent balance checks, some business systems do not tolerate such anomalies. Therefore, a higher isolation level must be employed. The Repeatable Read isolation level can prevent non-repeatable reads and read skew but cannot prevent phantom reads.

Serializable Serializable isolation level is the strictest level defined by ANSI SQL-92, capable of preventing all concurrency-related data anomalies. Serializable isolation ensures that the outcome of concurrent transactions is identical to their serialized execution. Despite being the safest isolation level compared to weaker levels, serializable isolation is not widely used due to its impact on concurrency performance. It is typically reserved for scenarios where strong isolation guarantees are critical.

Currently, databases implement serializable isolation using two main approaches:

1. Execute multiple transactions in a serial manner.
2. Concurrency control protocols include pessimistic concurrency control, optimistic concurrency control, and multi-version concurrency control protocols.

Concurrency control protocols will be discussed in detail in later chapters. Next, we will introduce a method to achieve the serializability isolation level by executing transactions in a serial manner.

The idea of achieving the serializability isolation level by serially executing multiple transactions is straightforward, but few database vendors implement it for a simple reason: the inefficiency of single-threaded serialization. Executing all transactions sequentially on a single thread, where only one transaction runs at a time, avoids many concurrency issues and eliminates the complexities associated with concurrency control. However, the efficiency of single-threaded execution cannot match that of multi-threaded execution. It was not until the past decade, with rapid advancements in hardware, that traditional database design thinking began to change. Papers like H-Store [62] truly demonstrated the effectiveness of single-threaded databases.

Implementing the transaction serializable isolation level in a serial execution manner has the following background characteristics:

- Memory. As memory capacities have increased, large Online Transaction Processing (OLTP) databases can now be entirely stored in memory. Traditional database designs were disk-oriented and unable to fully leverage the advantages of memory. However, modern database designs prioritize memory over disk, utilizing its benefits to optimize performance.
- Multi-threading. If all data resides in memory, transaction execution processes are typically short. In such scenarios, the advantages of multi-threading in terms of latency and throughput are not significant. Using a single thread can eliminate the need for concurrency control and concurrent data structures, making the overall architecture more reliable compared to multi-threaded designs. As a result, performance may even be better.

The memory database Redis, widely adopted in the industry [109], operates transactions in a serialized manner. Redis is a classic single-threaded database. In Redis, transactions are initiated with the MULTI command to start the transaction state. Subsequently, multiple commands are queued within the transaction. Finally, the EXEC command is used to trigger the transaction commit and execute all queued commands together. Let's explore how Redis avoids the previous issue of non-repeatable read anomalies.

Listing 4.2 Transaction add operation

```
127.0.0.1:6379> multi
OK
127.0.0.1:6379> get key1
QUEUED
127.0.0.1:6379> set key1 value1
QUEUED
```

Listing 4.3 Modify the value of key1

```
127.0.0.1:6379> set key1 value2
OK
127.0.0.1:6379> get key1
"value2"
```

Listing 4.4 Execute the transaction

```
127.0.0.1:6379> exec
1) "value2"
2) OK
3) "value1"
```

To illustrate, initiating a transaction with MULTI and performing write operations within the transaction is demonstrated in Listing 4.2.At this point, open another Redis client and modify the value corresponding to key1, as shown in Listing 4.3.Return to the previous client and commit the transaction, as shown in Listing 4.4.

At this point, you can observe a non-repeatable read anomaly occurring. The transaction's three operations will execute sequentially. The first read operation will fetch the value modified by the other client after the transaction began: key1 = value2. The second read operation within the transaction will then read the modified value by its own transaction: key1 = value1. This inconsistency arises because the transaction reads different values for the same key in two consecutive reads. Similarly, read skew anomalies can also occur.

So how does Redis prevent these data anomalies? Common strategies to avoid read skew or non-repeatable reads typically rely on locking mechanisms. Transactions acquire locks on keys they read during execution, restricting other transactions from accessing the same keys during that time. Redis also implements a similar solution.

First, use the WATCH command to monitor the keys involved in reads and writes before starting the transaction. Then, use the MULTI command to begin a transaction and perform write operations within it, as shown in Listing 4.5.

Listing 4.5 Monitor transaction execution using watch

```
127.0.0.1:6379> watch key1
OK
127.0.0.1:6379> multi
OK
127.0.0.1:6379> get key1
QUEUED
127.0.0.1:6379> set key1 value1
QUEUED
```

Listing 4.6 Modify the value of key1

```
127.0.0.1:6379> set key1 value2
OK
127.0.0.1:6379> get key1
"value2"
```

Listing 4.7 Execute the transaction

```
127.0.0.1:6379> exec
(nil)
127.0.0.1:6379> get msg
"value2"
```

At this point, open another Redis client and modify the content of key1, as shown in Listing 4.6. Return to the previous client and commit the transaction, as shown in Listing 4.7.

From the Redis transaction execution example above, it is evident that using the WATCH command helps transactions avoid non-repeatable reads and read skew anomalies. If a watched key is modified by another transaction during execution, the current transaction will be aborted. In Redis, the single-threaded nature of the main event loop ensures that no other commands are executed while a transaction is running. This guarantees that all transactions are truly serializable and do not violate isolation levels.

The isolation characteristics of the four isolation levels proposed by ANSI SQL-92 for data anomaly prevention are summarized in Table 4.1.

The serializable isolation level can prevent all data anomaly phenomena, but the four isolation levels proposed by ANSI SQL-92 are not exhaustive. The paper mentioned earlier [10] not only supplements data anomaly phenomena but also introduces Snapshot Isolation (SI) and Cursor Stability Isolation Level.

In Snapshot Isolation, each transaction reads from a consistent database snapshot. Initially, a transaction can only see data that has been committed and is the latest committed transaction, known as the snapshot of the database version. Even if data is subsequently modified by another transaction, Snapshot Isolation ensures that transactions see data that has already been committed. Currently, Snapshot Isolation has been implemented in multiple mainstream databases such as Oracle,

Table 4.1 Isolation levels and their handling of data anomalies

Transaction isolation levels	P0 dirty write	P1 dirty read	P2 fuzzy read	P3 phantom read	P4C cursor lost update	P4 lost update	A5A read skew	A5B write skew
Read uncommitted	Impossible	Possible	Possible	Possible	Possible	Possible	Possible	Possible
Read committed	Impossible	Impossible	Possible	Possible	Possible	Possible	Possible	Possible
Repeatable read	Impossible	Impossible	Impossible	Possible	Impossible	Impossible	Impossible	Impossible
Serializable	Impossible	Impossible	Impossible	Impossible	Impossible	Impossible	Impossible	Impossible

MySQL, PostgreSQL, etc. The implementation of Snapshot Isolation typically involves locks and multi-version concurrency control, which will be discussed in subsequent chapters.

At the start of a transaction (denoted as T_1), a timestamp called the Start Timestamp (ST) is obtained. Each historical version of every data item in the database records a corresponding timestamp called the Commit Timestamp (CT). The snapshot that T_1 reads consists of the most recent historical versions of all data items where the CT is less than the ST. Since these data items are historical versions and are not locked by write operations, there are no read-write conflicts, ensuring that read operations within the snapshot are never blocked.

Modifications by other transactions after ST are not visible to T_1. Upon committing, T_1 obtains a CT that ensures it is greater than any existing timestamp in the database at that moment (either ST or CT). This CT serves as the version timestamp for data items during persistence. The write operations of T_1 are also reflected in T_1's snapshot and can be read again by read operations within T_1. After T_1 commits, its modifications become visible to transactions holding ST greater than T_1's CT.

If another transaction (T_2) performs write operations on the same data items during T_1 's execution interval [ST, CT], T_1 will abort, and T_2 will commit successfully. This feature is known as First-Committer-Wins, ensuring no Lost Update occurs. In practice, some databases adjust this to First-Write-Wins, determining conflicts at the time of write operations to reduce the cost of conflicts and prevent Lost Update from occurring.

The Cursor Stability Isolation Level aims to avoid the phenomenon of Lost Update. Recall the Lost Update phenomenon: when transaction T_1 reads a data item, then transaction T_2 updates the data item (possibly based on the earlier read), and then T_1 updates the data item (based on its earlier read value) and commits, a Lost Update occurs. The time sequences of the two transactions are: $r_1[x] \ldots w_2[x] \ldots w_1[x] \ldots c_1$ (Lost Update). Cursor Stability extends the row-locking behavior for SQL cursors under the read committed isolation level. It introduces read operations on cursors (rc), which require holding a long read lock on the current data item of the cursor until the cursor moves or closes (possibly through commit). Moreover, transactions reading on cursors can update rows (wc), and even if the cursor moves on subsequent reads, the write lock remains on the row until the transaction commits.

The overall transaction isolation levels are depicted in Fig. 4.14, excerpted from the Jepsen official website [60] and the paper [8].

In Fig. 4.14, the solid line portion represents transaction isolation levels with lower availability, while the dashed line portion represents levels with higher isolation. Here, lower availability refers to higher constraints on the concurrency order of transactions, potentially affecting transaction execution efficiency, but ensuring higher data accuracy. Most isolation levels in the figure have been previously discussed. However, two additional levels not yet mentioned are *strict serializable* and *monotonic atomic view (MAV)*.

Strict serializable isolation ensures operations occur atomically: suboperations of a transaction do not interleave with suboperations from other transactions. MAV

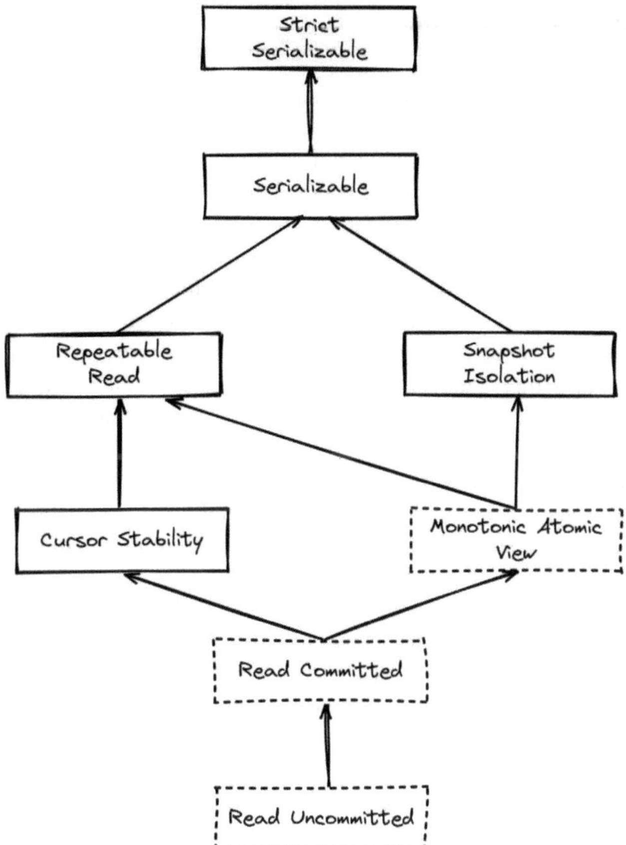

Fig. 4.14 Transaction isolation levels

isolation guarantees that if one transaction observes another transaction, it sees all transaction operations. MAV isolation is less commonly discussed, with minimal presence in the literature. According to the paper [8], MAV isolation is described as follows.

Under monotonic atomic view isolation, once some operations of transaction T_i are observed by another transaction T_j, all operations of T_i must be observed by T_j. Specifically, if T_j reads a version of an object written by T_i, subsequent reads by T_j cannot reflect newer versions of that object modified by T_i. MAV isolation provides a higher level of isolation compared to read-committed isolation.

4.1.2 Global Transactions

Global transactions are a concept contrasted with local transactions, involving transaction mechanisms that span multiple data sources. Returning to the earlier example of bank transfers, where account A transfers funds to account B, from an implementation perspective, this typically involves two steps: "deducting money from account A" and "depositing money into account B." In many cases, these two accounts would reside in different databases and might require two separate service calls. These operations require either both to succeed or both to fail simultaneously. On an e-commerce website, after a consumer clicks the purchase button, the backend processes involve inventory checks, order placement, inventory deduction, and updating order status through a series of service calls. Each operation corresponds to an independent service, often with its own database, thus involving multiple data sources in the transaction.

The scenario described above is a typical example of distributed transactions. Global transactions are not distinct from distributed transactions but can be viewed as a concept within distributed transactions. Compared to distributed transactions, global transactions are a narrower concept. Global transactions are part of the Distributed Transaction Processing (DTP) model defined by X/Open Distributed Transaction Processing Reference Model. The X/Open DTP model is a software architecture model proposed by "The Open Group" [126], which has become a de facto standard for transactional behavior in components of transaction models. It allows multiple applications to share resources provided by multiple resource managers and coordinates their work within global transactions. The X/Open DTP model identifies key entities in the distributed transaction processing environment and standardizes their roles and interactions. Entities in the DTP model include:

- Resource Manager (RM): A database management system or message server management system through which applications control resources.
- Application Program (AP): An application that utilizes the DTP model.
- Transaction Manager (TM): Responsible for coordinating and managing transactions, providing application programming interfaces (APIs) to applications, and managing resource managers.

The X/Open DTP model is illustrated in Fig. 4.15.

In the X/Open DTP model, two protocol interfaces are defined: the XA protocol and the TX protocol. The XA protocol serves as a communication bridge between a transaction manager and multiple resource managers, while the TX protocol acts as the interface for communication between the transaction manager and the application. The X/Open DTP specification, known as the XA specification, aims to enable multiple resources (such as databases, application servers, message queues, etc.) to be accessed within the same transaction. This ensures the ACID properties remain effective across different applications, maintaining transactional integrity and consistency. The XA specification employs the two-phase commit protocol to guarantee that all resources either commit or roll back any specific transaction

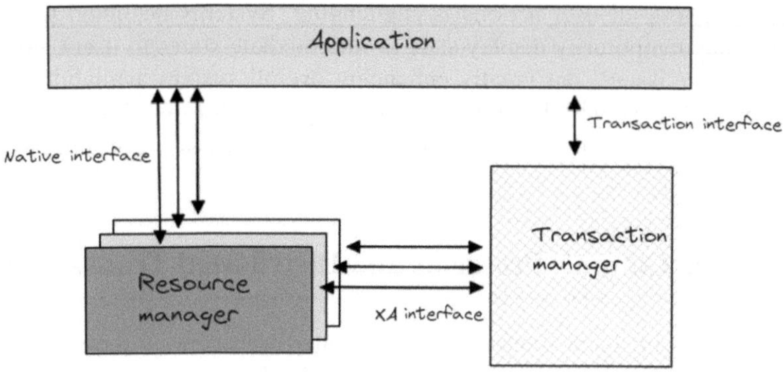

Fig. 4.15 X/Open DTP model [99]

simultaneously. Details of the two-phase commit protocol will be elaborated in subsequent chapters.

4.1.3 Distributed Transactions

In Sect. 4.1.2, we have explored some distributed transaction scenarios and recognized the XA specification as an industrial standard for distributed transaction processing. In this section, we aim to expand on the discussion of distributed transactions by integrating concepts from distributed systems with those of database transactions.

The CAP theorem (detailed in Chap. 1) is an axiom in the field of distributed systems. It states that consistency, availability, and partition tolerance cannot all be achieved simultaneously—only two out of three are possible. Given that partition tolerance is a fundamental requirement in distributed systems (unless ideal conditions of constant communication are assumed, which is impractical), real-world distributed systems typically face trade-offs between consistency and high availability.

Most modern distributed systems today are AP systems, prioritizing high availability at the expense of strong consistency. Therefore, architects at eBay proposed the BASE theory [106] as an extension of the CAP theorem for AP systems. BASE stands for Basically Available, Soft state, and Eventually Consistent, contrasting with the ACID properties mentioned earlier. While ACID ensures atomicity, consistency, isolation, and durability for safe data operations, BASE proposes a shift toward eventual consistency, which aligns better with many business requirements. The core idea of BASE is to maintain core functionality even during system failures, allowing noncritical functionality to be temporarily offline if necessary. BASE is considered the antithesis of ACID and is more compatible with large-scale distributed systems, such as online shopping platforms where achieving strong consistency across all user actions is challenging. Systems designed with BASE principles may temporarily display stale or intermediate states to users, potentially leading to some issues, but greatly enhancing overall system availability in most cases. In Sect. 4.2, we will delve into common solutions for distributed transactions and detail atomic commit protocols in distributed transaction processing.

4.2 Atomic Commit Protocols and Distributed Transactions

In comparison with local transactions, distributed transactions require additional constraints to maintain transactional properties. In this section, we will introduce commonly used methods for preserving transactional characteristics in distributed environments. First, we will discuss atomic commit protocols, which are widely used in the underlying architecture of distributed environments. Following that, we will explore higher-level distributed transaction solutions.

4.2.1 2PC Protocol

In Sect. 4.1.2, we mentioned the use of the XA specification and the two-phase commit (2PC) protocol [12] to ensure that all transactions in distributed transactions can either commit or roll back simultaneously. Phrases like "all transactions... simultaneously..." reflect the atomicity of these transactions as a whole, which prevents partially failed transactions from disrupting the system. Just imagine the scenario of bank transfer operations would be like without introducing the concept of transactions.

The atomicity of local transactions is well-ensured and typically managed by the database storage engine. When a client submits a transaction request to the database, the database first saves the transaction's writes to the log and then appends the commit record to the log. If an unexpected situation interrupts this process, the system determines the safety of the transaction by tracing the log (if the commit record was successfully written to disk before a system crash, the transaction is considered safely committed; otherwise, it rolls back the transaction), thus achieving the goal of "all operations fail/succeed."

Maintaining the atomicity of distributed transactions is much more complex than local transactions, as distributed transactions face more exceptional situations. For instance, due to network partitions, transactions submitted from remote nodes may fail to reach certain database nodes, while other transaction submissions may succeed. Additionally, some database nodes may experience local system crashes during transaction commits, which upon recovery may roll back transactions, whereas other nodes may successfully commit them. The 2PC protocol described in this section is precisely a protocol for atomic commitment of distributed transactions, aiming to maintain their atomicity. The 2PC protocol is a distributed

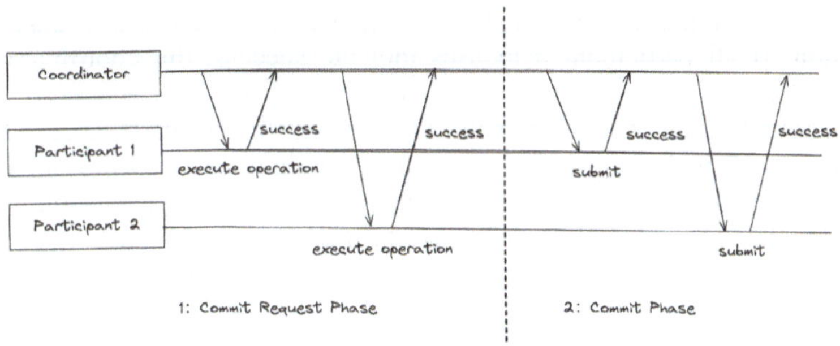

Fig. 4.16 The process of successfully executing a transaction using the 2PC

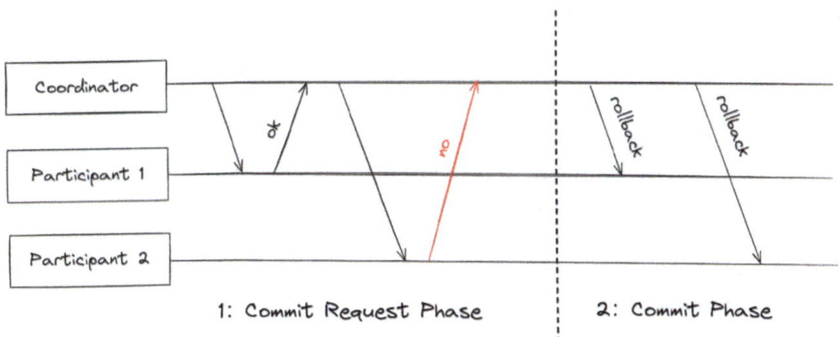

Fig. 4.17 The process of transaction failure in the 2PC protocol

coordination algorithm used for achieving transaction atomicity among multiple nodes. It ensures consistency among all nodes in a distributed system during transaction commits, ensuring either all nodes commit or all nodes abort.

The process of successful execution of the 2PC protocol is depicted in Fig. 4.16, while the rollback process in case of failure is shown in Fig. 4.17. The execution process of a 2PC transaction proceeds as follows.

Phase 1: Commit Request Coordinator Sends Commit Request: The coordinator node, initiating the 2PC transaction, sends a commit request message to all participant nodes. This message includes the transaction's global ID and the specific transactional operations to be executed. Upon receiving this commit request, each participant node checks if its resources meet the transaction requirements and locks these resources. Transactional operations are executed at this stage but are not yet committed. If a participant successfully replies to the coordinator, it indicates the node can ensure completion of the transactional operations under any circumstances and safely persist transaction data to disk.

Phase 2: Commit If Phase 1 fails (due to communication timeouts between the coordinator and participants or if the coordinator receives any execution failure messages), the coordinator notifies all participants to either cancel the transaction or roll it back, regardless of whether the participant successfully executed the transaction. If all participant responses indicate success, the coordinator makes the final decision to write the transaction to the disk's transaction log, known as the commit point, used for recovery in case of system crashes. Once the coordinator node makes the decision to commit and persist it, it will send a commit message to all participants. If any participant fails to receive this message, the coordinator retries until all participants confirm receipt. Participants that encounter failures during commit must retry after recovery until the transaction is successfully committed.

The 2PC transaction is a classic example of a "rigid transaction." The rationale for its rigidity can be observed from its execution process. In Phase 1 of the 2PC transaction, the coordinator offers two options to all participants: either execute or abandon the transaction. If all participants choose to execute the transaction, the coordinator proceeds with the commit and ensures all participants undergo the same process. The 2PC mandates that the purpose of executing the transaction must be achieved, regardless of how many times it fails; the ultimate result must be successful transaction completion. If some participants choose to abandon executing the transaction (due to resource reservation failures or other issues), the coordinator aborts the transaction and instructs all participants to do the same. Therefore, the 2PC can also be viewed as a consensus protocol, ensuring all nodes reach consensus on a transaction decision.

The execution of a 2PC transaction is highly stringent and relies on the following prerequisites for successful completion:

- Each node must be capable of reliably storing a write-ahead log.
- There are no scenarios where nodes remain permanently down.
- The write-ahead log is not susceptible to corruption due to unexpected circumstances.
- There are no situations of permanent network partitioning.

The 2PC protocol achieves node consensus on transaction states in a straightforward manner, but its simplicity comes with several drawbacks:

1. **Coordinator Node Single Point of Failure**. In the discussed execution process of the 2PC protocol, the handling of node failure is equivalent to waiting for the node to recover. However, if the coordinator itself crashes, it results in a serious consequence where all participant nodes await instructions from the coordinator. As depicted in Fig. 4.18, after the coordinator node crashes, participant node 2 remains indefinitely in a state of waiting for transaction execution commands. This is because individual transactions cannot determine whether to proceed with execution or rollback abandonment after the first phase. The 2PC protocol commonly handles coordinator node failures by substituting a standby coordinator node. This standby node queries the status of each participant node to determine whether the transaction should proceed, and sends messages to participants who have not yet executed the second phase. Some might consider allowing participant nodes to query each other about the first

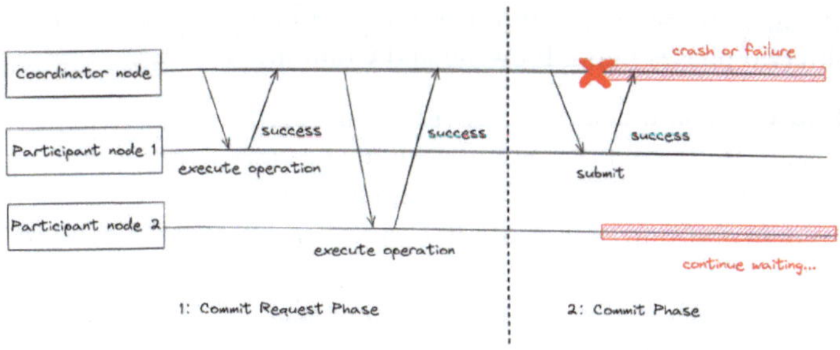

Fig. 4.18 In the event of coordinator node failure in the 2PC protocol

phase's transaction status and combine it with their own execution status to decide the second phase's path. While feasible, this approach extends beyond the scope of the 2PC protocol and intersects with discussions on consensus algorithms in subsequent sections.

2. **Blocking Issue**. After the first phase of the 2PC protocol, resources involved in the transaction remain locked until participant nodes receive commands from the coordinator node in the second phase.
3. **Coordinator and Any Number of Participant Nodes Crash in the Second Phase**. As mentioned earlier, a common approach to handling coordinator node single-point failures is to activate a standby coordinator node to query transaction execution status from other participants, ensuring continuity in the second phase. However, if the standby node cannot query all participant nodes due to partial participant failures, the standby coordinator node cannot ascertain the course of the second phase transaction. In such scenarios, the remaining active nodes in the entire distributed cluster must wait for all failed nodes to recover.

Classic Engineering Practice of 2PC: Percolator Percolator, introduced by Google in 2010 in their paper [102], is a distributed database. The transaction commit protocol used in the paper is an optimized version based on the 2PC protocol, commonly referred to as the Percolator protocol thereafter.

Percolator is a distributed database built on BigTable [24], Google's distributed structured data storage system. Percolator extends BigTable's single-row transactions to support distributed multi-row transactions using a global timestamp server. Next, we will detail how Percolator optimizes the 2PC protocol as a transaction processing protocol.

The Percolator transaction protocol consists of the following three components:

- Client: Equivalent to the coordinator in the 2PC protocol, it initiates the transaction.
- TSO (Timestamp Oracle): The cluster-wide global timestamp server provides globally unique and incrementing timestamps.
- Bigtable: The distributed storage system responsible for persisting data.

The Percolator transaction commit protocol belongs to the standard two-phase commit, with two phases being Prewrite and Commit.

Prewrite Phase Upon transaction initiation, the Client acquires a globally unique timestamp, start_ts, from the TSO. The Client sends a Prewrite request to the underlying distributed storage system, designating the first Key among the multiple rows involved in the transaction as its Primary Key; other Keys are Secondary Keys, each recording its corresponding Primary Key. Each row of data contains three columns for transaction management metadata: lock, write, and data. The lock column stores the locking status of the row's data, the write column stores transaction commit information (data version information), and the data column stores various versions of the row's data. Each row affected by the transaction writes its start_ts into its lock column and inserts new data into the data column along with start_ts. For instance, 5:"value2" denotes that data "value2" was written with start_ts 5. Among these rows, the lock in the locks column of the row corresponding to the Primary Key is chosen as the Primary Lock, while other locks are Secondary Locks, each containing a pointer to the Primary Lock. The transaction rolls back under two conditions: if the data row to be written already has a version in the data column with a larger start_ts or if a lock already exists in the lock column of the data row.

Commit Phase If the transaction is not terminated during the Prewrite phase, it will proceed to the Commit phase. The Client acquires another globally unique timestamp, commit_ts, from the TSO. It checks if the lock in the lock column corresponding to the Primary Key exists. If the lock has been cleared by another transaction, the transaction fails and terminates. If the lock exists, the Client writes commit_ts to the write column as the timestamp and start_ts as the value. Read operations first access the write column to retrieve start_ts, and then use start_ts to read the value from the data column. The Client deletes the lock from the lock column corresponding to the Primary Key. If these steps succeed, the Client can acknowledge a successful transaction commit. All lock columns in Secondary Keys undergo asynchronous deletion of locks and writing to the write column, repeating the above steps.

The specific execution flow of the Percolator protocol is detailed as follows:

1. Initial State. Before the transaction starts, the balances in two accounts, Bob and Joe, are 10 and 2, respectively. The data in the write column indicates that the effective data column corresponds to a timestamp of 5. The column family information is shown in Table 4.2.
2. Prewrite Phase. The transfer transaction begins by acquiring start_ts as 7 from the TSO. Subsequently, the Primary Lock is written into the lock column of Bob's row, which serves as the Primary Key. The column family information is illustrated in Table 4.3.
3. Prewrite Phase. Writing data into Secondary Keys begins. The transaction locks Joe's data, with the lock column designated as a Secondary Lock, containing a reference to Bob's Primary Lock. Simultaneously, data is written into the data column. The column family information is depicted in Table 4.4.

Table 4.2 Initial state

Key	Data	Lock	Write
Bob	5: 10	5:	5:
	6:	6:	6: data@5
Joe	5: 2	5:	5:
	6:	6:	6: data@5

Table 4.3 Prewrite phase-1

Key	Data	Lock	Write
Bob	5: 10	5:	5:
	6:	6:	6: data@5
	7: 3	7: Primary Lock	7:
Joe	5: 2	5:	5:
	6:	6:	6: data@5

Table 4.4 Prewrite phase-2

Key	Data	Lock	Write
Bob	5: 10	5:	5:
	6:	6:	6: data@5
	7: 3	7: Primary Lock	7:
Joe	5: 2	5:	5:
	6:	6:	6: data@5
	7: 9	7: Primary@Bob	7:

Table 4.5 Commit phase-1

Key	Data	Lock	Write
Bob	5: 10	5:	5:
	6:	6:	6: data@5
	7: 3	7:	7:
	8:	8:	8: data@7
Joe	5: 2	5:	5:
	6:	6:	6: data@5
	7: 9	7: Primary@Bob	7:

4. Commit Phase. Within the Primary Key, the Primary Lock is first removed, and a new entry is written into the write column with the new timestamp (specifically commit_ts = 8). Simultaneously, the data in the lock column is cleared. The column family information is represented in Table 4.5.
5. Commit Phase. The lock information within the Secondary Key is cleared. A new entry is written into the write column with the new timestamp, corresponding to the data at start_ts. The column family information is displayed in Table 4.6.

When performing data reads, the first step is to acquire the current read timestamp, ts. Subsequently, it checks whether there exists a lock within the data row being read that falls within the timestamp range [0, ts]:

- If there exists a lock with a timestamp within the range [0, ts], it indicates that the current data is locked by a transaction that started earlier than the current

Table 4.6 Commit phase-2

Key	Data	Lock	Write
Bob	5: 10	5:	5:
	6:	6:	6: data@5
	7: 3	7:	7:
	8:	8:	8: data@7
Joe	5: 2	5:	5:
	6:	6:	6: data@5
	7: 9	7:	7:
	8:	8:	8: data@7

reading transaction but has not yet committed. *Because it's currently impossible to determine whether the transaction locking this data will commit, the current read request cannot be fulfilled. It must wait until the lock is released before proceeding with data retrieval.*

- If there is no lock, or if the timestamp of the lock is greater than ts, then the read request can be fulfilled.

Retrieve the record with the maximum commit_ts within the range [0, ts] from the write column, and use this record to fetch data from the data column. For example, in Table 4.6, if a request is made to read Joe's balance data at ts = 9, first check if the lock within that Key exists. If no lock exists at that moment, the data with the maximum commit_ts from the write column will be read: data@7. This allows the read transaction to determine that the data with start_ts = 7 in the data column is the latest committed data, which is 9.

From the above example of a transfer business, the specific execution process of the Percolator protocol shows significant improvements over the 2PC protocol. Firstly, the coordinator no longer maintains persistent state; instead, it uses the concept of the Primary Key to record the state of the transaction involving the first row of data processed as the Primary Key. Other Keys record their corresponding Primary Key. Percolator introduces a method of multi-version control where each Key records the state of previous transactions, facilitating rollback in case of transaction failures. During transaction execution, if the coordinator fails, other participants can decide whether to roll back or commit their local modifications by querying the transaction state stored in the Primary Key (i.e., the write column).

However, Percolator still faces some limitations inherent in the original 2PC protocol, such as blocking issues. During the Commit phase, all Secondary Keys still need to wait for the Primary Key to complete the commit and persistent data operations, which unavoidably introduces blocking delays in this phase.

4.2.2 3PC Protocol

In the two-phase commit protocol, nodes respond to the coordinator in the first phase indicating successful preparation for the operation. Subsequently, their resources remain locked until the node receives a commit or rollback instruction from the coordinator. Therefore, the two-phase commit protocol is also known as a blocking atomic commit protocol. The three-phase commit (3PC) protocol [119] is an improvement over 2PC and offers better fault tolerance. 3PC divides the first phase of 2PC, known as the "Commit Request Phase," into two stages: the "Request to Prepare Phase" and the "Prepare to Commit Phase." Each phase incorporates timeout handling logic to address resource locking issues caused by single node failures in 2PC. The execution process of the 3PC protocol is depicted in Fig. 4.19.

The execution process of a 3PC transaction is as follows:

1. **Phase 1: Request to Prepare**. In this phase, the coordinator node inquiries all participant nodes to check whether the transaction operation can be executed. If all return "ok," the transaction can proceed smoothly; if any participant node cannot complete the preparation work, the transaction will terminate its execution.
2. **Phase 2: Prepare to Commit**. In the first phase, upon receiving successful preparation responses from all participant nodes, the coordinator can proceed to this stage. During this phase, all participant nodes perform actions similar to the first phase of the 2PC protocol: they write preparatory logs, lock resources, and execute the transaction without committing. After executing these steps, they reply with a message indicating successful execution to the coordinator node. If any participant node encounters a failure during these actions, the transaction execution will be aborted.
3. **Phase 3: Commit**. If all participant nodes successfully complete the second phase, the transaction proceeds to the third phase for committing operations. The coordinator node initiates a commit command to all participant nodes, instructing them to commit the transaction, release locked resources, and subsequently respond with a success message to the coordinator node. Upon receiving

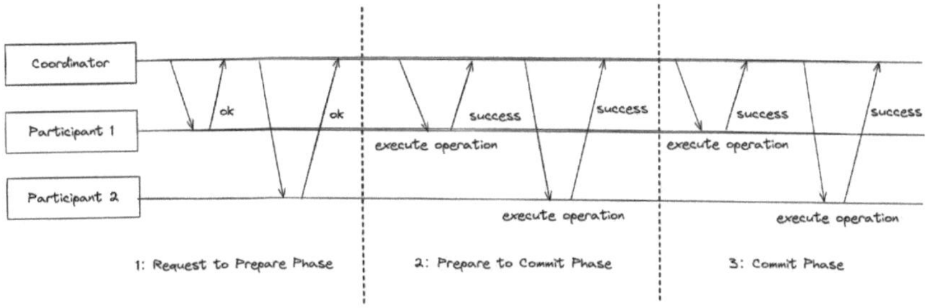

Fig. 4.19 The execution process of the 3PC protocol

confirmation messages from all participant nodes, the coordinator completes the transaction commit process. If any participant node crashes during this phase, the system will wait for recovery and ensure the completion of the transaction commit once the node has recovered.

The execution process of the 3PC protocol is very similar to the 2PC protocol, except that it adds an additional phase at the beginning of the 2PC's first phase to ask all nodes if they are ready to execute the transaction. What is the purpose of adding this preparation phase? In the 2PC protocol, the first phase has too much workload, and unexpected situations may cause nodes to fail during this phase. If some nodes have resources occupied by other transactions, the work done by other participating nodes would be in vain. Asking all nodes whether they can execute before the transaction phase can greatly reduce the probability of a participant node crashing and causing the entire cluster transaction to roll back.

3PC reduces the probability of transaction failure and rollback. Therefore, if the coordinator node crashes during the second phase, all participant nodes will default to committing the transaction based on the timeout period set in advance, instead of continuing to wait as in the 2PC protocol. In a sense, this alleviates the single-point-of-failure problem of the coordinator node in 2PC. Therefore, the 3PC protocol is also considered a *non-blocking protocol* that does not indefinitely lock resources and can continue to complete transaction commits after a single node failure.

However, 3PC also has significant drawbacks. As mentioned above, the default transaction commit operation increases the risk of data inconsistency in the 3PC protocol because *network partitioning issues cannot be avoided* in actual systems. Therefore, if the coordinator node's abort transaction command is delayed due to network latency and cannot be delivered to the participant nodes for a long time, some participant nodes will consider the coordinator node to have failed and commit the transaction in the default manner, while others may receive the abort transaction command normally. This inconsistency in transaction execution status ultimately leads to data inconsistency. In addition to data inconsistency issues, 3PC also increases transaction latency due to the additional communication flow compared to the 2PC protocol. Therefore, the widely used protocol currently is still the 2PC protocol.

4.2.3 Best Efforts 1PC Transaction

Previously, we introduced the 2PC and 3PC protocols, which operate on the underlying distributed transaction infrastructure and do not directly involve specific technologies used at the higher layers. In this section, we will introduce the *Best Efforts 1 Phase Commit* transaction, which represents a completely different transaction execution mode from the previously discussed 2PC and 3PC protocols, albeit sharing some similarities.

From the term "one-phase commit," it can be inferred that this is a simplified operation compared to the original 2PC. Under the Best Efforts 1 Phase Commit mode, there is no coordination among transaction managers of the participating resources; instead, each resource manager executes the commit operation in sequence. In this mode, if an error occurs during the business processing of a resource and no commit has been performed on any resource, all resources can be rolled back. This mode significantly reduces the likelihood of failures in system basic components (such as network and hardware). By optimizing the centralized coordination among resource managers in the two-phase commit, better concurrency and higher throughput performance can be achieved.

Best Efforts 1 Phase Commit can also be referred to as a reliable event queue or a final consistency solution based on reliable messaging. It primarily leverages the reliability mechanisms of message middleware (such as Kafka, RocketMQ, or RabbitMQ) to achieve data consistency. For instance, in the payment scenario of an e-commerce platform, once a user completes an order payment, they do not need to synchronously wait for the payment result but can proceed with other tasks. For the system, most actions occur after initiating the payment, awaiting asynchronous payment notifications from third-party payment platforms, and then adjusting the payment status of the order based on these results. Additionally, if the payment is successful, many e-commerce platforms will also add certain bonus points to the user's account based on marketing strategies. Therefore, when the system receives the payment result from the third party, it needs to update the payment status in the payment service and adjust the points balance in the account service, posing a data consistency issue between these two services. From this scenario, it is evident that real-time data consistency is not required, making it suitable to adopt a final consistency solution based on reliable messaging to ensure data consistency between the payment service and account service.

Figure 4.20 depicts a simple example of a transaction processing procedure using the best-effort one-phase commit mode. We assume microservice A represents user order business, and microservice B represents specific payment operations:

1. The application initiates a transaction with microservice A. The merchant creates a payment order.
2. Microservice A starts a local transaction and sends out a message. The merchant sends a payment-related message to the message queue.
3. The messaging middleware sends a request to another microservice B to execute another transaction. The message queue sends a payment execution message to a component responsible for payment operations. The payment component then creates a payment transaction for the merchant and records the payment status based on the user's payment result.
4. Return the execution result of microservice B's transaction. Upon completion of the payment, trigger a callback notification to the message queue.
5. The message queue dequeues the execution result of microservice B to microservice A. The message queue sends a callback notification to the merchant.

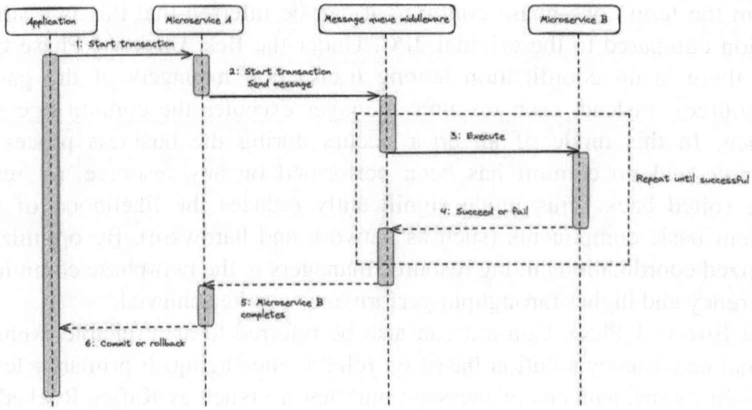

Fig. 4.20 Example of best efforts 1 phase commit execution

6. Microservice A decides whether to commit or roll back the entire transaction based on the execution result from microservice B. Upon receiving this notification, the merchant updates the status of the local payment order according to the result.

In this order processing scenario, ideally, the transactional state of the payment operation and the order status of the merchant should achieve eventual consistency after notification of completion. However, due to network uncertainties, notifications of payment results may fail or get lost, resulting in an unknown state for the merchant's payment orders. Therefore, the role of best-effort one-phase commit transactions becomes crucial. Suppose the merchant does not return a "SUCCESS" status code upon receiving the payment result notification. In that case, the callback request will be retried using an exponential backoff retry mechanism (increasing notification intervals), such as 1 minute, 5 minutes, 10 minutes, and 30 minutes, until reaching the maximum retry attempts. If the merchant still does not provide a definitive status after the specified attempts, default operations are typically executed or human intervention is required. For instance, microservice B may offer an additional transaction result query interface where the merchant can query the payment status based on the payment order number and update the payment order status accordingly. This process can be triggered by a timer or manually reconciled.

This pattern could lead to inconsistency if a database transaction succeeds at step 4 but fails at step 6. To mitigate such issues, implementing idempotent operations in business processes can prevent duplicate execution due to data inconsistency. Alternatively, transactional compensation mechanisms can ensure eventual consistency, resolving the aforementioned problems.

4.2.4 TCC Transaction

TCC (Try-Confirm/Cancel) transactions [55], also known as two-phase commit (2PC) at the application layer, consist of three phases. During each execution, not all three phases are necessarily performed; only one of the latter two phases is executed:

- Try. In the Try phase, two main operations are performed. First, business checks ensure the consistency of business resources, maintaining the state of resources before and after operations. Second, resource reservation isolates resources to be operated on from other transactions, akin to locking, preventing resource anomalies due to concurrent access by multiple transactions.
- Confirm. In the Confirm phase, no commit operations are executed. Instead, business operations are performed on resources prepared during the Try phase. TCC transactions assume that the Confirm phase will also succeed if the Try phase succeeds. Therefore, if the Confirm phase fails, it retries continuously. Hence, operations in the Confirm phase are idempotent.
- Cancel. In the Cancel phase, if errors occur during the Try phase, all operations performed in the Try phase are undone, and all business resources reserved in the Try phase are released. Similar to the Confirm phase, operations in the Cancel phase are assumed to always succeed. If Cancel fails, it retries continuously. Therefore, operations in the Cancel phase are also idempotent.

TCC transactions are not the original 2PC protocol based on the infrastructure level but operate at the user business code layer. Unlike traditional transaction mechanisms like X/Open XA Two-Phase-Commit, TCC transactions do not rely on Resource Managers (RM) supporting XA. Instead, they manage distributed transactions through scheduling of business logic provided by the business system.

For example, bank transfer operations, executing this business by using TCC transaction mechanism, are illustrated in Fig. 4.21.

The Detailed Execution Process of TCC Transactions Is As Follows

1. The transaction manager enters the Try phase by creating a transaction ID. Initially, it performs business checks to verify whether the transaction's execution adheres to business data consistency constraints. If the business checks pass without issues, it proceeds with resource locking operations. In the context of a bank transfer transaction, resource locking involves freezing the balance of account A.
2. Perform the Try operation on the receiver's status.
3. In the Confirm phase, the transaction manager completes the business process based on the results from the Try phase. This includes deducting 100 units from account A's balance and crediting 100 units to account B's balance. If there are network issues or other exceptions during this phase, the operation will be retried continuously until the Confirm phase successfully completes.

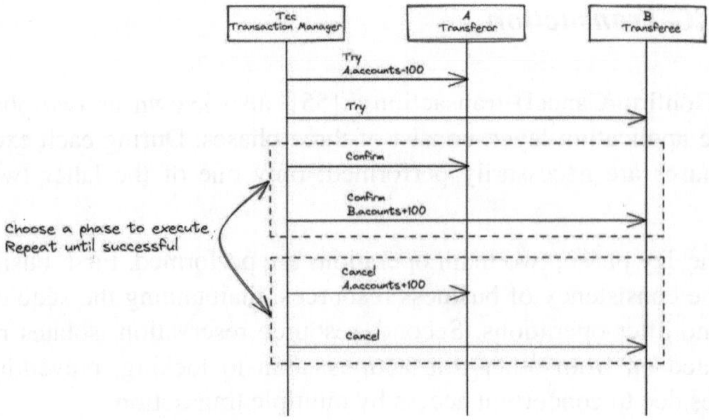

Fig. 4.21 TCC transaction process diagram

4. If the transaction manager cannot successfully pass the Try phase due to scenarios like network exceptions, violations of business data consistency, or unsuccessful resource locking, it will enter the Cancel phase. Specifically, for account A, the operation will revert the deduction of 100 units and release the lock on its resources. Similarly, for account B, the operation will cancel the transaction.

During the invocation of business operations, in the Try phase, the objective is not to complete the business logic immediately, but rather to observe if each service can operate normally and if necessary resources can be frozen. If the Try phase proceeds smoothly, it means databases, caches, message queues, etc., are writable and the required resources (such as frozen balances) are reserved.

Next, each service executes the Confirm phase. Due to the guarantees provided by the Try phase, the Confirm phase can largely ensure the completion of a distributed transaction. If a service fails during the Try phase, such as a database crash or Redis failure, each service's Cancel logic is automatically executed to roll back the operations performed in the Try phase, ensuring either success together or failure together. In case of unexpected incidents, like a sudden service crash followed by a restart, how does the TCC distributed transaction framework ensure the continuation of previously unfinished distributed transactions? TCC transaction frameworks generally incorporate a logging module, primarily used to record activities related to distributed transactions. These logs can be stored in log files on disk or in databases, capturing the state and relevant data of each phase of the distributed transaction's execution. If services encounter problems due to other reasons, the logs play a critical role.

Indeed, the TCC process bears a striking resemblance to the 2PC protocol. However, as outlined above, TCC transcends abstract transaction operations, delving into specific business aspects and allowing adjustments in the granularity of resource locking as per requirements. TCC transactions are commonly implemented using existing frameworks such as ByteTCC [18] and TCC-TRANSACTION [124].

4.2.5 SAGA Transaction

Although TCC transactions are widely used in the industry, they have a high degree of intrusion into business code. For certain distributed transaction scenarios where processes are complex, are lengthy, and may involve calling services from other companies, especially those that are uncontrollable and cannot adhere to the TCC development pattern, the first phase of TCC transactions may fail to complete successfully. In response to such business scenarios of distributed transaction processing, the SAGA transaction model has been proposed.

The SAGA transaction model was initially introduced in a paper by Hector Garcia-Molina and Kenneth Salem in 1987 [46]. This paper proposed improving the efficiency of "Long Lived Transactions," which occupy database resources for extended periods, hindering the completion of other transactions. Long-lived transactions are composed of multiple short transactions that can be seen as a collection of short transactions that can run interleaved. Each sub-transaction in a SAGA is a true transaction, mutually related and executed as a cohesive unit: if any sub-transaction in the SAGA fails to execute properly, compensatory actions must be performed.

Compensation actions are analogous to the Cancel phase in TCC transactions discussed earlier. For instance, if reducing inventory is a short transaction, then restoring inventory would be its compensatory transaction. Breaking down a long-lived transaction T into multiple short transactions $T_1, T_2, \ldots, T_n$, each short transaction has a corresponding compensatory transaction $C_1, C_2, \ldots, C_n$. The SAGA transaction manager generates a directed acyclic graph (DAG) based on the program's execution process. If a rollback operation is needed, compensatory transactions are invoked in the reverse order of the DAG:

- Sub-transaction T_i and its compensatory transaction C_i should be idempotent, meaning executing them multiple times produces the same result as executing them only once.
- The normal outcome of distributed transaction T should be consistent with the successful completion of all its sub-transactions T_i.

The SAGA transaction has two execution paths: one is the sequential execution of $T_1, T_2, \ldots, T_n$, resulting in a successful commit of T; the other is sequential execution of sub-transactions where an error occurs, triggering the execution of compensatory transactions in reverse order $T_1, T_2, \ldots, T_n, C_n, \ldots, C_2, C_1$, resulting in the cancellation of T. These two execution processes are illustrated in Fig. 4.22. If all sub-transactions can be successfully committed, then transaction T is successfully committed. Otherwise, compensatory actions are required. In SAGA transactions, there are primarily two types of compensation strategies: forward recovery and backward recovery.

Forward Recovery For forward recovery, the SAGA transaction execution component requires reliable code copies and a checkpoint for all missing transactions. This checkpoint is used by the application or system depending on which one halted

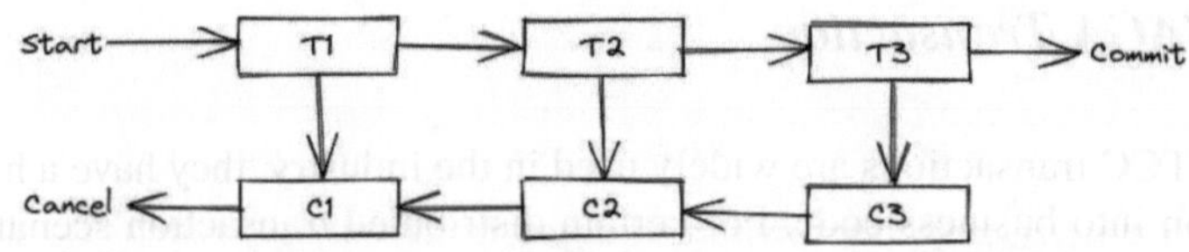

Fig. 4.22 SAGA transaction execution process

the SAGA transaction execution. In a system crash, the recovery component can designate the most recent checkpoint for each SAGA transaction. If a transaction execution fails, retries are performed continuously until the transaction can be executed successfully, similar to the best-effort approach mentioned earlier. Forward recovery does not necessitate the provision of compensatory transactions. In practical scenarios where sub-transactions ultimately succeed or defining compensatory transactions is difficult or impossible, forward recovery aligns better with business requirements. The sequence of forward recovery transaction execution is as follows: $T_1, T_2, \ldots, T_j$(fail), T_j(redo), $\ldots, T_n$, where no compensatory transactions are involved.

Backward Recovery In backward recovery, all completed transactions are compensated if any sub-transaction fails. This involves executing the sequence $T_1, T_2, \ldots, T_j, C_j, \ldots, C_2, C_1$, where T_j is the sub-transaction that encountered an error. This approach aims to undo all previously successful sub-transactions, thereby canceling the entire SAGA transaction execution. In theory, compensatory transactions should never fail. However, in a distributed environment, servers may crash, networks may fail, or even data centers may experience power outages. In such cases, the last resort is to provide fallback measures, such as manual intervention.

For services, implementing SAGA involves the following requirements:

1. T_i and C_i are idempotent.
2. C_i must be capable of succeeding; if it fails, manual intervention is required.
3. The execution results of $T_i - C_i$ and $C_i - T_i$ must be the same: rollback of the sub-transaction.

The first requirement is that T_i and C_i must be idempotent. For example, if T_i times out during execution, the outcome is uncertain. Applying a forward recovery strategy would involve resending T_i, potentially resulting in T_i being executed twice. Hence, T_i must be idempotent. Similarly, with a backward recovery strategy, C_i would be sent in response. If C_i also times out and is resent, C_i could potentially be executed twice, necessitating C_i to be idempotent.

The second requirement mandates that C_i must succeed. This is crucial because if C_i fails to execute successfully, it implies the Saga cannot be fully rolled back, which is unacceptable. However, exceptional situations such as bugs in C_i's code or prolonged service crashes may necessitate manual intervention.

The third requirement, though initially unconventional, can be illustrated as follows: consider the scenario where T_i times out, and a backward recovery approach is adopted, resulting in the transmission of C_i. In this case, three potential outcomes arise:

1. T_i's request is lost; consequently, T_i will not be executed before or after this event.
2. T_i is executed before C_i.
3. $C_i$is executed beforeT_i.

For the first case, it is easily manageable. For the second and third cases, however, it is required that T_i and C_i are commutative, and their final results both involve the cancellation of the sub-transaction.

The execution process of SAGA transactions is notably different from TCC transactions in that it does not involve resource locking or reservation operations. The benefit of not having resource reservation operations is clear: some business processes are straightforward, and adapting them to TCC would require modifying the original business logic, whereas SAGA only requires adding a compensatory action. Additionally, without reservation actions, there is no concern about resource release, making exception handling simpler.

TCC is designed for short transactions with high consistency requirements. Transactions with high consistency requirements are generally short; if a transaction takes a long time to complete, the consistency appears poor to users, making TCC unnecessary for such highly consistent designs. Therefore, the orchestration of TCC transaction branches is embedded in the program code, allowing flexible invocation based on the results of each branch. SAGA, on the other hand, targets transactions with lower consistency requirements, whether short or long. For scenarios like booking airline tickets, where transactions may last from several minutes to a day or two, the entire transaction orchestration needs to be stored on servers to prevent loss of transaction orchestration information due to application upgrades, failures, and similar reasons affecting applications initiating global transactions.

4.3 Concurrency Control Protocols

Concurrency control protocols are proposed to address the data anomaly issues caused by multiple concurrent transactions, primarily maintaining the isolation of transactions. In this chapter, we will introduce three concurrency control protocols: pessimistic concurrency control protocol, optimistic concurrency control protocol, and multi-version concurrency control protocol. There is no single best concurrency control protocol, as each has its advantages and disadvantages.

4.3.1 Pessimistic Concurrency Control Protocol

The term "pessimistic" in pessimistic concurrency control (PCC) represents a conservative approach: if some operations might conflict with other operations, directly abandon executing those operations and wait until conflicts can't occur before executing them. Similar to the concept of locks in concurrent programming, in practical schemes, pessimistic concurrency control also uses locks and timestamps to control the order of transaction execution, which inevitably incurs additional overhead. This chapter introduces the pessimistic concurrency control protocol from the perspectives of locks and timestamps.

In the database transaction concurrency control mechanism, there are mainly two types of locks: one is the write lock (X-Lock), also known as an exclusive lock or mutex; the other is the read lock (S-Lock), also known as a shared lock. Data resources occupied by a mutex cannot be shared with other transactions, but a shared lock can share critical resources with another transaction using a shared lock. The access conditions of resources by shared locks and mutexes are shown in Table 4.7.

The typical operation process of lock algorithms is roughly as follows:

1. Before using the data item x, the transaction needs to request a lock from the lock manager.
2. If x is already locked and the current lock is incompatible with the requested lock, the transaction will be delayed, waiting for the lock to be released.
3. Otherwise, the lock manager will grant the lock to the transaction.

Figure 4.23 shows the execution process of two transactions accessing the same data item. As can be seen, simply using mutexes and shared locks cannot achieve transaction isolation, leading to inconsistent results of transaction scheduling (executing these two transactions multiple times randomly, the final value of A read by transaction T_1 cannot remain consistent).

Two-Phase Locking Protocol (2PL) is a widely used protocol in databases. It defines how a transaction coordinates locks with other transactions during its operation to achieve serializable isolation levels. If two transactions simultaneously attempt to execute write operations on the same data, designers can straightforwardly think of using locks to implement concurrency control. 2PL is similar to this naive use of locks but is slightly more complex. 2PL is divided into two phases: the locking phase and the unlocking phase. In the locking phase, locks can only be acquired and not released; in the unlocking phase, locks can only be released and not acquired, as shown in Fig. 4.24.

Because of the existence of these two phases, the situation where two transactions are interleaved as shown in Fig. 4.23 will no longer exist. Under the 2PL protocol,

Table 4.7 Access conditions of shared locks and mutexes

	X-Lock	S-Lock
X-Lock	Conflict	Conflict
S-Lock	Conflict	Conflict-free

Fig. 4.23 Transaction scheduling scheme based on X-Lock and S-Lock

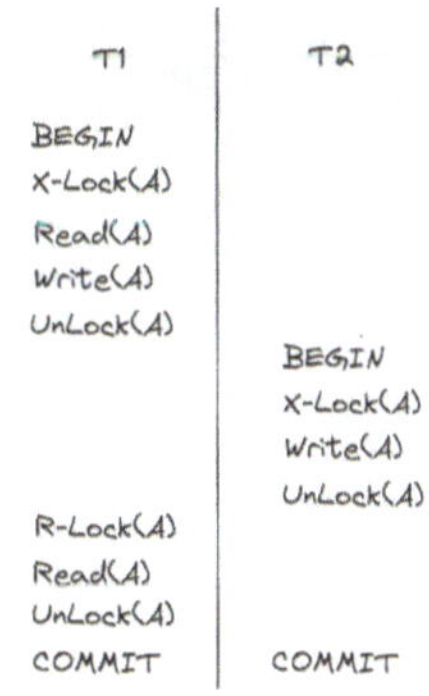

Fig. 4.24 2PL protocol [48]

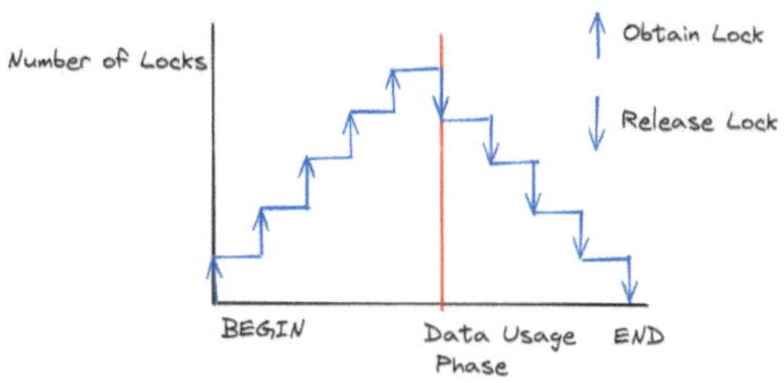

after T_1 executes the *Write*(A) operation, it will not release the lock because 2PL requires the lock acquisition operations to be gathered in one phase and then released uniformly at the end. T_1 will not release the lock until it has finished executing *Read*(A), and before releasing the lock, transaction T_2 naturally cannot obtain the lock to perform transaction operations.

2PL can ensure the isolation of transactions and guarantee serializable isolation levels. However, as you can see from the above description, the throughput and response time of 2PL transactions are much lower compared to other weaker isolation-level algorithms. This is a typical "pessimistic approach" where any operations that could potentially lead to data contention between two concurrent transactions are prohibited, and one transaction must wait for the other to complete before starting its operations. But even disregarding performance issues, 2PL is still unsatisfactory because it can lead to *Cascading Aborts* issues, as shown in Fig. 4.25. Transaction T_1 completes all lock acquisition steps and enters the lock release phase in 2PL after executing *Write*(A). At this point, transaction T_2 can acquire the relevant locks on data A. If transaction T_1 can commit, there will be no further issues. However, if T_1 encounters an error in subsequent operations and needs to roll back, then transaction T_2 must also roll back its operations on data A. This is the cascading aborts problem. The read operation performed by transaction T_2 is also known as a dirty read [52], where a transaction reads the results of an uncommitted operation from another transaction.

The *Strict Two-Phase Locking Protocol (S2PL)* can solve the cascading aborts problem. In S2PL, a transaction holds on to the write locks it has acquired until the transaction terminates, as shown in Fig. 4.26. In the scenario mentioned above, if the

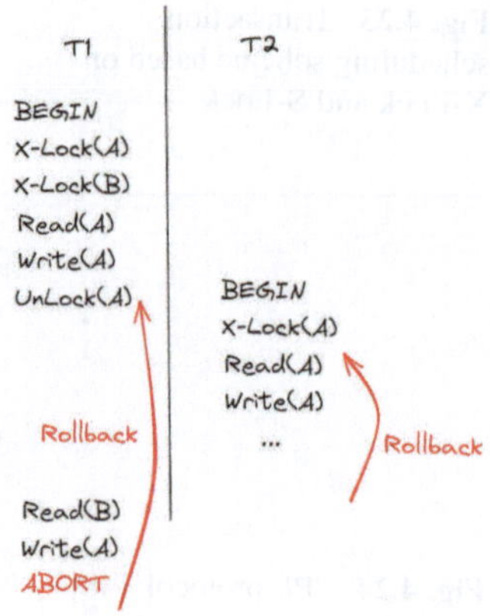

Fig. 4.25 Cascading aborts problem in 2PL

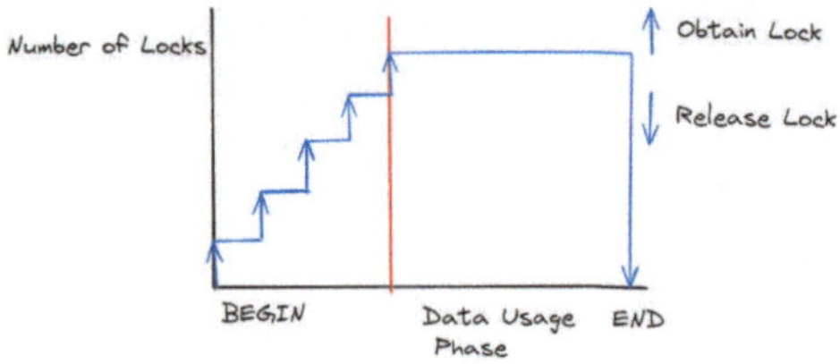

Fig. 4.26 Changes in the number of write locks in S2PL protocol [48]

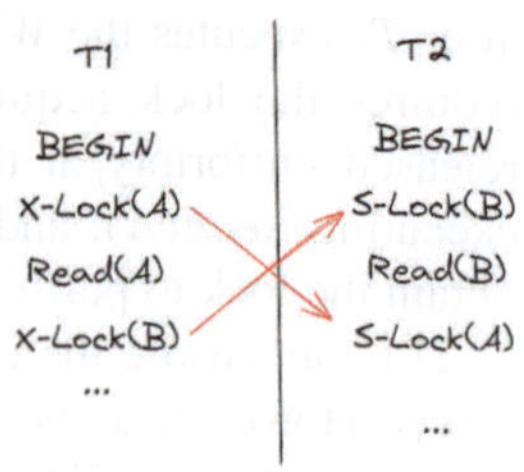

Fig. 4.27 Deadlock problem

concurrency control protocol is replaced with S2PL, transaction T_2 cannot obtain the mutex or shared lock on data A before transaction T_1 completes. Therefore, T_2 will be delayed, preventing the occurrence of cascading aborts and avoiding dirty reads.

In addition to S2PL, there is also the *Strong Strict Two-Phase Locking Protocol (SS2PL)*, which not only retains write locks but also retains read locks until the transaction ends. Since the abovementioned concurrency control protocols are based on locks, deadlock issues naturally need to be considered. Unfortunately, the 2PL protocol and its derivatives cannot avoid deadlocks. Consider the scenario shown in Fig. 4.27. Transaction T_1 first acquires an exclusive lock on A and then attempts to acquire an exclusive lock on B. Meanwhile, transaction T_2 first acquires a shared lock on B and then attempts to acquire a shared lock on A. Deadlock occurs at this point: transaction T_1 needs to acquire an exclusive lock on B to continue, but since transaction T_2 already holds a shared lock on B, T_1 cannot obtain the exclusive lock on B. Similarly, transaction T_2 cannot acquire a shared lock on A.

To solve the deadlock problem, database systems often use the following two approaches:

- **Deadlock Detection**. The database maintains a directed graph representing the state of lock requests by various transactions, known as the "wait-for graph." Each node in the graph represents a transaction, and each directed edge represents a lock-waiting relationship. An edge from node A to node B means that transaction A is waiting for transaction B to release a lock. The database periodically checks the wait-for graph, and if it finds a cycle, it will break the cycle by rolling back one or more of the involved transactions.
- **Deadlock Prevention**. Assign each transaction a priority based on a timestamp, whereas transactions with smaller timestamps have higher priority. There are generally two schemes:
  - Higher Priority Waits for Lower Priority (Wait-Die): When a lower-priority transaction attempts to acquire a lock held by a higher-priority transaction, it triggers the higher-priority transaction to roll back and release the lock. Conversely, when a higher-priority transaction seeks to acquire a lock held by a lower-priority transaction, it must wait for the lower-priority transaction to complete and release the lock naturally. This asymmetric behavior ensures system stability while managing resource conflicts between transactions of different priorities.
  - Lower Priority Waits for Higher Priority (Wound-Wait): In priority-based locking schemes, a higher-priority transaction can compel a lower-priority transaction to roll back and relinquish its lock. Conversely, a lower-priority transaction must wait for a higher-priority transaction to release its lock voluntarily. This asymmetric approach ensures that high-priority operations can proceed without undue delay while maintaining system integrity.

Under the 2PL and its derivative protocols, if transactions occur very frequently, deadlocks may also occur very frequently, leading to severe performance issues. If deadlock detection is used to solve the problem, the appropriate period for checking the wait-for graph needs to be considered. If the checking frequency is too high, it will impact performance; if the checking frequency is too low, the duration of deadlocks will be too long, which will also affect performance. Deadlock prevention has similar issues, such as the selection of transactions to roll back. Simply rolling back transactions based on timestamp alone is unsuitable because some transactions may have already been executed for a long time, making it inappropriate to roll them back. Thus, the basis for determining priority also needs to be considered.

To summarize, the 2PL protocol separates the locking and unlocking phases, where each phase can only perform the specified steps for that phase. We introduced two variants of 2PL: S2PL and SS2PL. In S2PL, transactions hold write locks until the transaction terminates, while in SS2PL, transactions hold both read and write locks until the transaction ends. Besides these two variants, there is also the *Conservative Two-Phase Locking Protocol (C2PL)*, where transactions acquire all necessary locks at the start.

Some distributed transaction models reflect these concurrency control protocol ideas, such as Percolator, an improved protocol of 2PC. Percolator stipulates that all secondary locks will not be released before the primary lock in multi-row transactions is released. After the primary lock is released, the secondary locks will also be released, which aligns with the S2PL protocol. Distributed databases generally use S2PL or SS2PL to achieve serializable isolation levels. However, these protocols require transactions to hold locks for extended periods, which can severely impact transaction concurrency performance, especially for long transactions. Therefore, most database users do not configure the isolation level to be serializable.

The *Serialization Graph Testing (SGT)* concurrency control protocol is based on the Multi-Version Serialization Graph (MVSG) proposed in the paper [2], often simply referred to as the Serialization Graph. This graph serves as a tool for analyzing conflict scenarios in database transactions. Each transaction is represented as a vertex within the graph, while the relationships between transactions are depicted as edges connecting these vertices. This visual representation facilitates the identification and analysis of transaction interdependencies and potential conflicts. Transactions are vertices, and when an operation in one transaction conflicts with an operation in another transaction, a directed edge can be drawn between the two vertices. An edge from transaction T_1 to transaction T_2 (requiring T_1 to precede T_2 in the execution order) can be created by three types of dependencies:

1. Write-Read Dependencies. If T_1 writes a version of an object and T_2 reads that version, T_1 appears to execute before T_2.
2. Write-Write Dependencies. If T_1 writes a version of an object and T_2 replaces that version with the next version, T_1 appears to execute before T_2.
3. Read-Write Dependencies. If T_1 writes a version of an object and T_2 reads the previous version of that object, T_1 appears to execute after T_2 because T_2 does not see T_1's update. Sometimes, these are also called read-write conflicts.

If a cycle emerges in the graph created by these three types of dependencies, the execution fails to conform to any serial order. This indicates a serializability conflict arising from Snapshot Isolation anomalies. Figure 4.28 shows a timing diagram of three transactions executing serially.

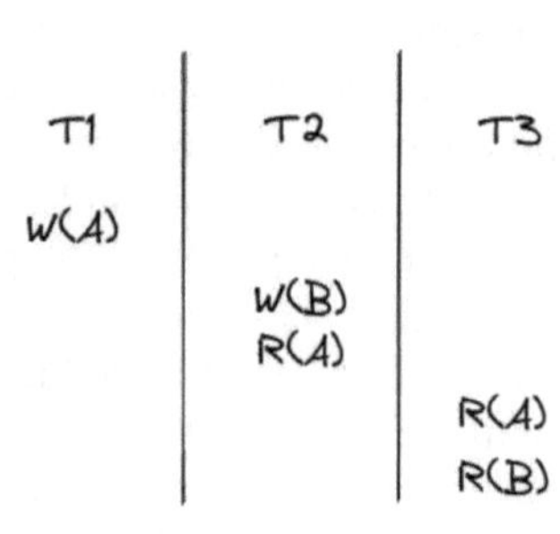

Fig. 4.28 Serial execution timing of transactions

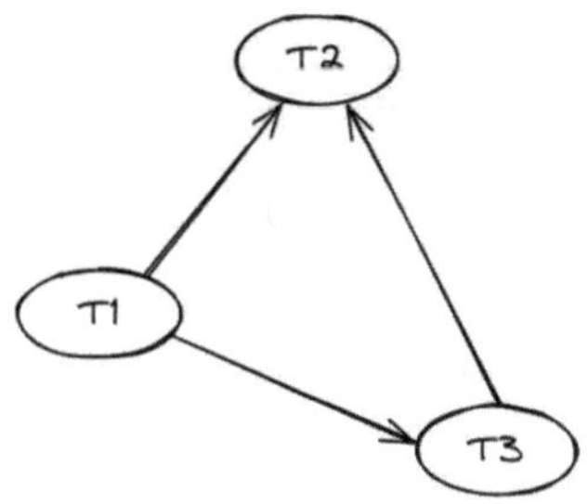

Fig. 4.29 Serialization graph of transactions

Fig. 4.30 Execution timing of transactions with deadlock

T1	T2
W(A)	W(B)
W(B)	W(A)

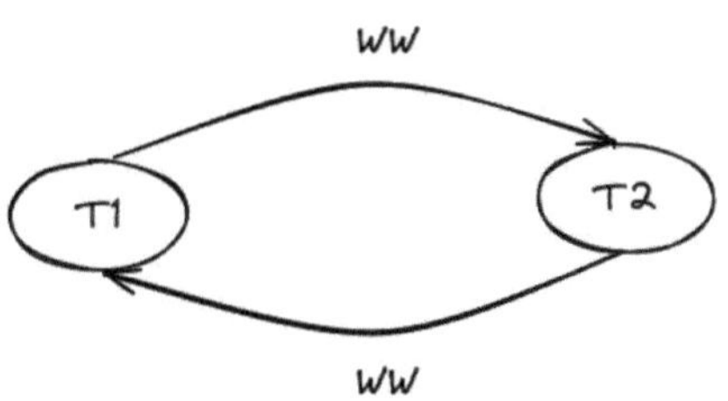

Fig. 4.31 Serialization graph of deadlock

In Fig. 4.28, three transactions are executed sequentially. Transaction T_1 first executes W(A); then T_2 executes R(A), creating a write-read dependency between T_1 and T_2, forming a directed edge from transaction T_1 to transaction T_2. Similarly, the W(B) operation of transaction T_2 and the R(B) operation of transaction T_3 create a read-write dependency. The W(A) operation of transaction T_1 and the R(A) operation of transaction T_3 also create a read-write dependency, forming two more directed edges from vertex T_2 to vertex T_3 and from vertex T_1 to vertex T_3, respectively. The resulting serialization graph is shown in Fig. 4.29.

Figure 4.29 is evidently an acyclic graph, indicating that the execution of the above three transactions can achieve serializable scheduling. What does the serialization graph look like if we construct two transactions that will deadlock, as shown in Fig. 4.30?

In Fig. 4.30, transactions T_1 and T_2 each perform write operations on different data items and then request locks from each other, causing a deadlock. The corresponding serialization graph is shown in Fig. 4.31.

The write-read dependencies and write-write dependencies introduced earlier align with our intuition about inter-transaction dependencies, but read-write dependencies might not be as intuitive. The paper [104] describes a scenario involving read-write dependencies among three transactions: there are two tables, a receipts table that records the day's income with each row containing a batch number, and a control table that maintains only the current batch number. There are three types of transactions:

- T_2 NEW-RECEIPT. Reads the current batch number from the control table and inserts a new entry with that batch number in the receipts table.
- T_3 CLOSE-BATCH. Closes the current batch by incrementing the current batch number in the control table, so subsequent transactions are assigned to the next batch.
- T_1 REPORT. Reads the current batch number from the control table and then reads all entries from the receipts table with the previous batch number (i.e., displays the total receipts for the previous day).

Once the T_1 REPORT transaction displays the total for a specific batch, subsequent transactions cannot alter the total for that batch. This is because the T_1 REPORT transaction displays the total for the previous batch and must follow the CLOSE-BATCH transaction. Each T_2 NEW-RECEIPT transaction must either precede both transactions, making its receipts visible to T_1 REPORT, or follow T_3 CLOSE-BATCH, in which case it will be assigned the next batch number. T_1 needs to execute after T_3, and T_2 needs to execute before T_3, so that receipts appear in the report for the current batch. The anomalous execution timing is shown in Fig. 4.32.

1. T_2 executes first to process a new receipt, acquiring the current batch number x from the control table.

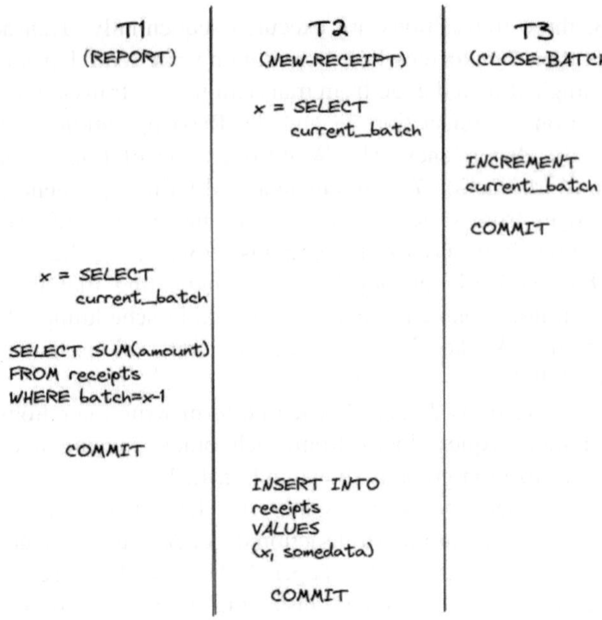

Fig. 4.32 Example of read-write dependency anomaly

Fig. 4.33 Serialization graph

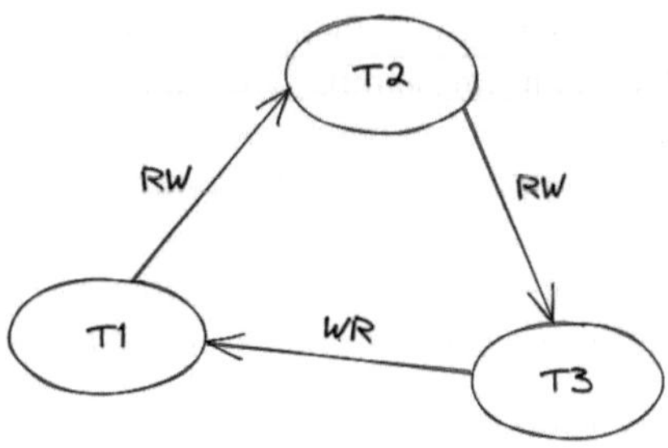

2. T_3 executes next, updating the batch number. The batch number obtained by T_2 is now outdated, but T_2 is still using x to record the receipt.
3. T_1 executes after T_3, but T_2's transaction has not yet been committed, so the receipt processed by T_2 is missing from T_1's report. Furthermore, this receipt will not appear in any subsequent report, leaving it without a trace.

Next, we use a serialization graph to illustrate this anomaly. First, determine the operations each transaction performs on the data items: T_1 reads the batch number from the control table and then reads the receipts table (R(control), R(receipts)); T_2 reads the batch number from the control table and then writes to the receipts table (R(control), W(receipts)); T_3 writes to the batch number in the control table (W(control)). Clearly, vertex T_2 and vertex T_3 form a read-write dependency edge on control, T_3 and T_1 form a write-read dependency edge on receipts, and vertex T_1 and vertex T_2 form a read-write dependency edge on receipts, as shown in the serialization graph in Fig. 4.33. Therefore, the three transactions described above cannot be executed under serializable isolation. Although T_1 is a read-only transaction, it creates a cycle involving the three transactions, proving that the intuition that read-only transactions do not affect concurrent execution is incorrect.

The *timestamp-ordering protocol* avoids the overhead of locks by predefining a sequence among transactions. A transaction's timestamp can be understood as its ID, derived either from the actual physical time or a logical counter. Transaction T_i is assigned a globally unique timestamp $TS(T_i)$, which is a time marker assigned by the DBMS's concurrency control manager before transaction T starts. Timestamp ordering ensures that $TS(T_i) < TS(T_j)$ if transaction T_i enters the database system before transaction T_j. The main idea is to order transactions by their timestamps. A schedule of transactions is serializable if the only allowed serial schedule follows the timestamp order. In simple terms, the schedule corresponds to a specific serial order based on the transactions' timestamps. The protocol ensures that for each data item accessed by conflicting operations, the access order does not violate the timestamp order. Two timestamp values are associated with each data item X to ensure this:

- $W\_TS(X)$. The largest timestamp of any transaction that successfully executed write(X).
- $R\_TS(X)$. The largest timestamp of any transaction that successfully executed read(X).

This protocol manages concurrent execution to determine a serializable order based on timestamps. The timestamp-ordering protocol ensures that any conflicting read and write operations are executed in timestamp order. Whenever a transaction T attempts to issue $R\_item(X)$ or $W\_item(X)$, the basic timestamp-ordering protocol compares T's timestamp with $R\_TS(X)$ and $W\_TS(X)$ to ensure that the timestamp order is not violated. The basic timestamp-ordering protocol for the following two cases is described as follows:

1. Whenever transaction T issues a $W\_item(X)$ operation, the following occurs:
   - If $R\_TS(X) > TS(T)$ or $W\_TS(X) > TS(T)$, then the transaction T is aborted and rolled back, and the operation is rejected.
   - Otherwise, the $W\_item(X)$ operation of T is executed, and $W\_TS(X)$ is set to $TS(T)$.
2. Whenever transaction T issues a $R\_item(X)$ operation, the following occurs:
   - If $W\_TS(X) > TS(T)$, then the transaction T is aborted and the operation is rejected.
   - Otherwise, if $W\_TS(X) <= TS(T)$, the $R\_item(X)$ operation of T is executed, and $R\_TS(X)$ is set to the larger of $TS(T)$ and the current $R\_TS(X)$.

An example of successfully committing transactions based on the timestamp-ordering protocol is shown in Fig. 4.34. Initially, the read timestamp R_TS and write timestamp W_TS of data items A and B are both 0. Transaction T_1 starts first, with a timestamp TS of 1. T_1 first performs the read operation on data item B. According to the timestamp-ordering protocol, since $W\_TS(B) <= TS(T_1)$, the operation can be executed, and the read timestamp $R\_TS(B)$ is set to the larger of $TS(T_1)$ and the current $R\_TS(B)$, which is $TS(T_1)$, thus $R\_TS(B) = TS(T_1) = 1$.

Transaction T_2 starts execution with a timestamp TS of 2. T_2 performs the read operation on data item B, similar to T_1's read operation, comparing with $W\_TS(B)$, and satisfying the condition, setting $R\_TS(B)$ to 2. T_2 then proceeds to update data item B. According to the timestamp-ordering protocol, since $R\_TS(B) = TS(T_2)$ and $W\_TS(B) < TS(T_2)$, the operation can be executed, and the write timestamp

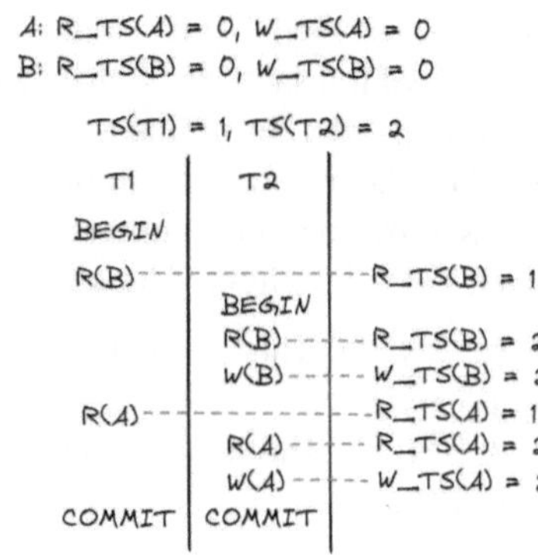

Fig. 4.34 Successful execution of transactions with timestamp-ordering protocol

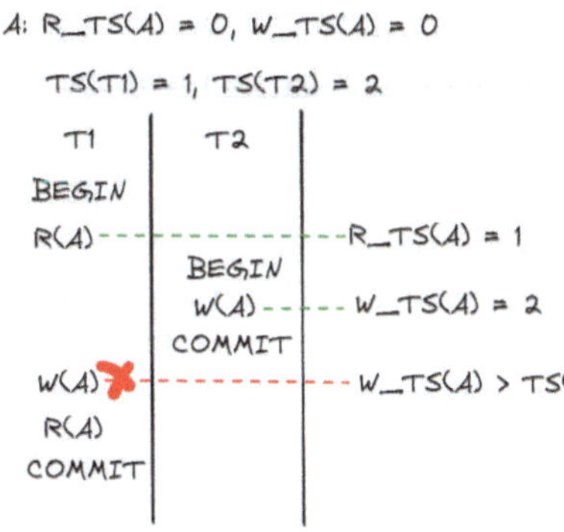

Fig. 4.35 Failed execution of transactions with timestamp-ordering protocol

$W\_TS(B)$ is set to 2. T_1 and T_2 continue execution with similar operations on data item A, with no read-write conflicts, allowing both transactions to commit successfully.

An example of a failed execution based on the timestamp-ordering protocol is shown in Fig. 4.35. Initially, the read timestamp R_TS and write timestamp W_TS of data item A are both 0. Transaction T_1 starts first, with a timestamp TS of 1. T_1 first performs the read operation on data item A. According to the timestamp-ordering protocol, since $W\_TS(A) <= TS(T_1)$, the operation can be executed, and the read timestamp $R\_TS(A)$ is set to the larger of $TS(T_1)$ and the current $R\_TS(A)$, which is $TS(T_1)$; thus $R\_TS(A) = TS(T_1) = 1$.

Transaction T_2 starts execution with a timestamp TS of 2. T_2 performs the update operation on data item A. According to the timestamp-ordering protocol, since $R\_TS(A) = TS(T_2)$ and $W\_TS(A) < TS(T_2)$, the operation can be executed, and the write timestamp $W\_TS(A)$ is set to 2, and transaction T_2 commits. T_1 continues execution, attempting to update data item A. According to the timestamp-ordering protocol, since $W\_TS(A) < TS(T_1)$, T_1 cannot complete the operation, and transaction T_1 is aborted. Although the timestamp-ordering protocol requires T_1 to roll back, it is unnecessary. Why? Because T_2 has already written to A, and the value T_1 wants to write to A will never be read. The reasons are as follows:

- Any transaction T_i satisfying $TS(T_i) < W\_TS(A) = TS(T_2)$ attempting to read(A) will be rolled back because it does not meet the timestamp-ordering protocol.
- Any transaction T_j satisfying $TS(T_j) > W\_TS(A) = TS(T_2)$ must read the value written by T_2 and not T_1.

Thus, the conclusion is that the write(Q) operation of T_3 is outdated and can be ignored. The execution result is equivalent to the serial execution of T_1 and T_2. To reduce the transaction conflict rate, T_1 can sometimes avoid aborting, simply not executing the operation. As long as the final state of the database is the same, it is sufficient. If X has already been read by a future transaction, T_1 must roll back because the future transaction reads an incorrect value; if X has been written by a future transaction, the database can ignore T_1's write.

The *Thomas Write Rule* ensures serializability by removing outdated write operations issued by transactions. When a transaction T_i issues a write operation W(A):

- If $TS(T_i) < R\_TS(A)$, then the value of A generated by T_i is needed earlier but has been assumed not to be generated by the system. Hence, the write operation is rejected, and T_i is rolled back.
- If $TS(T_i) < W\_TS(A)$, then the value of A that T_i attempts to write is outdated, and the write operation can be ignored.

The Thomas Write Rule ensures serializability by deleting outdated write operations issued by transactions. Whenever the timestamp-ordering protocol detects two conflicting operations occurring in the wrong order, it rejects the later operation by aborting the transaction that issued it. The schedules generated by the timestamp-ordering protocol are guaranteed to be serializable and deadlock-free, as conflicting transactions are rolled back, restarted, and assigned new timestamps rather than waiting. However, the protocol does not guarantee that the schedules are recoverable, and thus needs improvement to ensure recoverable and cascadeless schedules. The basic timestamp-ordering protocol may lead to cascading rollbacks. Suppose we have transaction T_1, and T_2 uses the value written by T_1. If T_1 is aborted and resubmitted to the system, T_2 must also be aborted and rolled back.

Summary of the timestamp-ordering protocol:

- The timestamp-ordering protocol ensures serializable isolation levels.
- The timestamp-ordering protocol ensures no deadlocks and no transaction waits.
- The schedules may not be cascadeless or even recoverable.

A variant of the basic timestamp-ordering protocol is called the *Strict Timestamp Ordering* protocol, which ensures strictly conflict-serializable schedules. In this variant protocol, a transaction T_2 issuing R(X) or W(X) will have its read or write operations delayed until the transaction T_1 that wrote the value of X has committed or aborted if $TS(T_2) > W\_TS(X)$.

4.3.2 Optimistic Concurrency Control Protocol

In the previous section, we introduced pessimistic concurrency control protocols, which achieve transaction concurrency control using locks or timestamps. Pessimistic concurrency control reduces overall efficiency to avoid low-probability bad events. To address the inefficiency of pessimistic concurrency control, Professor H.T. KUNG from Carnegie Mellon University proposed the optimistic concurrency control (OCC) protocol in the paper [67]. At the time of this paper's publication, the mainstream concurrency control protocols were primarily the lock-based 2PL protocols. The authors pointed out five disadvantages of lock-based concurrency control protocols:

1. High overhead of lock management, especially for read-only operations that still require locking and the cost of deadlock detection.
2. No universal deadlock-free protocol exists, so complex lock protocols need to be customized to avoid deadlock issues.
3. Data objects locked and waiting for disk I/O operations significantly reduce system concurrency throughput.
4. Locks cannot be released at any time but only at the end of the transaction, reducing concurrency throughput.
5. Locks are meant to solve low-probability bad situations and should not become the norm.

Optimistic concurrency control requires transactions to operate in a private workspace, so their modifications are invisible to others until they commit. When a transaction is ready to commit, it will validate all involved data items to check for conflicts with other transactions. If validation fails, the transaction must be aborted and restarted later. Optimistic concurrency control protocols clearly overcome the deadlock problem.

Compared to pessimistic concurrency control protocols, optimistic concurrency control protocols have the following advantages: no deadlocks and avoidance of any time-consuming node-locking scenarios. This approach is general in the sense that if transactions are primarily query-based (read operations), the overhead of optimistic concurrency control is almost negligible. In this method, *read operations are completely unrestricted, while write operations of transactions are strictly restricted*.

Optimistic concurrency control handles transactions in three phases:

- **Read Phase**. Transactions maintain a read set and a write set. The read set records data involved in read operations, and the write set records data involved in write operations. All read operations first access the transaction's local memory space. If the data does not exist, it is read from the database and cached in the transaction's private memory space. All write operations' cache is kept in a private space exclusive to the transaction and is not exposed to other transactions.
- **Validation Phase**. The transaction scheduler will validate the above operations to check if they meet a serializability standard and if the result can be replicated to the database. If validation fails, the transaction will be terminated; otherwise, it proceeds to the next phase.
- **Write Phase** (optional). The cached data in the transaction's private space is written to the database, making it globally visible.

Serialization Equivalence Validation is used in the validation phase to check if the transaction execution meets the serializability standard. During the validation phase, each transaction is assigned a transaction number T_i at the end of the READ phase, allocated in sequence. If the transaction passes the validation phase and completes successfully, it retains this number. If validation fails and the transaction is aborted or if the transaction is read-only, it releases this number for reassignment. Transaction numbers are assigned in ascending order. The transaction number

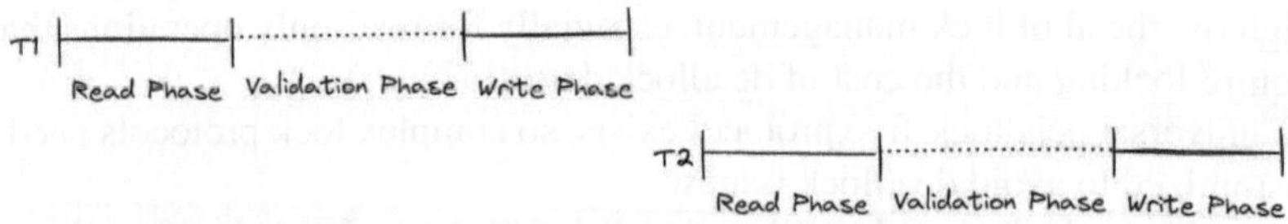

Fig. 4.36 The T_2 read phase occurs after the T_1 write phase is completed

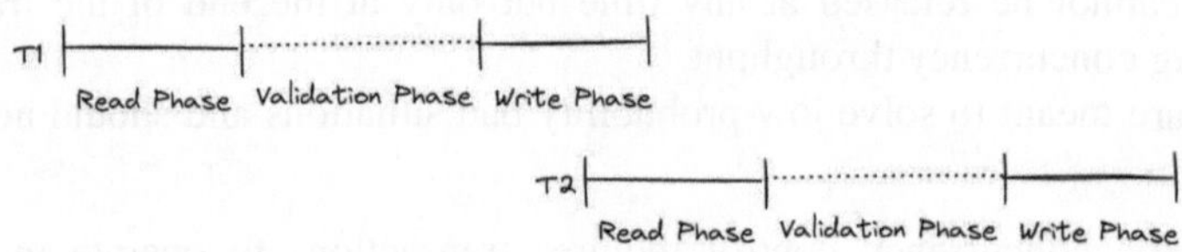

Fig. 4.37 The T_1 write phase and the T_2 read phase overlap

Fig. 4.38 T_1 and T_2 overlap in most phases

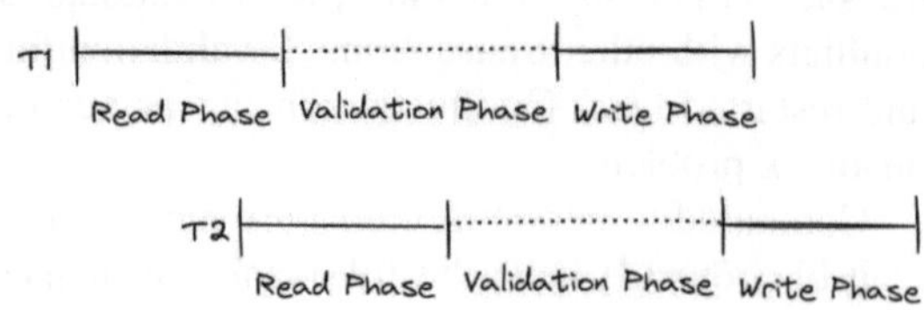

defines its position in time, and T_{id} satisfies the following properties: if $T_i < T_j$, transaction T_i occurs earlier. To check if T_j meets the serializable isolation level (i.e., T_i completes before T_j), the operations meet one of the following three validation conditions:

- As shown in Fig. 4.36, T_i completes the write phase before T_j starts the read phase. In this case, the execution time of the two transactions does not overlap, and T_i commits before T_j, ensuring data integrity.
- As shown in Fig. 4.37, T_i's write set does not intersect with T_j's read set, and T_i completes the write phase before T_j starts the write phase.
- As shown in Fig. 4.38, T_i's write set does not intersect with T_j's read or write sets, and T_i completes the read phase before T_j completes the read phase.

During the validation phase, there are two methods of validation: backward validation and forward validation. However, all transactions must use the same validation strategy. Backward validation compares with previously committed transactions. Forward validation compares with transactions that are currently executing but have not yet committed.

Optimistic concurrency control avoids the overhead of using locks when transaction conflicts are infrequent, potentially leading to better performance for database transactions. However, when transaction conflicts are frequent, it can result in frequent rollbacks, reducing efficiency.

4.3.3 Multi-version Concurrency Control Protocol

Multi-version concurrency control (MVCC) maintains multiple versions of data for each data item. When a transaction updates data, it creates a new version rather than modifying the previous version. MVCC was first proposed in the 1978 paper [111], and the detailed implementation was described in the 1981 paper [11]. The advantages of MVCC are as follows:

- Write operations do not block read operations, and read operations do not block write operations.
- No need for locks.
- Supports time-travel queries.

MVCC is not part of the previously mentioned optimistic or pessimistic concurrency control; it can be combined with either of these concurrency control protocols.

MVCC can be broken down into the following components:

1. Concurrency control protocol, which can be either a pessimistic concurrency control protocol or an optimistic concurrency control protocol.
2. Data version storage.
3. Garbage collection, which discards data versions that are no longer needed.

Data Version Storage

In MVCC, a data item can have multiple versions. Database management systems usually manage these multiple versions using a version chain. As the name implies, multiple versions of the same data item form a chain. There are three storage methods for version chains:

- Append-Only Storage. As shown in Fig. 4.39, all versions of the data are stored in the same table, with new data directly appended at the end of the chain.
- Time-Travel Storage. The latest data versions and historical data versions are stored separately. The latest data version has a pointer linked to the table of historical data versions. As shown in Fig. 4.40, the Main Table stores the latest version of the data, and the Time-Travel table stores historical versions of the data in chronological order, with the latest version pointing to the old versions.
- Delta Storage. As shown in Fig. 4.41, each time a version update occurs, only the changed fields are stored in the Delta Table, not all fields. The chain structure is similar to Time-Travel storage, with the Main Table, but the Delta Table does not store all field information of the old version data.

Garbage Collection

Over time, the number of data versions increases, so a garbage collection strategy is needed to remove outdated data. For example, some versions are not observed by active transactions, or others are created by aborted transactions.

- Clearing based on timestamps. Typically, a daemon thread periodically checks the versions of each data item. It will be deleted if the version's timestamp is less

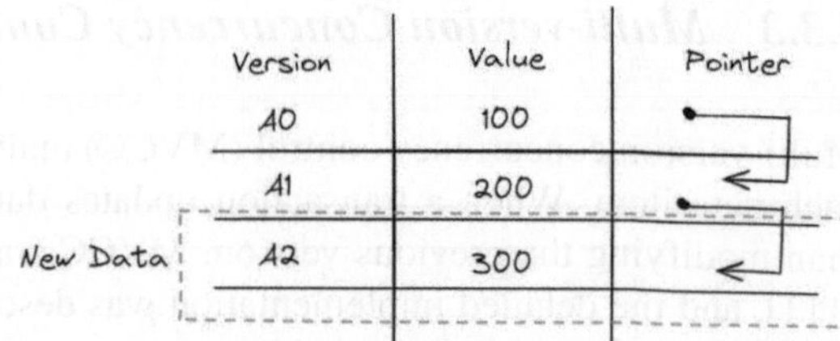

Fig. 4.39 Append-only storage

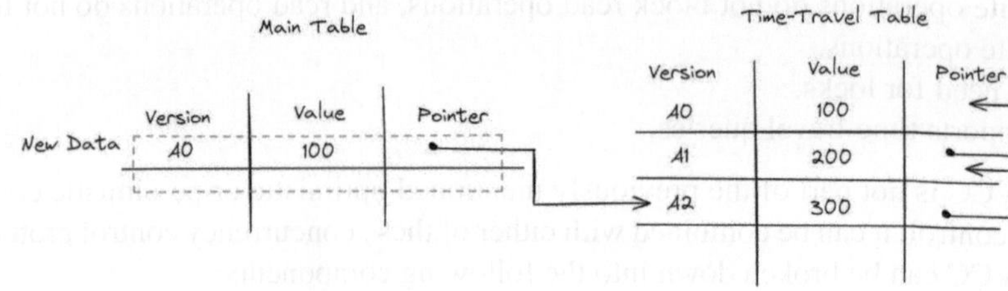

Fig. 4.40 Time-travel storage

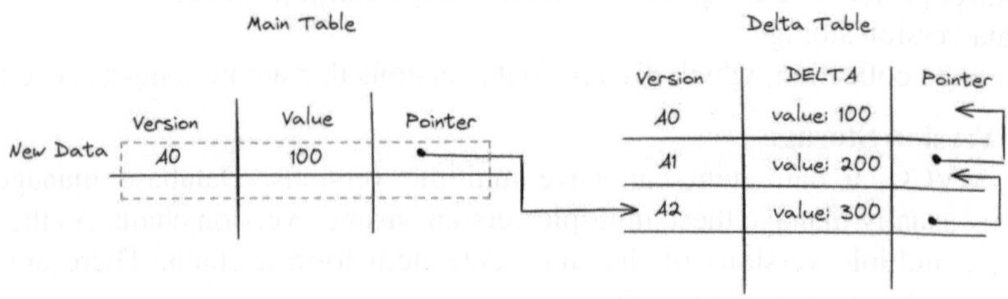

Fig. 4.41 Delta storage

than the minimum timestamp of the current active transactions. An optimized version of this method uses an additional bitmap to record modified data, so the daemon thread only checks modified data, improving garbage collection efficiency.

- Clearing based on transaction read/write sets. Each transaction has a read/write dataset. When the database decides that the versions created by the transaction can be cleared, it clears the data versions based on the read/write set.

4.4 Summary

In this chapter, we primarily introduce transaction processing in a distributed environment, using a common example of geographically distributed banking services to highlight the important role that transactions play in distributed data services. We start from the perspective of data service calls to different data sources and introduce three types of transactions: local, global, and distributed. In local transactions, we explain the ACID properties of transactions, transaction states, their transitions, and transaction isolation levels. Transaction isolation levels are a crucial concept in transactions, often used to determine the data quality of data services. Alongside introducing transaction isolation levels, we also discuss data anomalies, such as dirty reads, dirty writes, and phantom reads. These data anomalies can occur in actual data service calls, and only higher transaction isolation levels can prevent them from happening. Global and distributed transactions can be understood as single or multiple data services involving multiple data sources. We introduce the X/Open DTP model, which is currently the industrial standard for distributed transaction processing, followed by a discussion on transaction solutions in a distributed transaction environment.

In the distributed transaction solutions section, we explore atomic commit protocols frequently employed in the underlying architecture of distributed environments, such as the 2PC and 3PC protocols. We analyze the strengths and limitations of these atomic commit protocols. Subsequently, we introduce higher-level distributed transaction solutions that build upon these atomic commit protocols in distributed system architectures. These include TCC transactions and SAGA transactions. These advanced transaction models offer diverse approaches to rollback and fault tolerance in distributed transactions, enhancing the preservation of ACID properties in distributed environments.

Concurrency control protocols maintain transaction isolation. We introduce concurrency control protocols from three categories: pessimistic concurrency control protocols, optimistic concurrency control protocols, and multi-version concurrency control protocols. Pessimistic concurrency control protocols embody a conservative approach, assuming that operations will conflict. Therefore, these protocols strictly adhere to certain specifications throughout, either prohibiting operations that might conflict from executing or following a strict rule set. We introduce two-phase locking protocols, strict two-phase locking protocols, and other pessimistic concurrency control protocols. In contrast, optimistic concurrency control protocols operate on the principle of correcting mistakes if they occur. They mainly consist of three phases: the read phase, the validation phase, and the write phase. The degree of overlap between different transactions in these phases determines whether the transactions meet the serializable isolation level. Multi-version concurrency control protocols can be combined with pessimistic or optimistic concurrency control protocols and introduce a mechanism for data version storage, allowing multiple versions of the same data item to be stored. This supports time-travel queries, making MVCC a commonly used concurrency control protocol in actual data services today.

Chapter 5
Consistency of Distributed Data Services

In this chapter, we will introduce the concept of consistency in distributed data services, which is one of the most important concepts in this field. In distributed data services, clients are located in different geographical locations. If there is only one data source, the single-point load pressure and long-distance communication overhead are unbearable. Naturally, people have considered using multiple data sources and load balancing mechanisms to solve these problems. However, this brings new challenges, as multiple data sources can lead to data inconsistency due to the "replication lag" problem. This is because a client's write request cannot reach different data sources simultaneously. Even if it could reach different data sources simultaneously, it is impossible to ensure that all data sources can successfully execute the write request. Therefore, users may encounter the following situation: client C_1 requests the data service to update the value V_1 to V_2, but another client C_2, located in a different geographical location, still sees the value as V_1. Although clients do not directly communicate with each other and thus will not discover this data anomaly, it indeed happens and can seriously affect the quality of the data service. In this chapter, we will introduce some data synchronization methods, data consistency levels, and data consistency/consensus algorithms to see how modern distributed systems solve these issues.

5.1 Data Synchronization Methods

The foundation for maintaining data consistency is data synchronization methods, which implement data operations and state synchronization at the database level to achieve data consistency. In the following chapters, we will introduce three common data synchronization methods: master-slave replication, multi-master replication, and masterless replication.

Z. Xu, *Distributed Data Services*, https://doi.org/10.1007/978-981-96-9392-4_5

5.1.1 Master-Slave Replication

The master-slave architecture is one of the most widely used architectures in distributed systems. The term "master-slave" indicates that in a distributed system, the roles and statuses of each node are unequal. If node A issues commands to node B, directing B's subsequent operations, then node A is the master node, and node B is the slave node. In a master-slave architecture, it typically refers to a single-master, multiple-slave setup, distinguishing it from multi-master architectures. The common client/server architecture is also a type of master-slave architecture.

Master-slave replication is based on the master-slave architecture. It allows data from one database server (the master server) to be replicated to one or more other database servers (the slave servers). The master server records the write operation updates and then propagates these operations to all the slave servers. The operation mechanism of master-slave replication is as follows:

- The master node receives the client's write request, executes it locally, and then sends the operation instruction to the other nodes. Write operation requests can only be sent to the master node; otherwise, data consistency cannot be achieved.
- The slave nodes receive the write operation instructions from the master node, execute these operations, and ensure they are executed in the same order as on the master node.
- Both master and slave nodes can receive and execute read operation instructions.

Master-slave replication can be either synchronous or asynchronous. The difference lies only in the timing of change propagation. If the master server waits for the slave servers to execute the update operations after receiving a client's write operation, and only returns the write operation result to the client after all data synchronization is complete, it is synchronous replication. If the master node returns the execution result to the client without waiting for the slave nodes to complete the write operations, it is asynchronous replication. The execution processes of synchronous and asynchronous replication based on the master-slave architecture are shown in Figs. 5.1 and 5.2, respectively. Synchronous replication ensures that the data accessed by clients is always the latest version, but it has a lower throughput compared to asynchronous replication because each client write operation takes

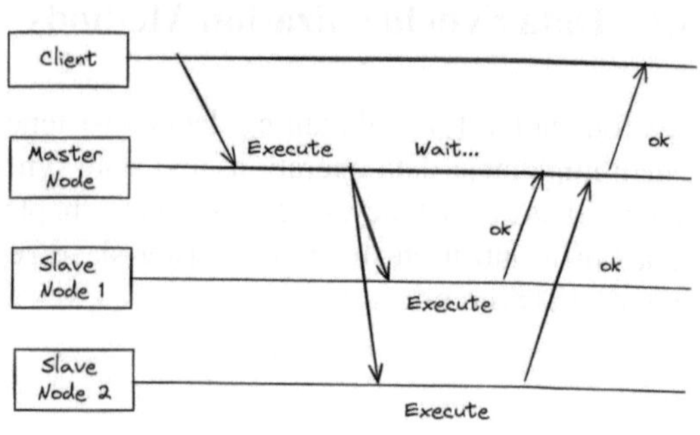

Fig. 5.1 Master-slave synchronous replication

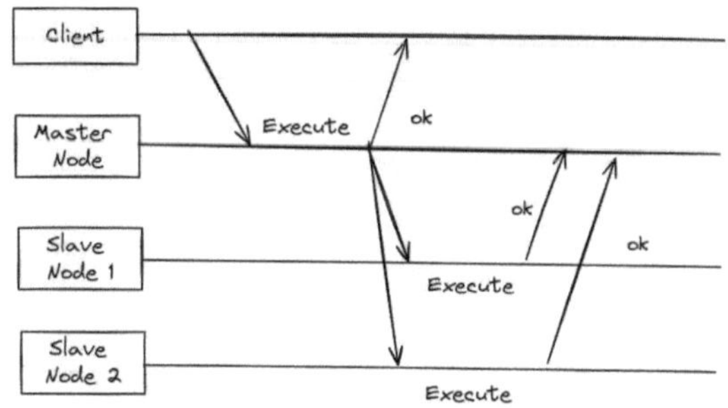

Fig. 5.2 Master-slave asynchronous replication

longer. Additionally, because it needs to wait for the execution results of the slave nodes, the overall efficiency can be affected by the slow execution speed of a single slave node, causing both the client and the master node to wait. The advantage of asynchronous replication is higher efficiency of the data service, but there will be a window period of data inconsistency.

Since both synchronous and asynchronous replication have their own advantages and disadvantages, Facebook proposed "semi-synchronous replication" [114], where one slave node in the cluster uses synchronous replication while the other slave nodes use asynchronous replication. This means that the master node only needs to wait for the execution result of the slave node using synchronous replication before returning the result to the client.

5.1.2 *Multi-master Replication*

As mentioned above, in the master-slave replication architecture, only the master node can handle client write operation requests. Due to the single-master and multiple-slave nature, the distributed system cannot increase the concurrency of handling write requests as the cluster scales. The entire cluster is limited by the performance of the master node, and if there is a network interruption between the master node and the slave nodes, master-slave replication will affect the client's write operations.

Multi-master replication, compared to the previously mentioned master-slave replication mode, allows distributed systems to have more master nodes to handle client write operation requests. Each master node that processes a write request must propagate the write operation to all other nodes. From the perspective of each master node handling write operations, all other nodes are its slave nodes, and it is also a slave node for other master nodes.

The usage scenario for multi-master replication is typically a geographically distributed multi-data center data service system. The structure of a multi-master replication cluster is shown in Fig. 5.3. A master-slave replication model is used within each data center, where each has one master node to handle the client write requests for that data center. The model operates according to the master-

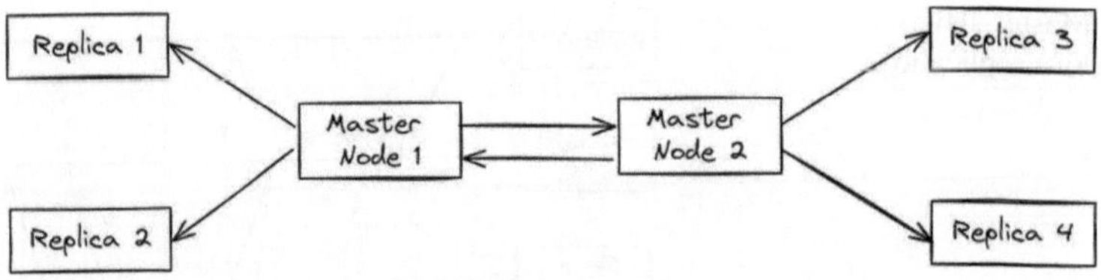

Fig. 5.3 Multi-master replication cluster structure

slave replication mode. Among multiple geographically distributed data centers, the master nodes are responsible for exchanging update operations to achieve synchronization across data centers.

The advantages of the multi-master replication model are as follows:

- **Load Balancing**. The most common reason for using multi-master replication is to allow a database cluster to handle more write traffic than a single node can. To truly scale a database cluster, we must scale read operations by adding read replicas and handling write operations with multiple machines. Therefore, when write operations on the database cluster become a bottleneck, multiple-master nodes should be used instead of a single master node to receive incoming write operations. This allows the cluster to distribute the load and handle multiple write requests. Typically, clients will choose one of the multiple-master nodes to send write requests to. These updates are then propagated asynchronously to the other master nodes, keeping them synchronized with the changes and maintaining eventual consistency in the system.
- **Avoiding Single Points of Failure**. Like any other node in the database cluster, the master node can also crash. If a cluster has only one master node handling all the write requests, the crash of the master node will render the entire distributed system unable to provide data services, leading to prolonged downtime. This is a typical case of a single point of failure. Since node crashes are inevitable, it makes sense to run multiple-master nodes in a cluster and have all master nodes handle write requests. With this approach, if one master node crashes, another master node can seamlessly continue handling write requests, allowing the cluster to keep functioning.
- **Low Latency for Geographically Distributed Data Services**. The most significant difference between distributed systems and single-machine systems is geographical distribution, which means not just having multiple nodes but also having nodes distributed across different geographical locations. If the master-slave replication model is used, the write latency will increase dramatically when database clients are distributed across different geographical locations, because all write operations from clients in all geographical locations must reach the region where the master node is located. To minimize cross-geographical write latency, a multi-master replication model can be used, with a master node residing in a region closer to the client. When a client issues a write request, it can be served by the nearest master node, providing good data service.

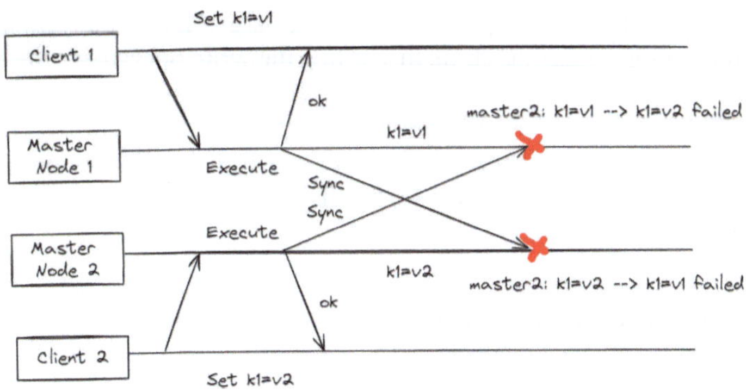

Fig. 5.4 Write conflicts in multi-master replication

Although the multi-master replication model has many advantages, it also has some significant issues, the most notable being write conflicts.

In the multi-master replication model, if multiple master nodes simultaneously handle write requests for the same data item, it will lead to write conflicts, as shown in Fig. 5.4. Client1 and Master1 belong to the same data center, and Client2 and Master2 belong to another data center. Client1 and Client2 simultaneously send update requests for the same data item k_1 to their respective data center's master nodes. The initial value of k_1 is v_0. Client1's update operation changes v_0 to v_1, and Client2's update operation changes v_0 to v_2. The master nodes in each data center complete the write operations from their clients and then send the write requests to the other data center's master nodes to synchronize the data. At this point, a conflict occurs: Master1 cannot execute the instruction from Master2 to change k_1 from v_0 to v_2 because the value of k_1 on Master1 is v_1. Similarly, Master2 cannot execute the write request from Master1. Generally, there are two main strategies to resolve write conflicts in multi-master replication:

- **Eliminate conflicts from the source**. The core of this conflict resolution strategy is to distribute data according to its attributes, with different segments stored in distinct data centers. This approach avoids the need for cross-data center synchronization while still allowing the benefits of multi-master replication to enhance distributed data services. For instance, data for various users can be kept in separate data centers: write operations for user A's information are directed to the primary node where user A's data is located, and similarly, write operations for user B's information are sent to the primary node holding user B's data. To the client, this method functions much like a master-slave replication setup. However, it does not address the problem of data center outages, which would necessitate an additional conflict resolution mechanism.
- **Use strategies to achieve data consistency after conflicts occur**. There are several strategies to resolve conflicts from this perspective. "Last-Writer-Wins" is one such strategy where each write request is assigned a globally unique, sequentially ordered ID, usually a timestamp or a random number. When a

conflict occurs, the write operation with the larger ID is the winner, and the master node adopts this operation, discarding the write operation with the smaller ID. Multiple-master nodes execute the operation with the largest ID, eventually achieving data consistency. Besides assigning IDs to operations, IDs can also be assigned to nodes, where conflicts are resolved by always prioritizing the operations from the node with the higher ID.

5.1.3 Masterless Replication

The main difference between masterless replication and the two previously mentioned replication schemes (master-slave replication and multi-master replication) is that in the masterless replication scheme, all nodes in the distributed cluster have the same role level. There is no distinction between master and slave nodes. Hence, it is also referred to as P2P replication (peer-to-peer replication). The rise of the masterless replication model mainly originated from Amazon Dynamo [34]. Before Dynamo, research hotspots were mostly concentrated on centralized replication, namely, the master-slave and multi-master replication schemes mentioned above. After the Dynamo paper was published, many databases inspired by Dynamo and based on masterless replication emerged, such as Riak [113] and Cassandra [21].

In a distributed environment, a common question is: when to resolve data conflicts (data inconsistencies)? The previously mentioned master-slave replication and multi-master replication modes resolve data conflicts when processing write requests, that is, solving the problem at the source. For example, in the master-slave synchronous replication scheme, to avoid data conflicts, the master node will always wait for the slave node to acknowledge the completion of the write request. If there is a network partition between the master and slave nodes, the write request is likely to fail, and the same applies to multi-master replication. This leads to a problem: the quality of data service that users receive will be very poor. This quality does not refer to data correctness, which is actually very high with this scheme that resolves data conflicts when processing write requests. The poor quality of data service here means that the user has to wait a long time for the system to respond. Clients and servers define a "Service-Level Agreement" (SLA), which is the service quality commitment from the service provider to the user. In a data service system, service quality can be understood as the upper limit of time taken to complete a user's request under certain conditions. For example, a service may guarantee to the client that under a 500 QPS load, 99.9% of requests will be processed within 300 ms. Different types of data service systems may have vastly different SLAs; some systems may emphasize data correctness more, while others may allow short-term data errors but prioritize data service efficiency. The master-slave synchronous replication model mentioned above is often used in systems that prioritize data correctness. However, most systems do not require such high correctness, and users care more about their experience.

Let's return to the question: when should data conflicts be resolved? Since resolving data conflicts when processing write requests affects the user experience of write operations, what about resolving data conflicts when processing read requests? The masterless replication mode adopts this scheme. In masterless replication, there is no central node like a master to coordinate client write requests. Instead, masterless replication uses the Quorum protocol to handle read and write requests:

1. The Quorum protocol has two configuration parameters: the minimum number of participants required for a read operation R and the minimum number of participants required for a write operation W. The number of nodes in the cluster system is N.
2. The Quorum protocol requires: $R + W > N$.

In a distributed system, N generally refers to the total number of nodes used for data backup, which is usually not very large and typically an odd number, such as 3 or 5. The masterless replication scheme generally requires the number of nodes participating in write requests and read requests to be $W = R = (N/2)+1$ (rounded up). This reduces the time required to process write requests to only waiting for the majority of nodes' acknowledgments, shifting some of the waiting overhead to the read request processing time. Figure 5.5 shows an example of the masterless replication scheme.

In Fig. 5.5, the client sends a write request to Replica 1, which then sends synchronization requests to the other two replicas. Replica 2 replies with "ok," and at this point, Replica 1 does not need to wait for Replica 3's response to inform the client that the write request has been processed. When Replica 1 processes read requests, it follows the same process as handling write requests. Even if Replica 3 crashes at this time, it can still respond to the client's read request. At this point, $R = 2, W = 2, N = 3, R + W > N$, satisfying the Quorum protocol operation requirements.

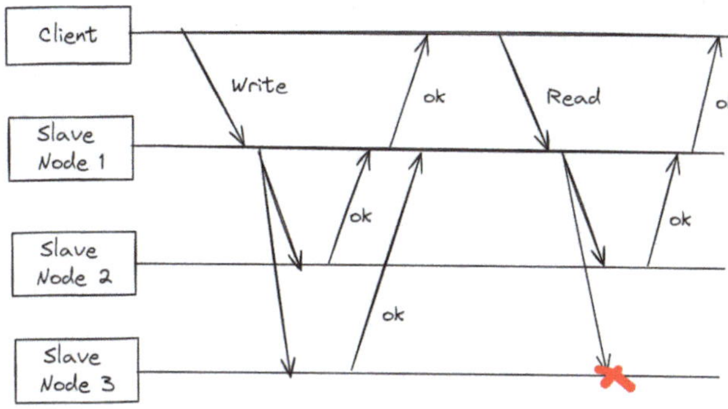

Fig. 5.5 Quorum protocol operation process in masterless replication

5.2 Distributed Data Consistency Models

In the previous section, we introduced several data synchronization methods, each with its own advantages and disadvantages. The most apparent differences lie in the window of data inconsistency and the read/write efficiency of the data services provided. The fundamental reason for these differences is the varying levels of data consistency applied within these synchronization methods. Different levels of data consistency result in different data service characteristics. Some levels of data consistency prioritize read/write efficiency without emphasizing the timeliness of the data, while others stress that the data obtained "everywhere" is consistent, even if this means lower read/write efficiency. Business systems with different requirements will adopt the corresponding level of data consistency. This section will provide a comprehensive introduction to the various levels of data consistency, giving readers a clear and layered understanding of data consistency.

5.2.1 Linearizability

Linearizability holds the highest position among all data consistency models and is sometimes referred to as strict consistency, atomic consistency, or strong consistency. In linearizability, all client accesses to the data service system will yield a uniform result, with no issues of replication lag. Each operation in linearizability seems to occur atomically and in real-time order: for instance, if operation A completes before operation B starts, then logically, B should take effect after A. The linearizability model is a single data object model, but the scope of "one object" can vary. Some systems provide linearizability for individual keys in key-value stores; others may offer linearizable operations for multiple keys within a table or across multiple tables in a database, but not between different tables or databases. A common and general description of linearizability is that the entire distributed system appears to have only one data replica, with all operations being atomic.

Linearizability was first proposed in the paper [56], which is rather abstract and difficult to understand. However, we will reference a few concepts from the paper to explain linearizability as clearly as possible. The paper introduces the concept of history (H), which is a finite sequence of invocations and responses. A subhistory of H is a subsequence of all events in H:

- Invocation. Denoted as < x op(args*) A>, where x is the object name, op is the operation name, args* represents the parameter values, and A is the process name.
- Response. Denoted as < x term(res*) A>, where term (termination) represents the ending condition, and res* is the return result.

If history H is ordered, then H has the following properties:

- The first event in H is the first invocation.

- Every invocation, except the last one, is immediately followed by a matching response. Every response is also immediately followed by an invocation.

If a History Is Not Ordered, Then the History Is Concurrent There exists a nonreflexive partial order $<_H$: $e_0 \quad <_H \quad e_1$ *if* $res(e_0)$ *precedes* $inv(e_1) in H$, where res represents response and inv represents invocation. In simple terms, this means that if the response of e_0 happens before the invocation of e_1 in H, then we can say $e_0 <_H e_1$. Some readers might already be thinking of the happens-before relationship, which originated from Leslie Lamport's 1978 paper "Time, Clocks, and the Ordering of Events in a Distributed System" [73].

If one event happens before another, this must be reflected in the result. If event a affects event b, a happens before b, or b depends on a, we define $a \rightarrow b$. The "happens-before" relationship is transitive; if $a \rightarrow b \wedge b \rightarrow c$, then $a \rightarrow c$. Events that occur sequentially within the same process or the sending and receiving of the same message have this relationship.

This partial order $<_H$ is essentially the happens-before relationship. A history H is linearizable if it can be extended by adding response events to become H' and satisfies the following two conditions:

- L1: complete(H') is equivalent to some legal sequential history S.
- L2: $<_H \subseteq <_S$.

complete(H') indicates that processes interact with complete operations, and L2 indicates that if operation 1 precedes operation 2 in H (where "precedes" emphasizes real-time order), then it is also the case in S. We call S the linearization point of H. Linearizability has the following characteristics:

- **Locality**. If each object in a concurrent system satisfies a property P, then the concurrent system also satisfies property P. Linearizability is a local property: H is linearizable if and only if each object x in H is linearizable.
- **Non-blocking**. A fully operational yet unfinished invocation need not wait for another unfinished invocation to complete; invocation events do not have to wait for the corresponding response events.

A linearizable history H has an equivalent sequential history S, and the real-time order of operations in H is consistent with the order in S, preserving single-threaded semantics.

In the 2014 paper by Sebastian Burckhardt et al. [17], linearizability was broken down into the following three properties:

1. SingleOrder. There exists a global order in the distributed system.
2. RealTime. The global order follows the physical time order of operations.
3. RVal. Consistency of return values is constrained.

Let us illustrate linearizability with an example. Figure 5.6 simulates four client processes accessing the same data item. Time progresses from left to right, and all operations reach the replicas at different times. Process P_1 updates the data item x

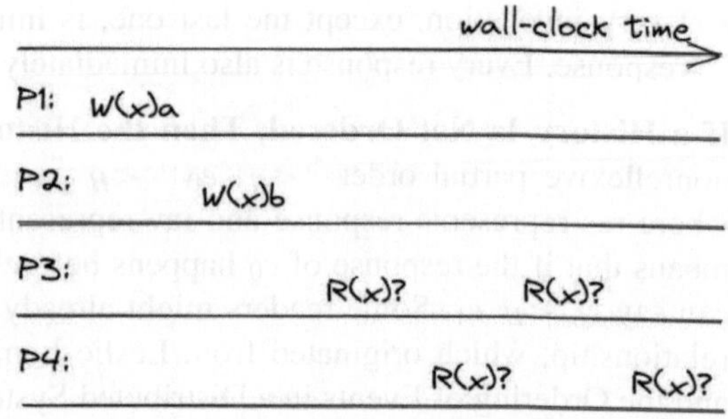

Fig. 5.6 Execution timing diagram of data consistency

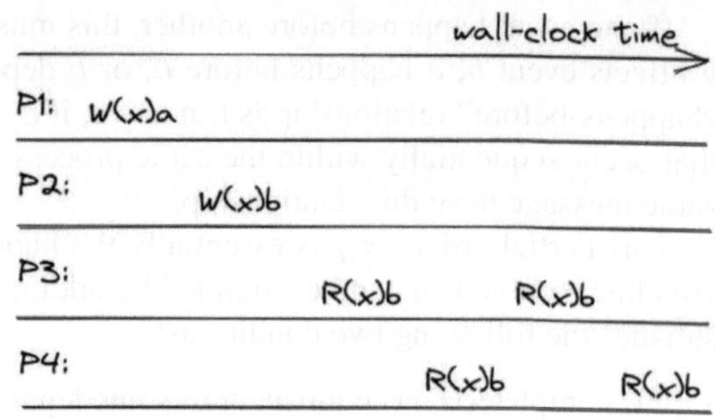

Fig. 5.7 Execution timing diagram of data consistency

to a, followed by process P_2 updating the data item x to b. What should P_3 and P_4 obtain?

Under linearizability, the execution of any operation is the same as all read/write operations initiated, in the order of physical time. Therefore, **the value read is the latest, and all replicas follow the write operation order of the physical clock.** Under the linearizability consistency level, the actual execution should be as shown in Fig. 5.7.

The execution results of P_3 and P_4 conform to the execution order of the write operations of P_1 and P_2, i.e., the result of the latest write operation.

To summarize the characteristics of linearizability:

- There exists a legal total order of operations. **Legal: Read operations always see the most recent write operation.**
- All operations follow the real-time order; if operation A completes before operation B starts, then A must precede B in the total order.
- A completed write operation will be observed by all subsequent read operations.
- Once a read operation sees a value, all other read operations after that must return the same value (until a new write operation occurs).

The advantage of linearizability is that it ensures the correctness of the read data, avoiding data anomalies. The time boundary of linearizability ensures that after an operation is completed; all changes are visible to other participants, thus preventing stale reads. Each read reads a state that is somewhere between the invocation time and the completion time, but never before the read request was invoked. Linearizability also prevents non-monotonic reads, such as a read request first seeing a new value, then an old value. Many concurrent programming models are built on linearizability. In JavaScript, all variables are linearizable; others include volatile variables in Java, atoms in Clojure, and independent processes in Erlang. Most programming languages implement mutexes and semaphores, which are also linearizable. Strong assumptions often lead to strong guarantees.

The disadvantage of linearizability is also evident: achieving these strong assumptions is not easy, and the obvious cost is high latency in data service read and write operations. The paper by Seth Gilbert and Nancy Lynch published in 2002 [47] formally proves the CAP theorem, which states that the relationship between strong consistency and high availability is confirmed: in the presence of network partitions, a distributed storage system must sacrifice either availability or linearizability. Therefore, many data service systems use consistency levels weaker than linearizability, which we will introduce later.

5.2.2 Sequential Consistency and PRAM Consistency

Before introducing sequential consistency, we first want to introduce PRAM consistency. PRAM is the abbreviation for "Pipeline Random Access Memory," which was first proposed in an academic report in 1988 [105]. It requires that all processes see the sequence of write operations issued by a given process in the same order as that process's invocations. On the other hand, processes may observe writes from different processes in different orders. Therefore, PRAM consistency does not require a global order. However, writes from any given process (session) must be serialized in the order they were issued, like they are in a pipeline, hence the name.

Jerzy Brzezinski and others proved in a 2003 paper [16] that PRAM consistency is equivalent to simultaneously achieving the following three consistency levels: "read your writes," "monotonic writes," and "monotonic reads." These three consistency levels are quite straightforward:

- "read your writes" means that if a process performs a write operation *w* and then subsequently performs a read operation *r*, *r* must observe the effect of *w*.
- "monotonic writes" means that if a process performs write operation *w1* and then performs write operation *w2*, all processes will observe *w1* before *w2*.
- "monotonic reads" means that if a process performs read operation *r1* and then performs read operation *r2*, *r2* cannot observe a state prior to the writes reflected in *r1*; intuitively, read operations cannot go backward.

Sequential consistency is essentially **PRAM consistency with a global order**. In a storage system that implements sequential consistency, all operations are serialized in the same order on all replicas, while preserving the order of operations determined by each process. Following the approach used to break down the properties of linearizability, we can decompose sequential consistency into the following attributes:

- SingleOrder. There exists a global order in the distributed system.
- PRAM. Preserves the real-time order of operations invoked by the same process.
- RVal. Constrains the consistency of return values.

By comparing these attributes with those of linearizability, we can see that sequential consistency mainly degrades the RealTime attribute to PRAM consistency. In simpler terms, sequential consistency does not guarantee a global order across processes (sessions) but ensures the order within a single process (session).

Sequential consistency can be traced back to Leslie Lamport's 1979 paper [69]. Lamport proposed sequential consistency based on shared memory multiprocessor systems, with two constraints for accessing shared objects:

- From the perspective of a single processor (thread or process), its instructions are executed in the order specified by the program.
- From the perspective of all processors (threads or processes), instructions maintain a single execution order.

If a system meets these two constraints, then it satisfies sequential consistency. We can understand such a system as a synchronous distributed model, extending this definition to the field of distributed systems. Let's illustrate the difference between sequential consistency and linearizability with an example, as shown in Fig. 5.8. Figure 5.8 shows an execution process where two processes perform write operations on a shared object. Let's assume that these two processes also perform read operations consecutively. Each process will observe a specific order of write operations. If we assume the system follows PRAM consistency, then these two processes might observe the following sequences of write operations:

- S_{PA}. $W1 \quad W2 \quad W3 \quad W5 \quad W4 \quad W7 \quad W6 \quad W8$
- S_{PB}. $W1 \quad W3 \quad W5 \quad W7 \quad W2 \quad W4 \quad W6 \quad W8$

PRAM consistency requires that within each process, the order of observed writes ensures that the writes from the same process are executed in sequence. As mentioned earlier, sequential consistency includes PRAM consistency and also has a unique global order. If the system further follows sequential consistency, then S_{PA} will be equal to S_{PB}, so S_{PA} or S_{PB} are both acceptable for sequential consistency, but the entire system must adopt a unified order, not different orders observed in different processes.

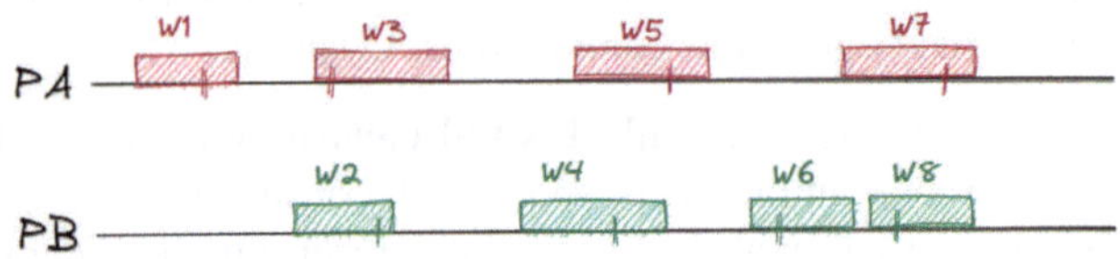

Fig. 5.8 Processes performing write operations on a shared data object, with vertical lines indicating chosen linearization points

If we further impose higher requirements on the system to satisfy linearizability, following the linearization points allocated in Fig. 5.8, the order of execution in the system can only follow the sequence:

$S_{Linearizability}$. $W1 \quad W3 \quad W2 \quad W4 \quad W5 \quad W6 \quad W8 \quad W7$

Sequential consistency is less stringent than linearizability and is easier to achieve. Although sequential consistency has a global order, the execution order of local nodes does not necessarily need to strictly follow the global time order and **does not require the execution order to strictly follow real-time sequence**. Additionally, sequential consistency **does not specify the order of operations for different processes (nodes)**. For example, *PA*'s $W3$ operation and *PB*'s $W2$ operation are executed in opposite orders in S_{PA} and S_{PB}, and sequential consistency does not specify which order is correct, as long as all processes (nodes) execute in a unified order. Generally, sequential consistency requires that the execution order of operations within a single process follows the program order, while the read/write order of different processes has no requirements.

5.2.3 Causal Consistency

Causal consistency captures potential causal relationships between operations. Operations related by causal relationships should appear in the same order across all processes, but the order of operations without causal relationships may not be consistent. In the previous section, we discussed PRAM consistency. The PRAM consistency model can only be used in scenarios where consistency requirements are relatively weak and there are no causal relationships between processes. However, if we consider real-life scenarios and business scenarios in distributed systems, in practice, many operations of different processes do have causal relationships. These causal relationships are often assigned by the system to make the entire system run more normally and not confuse users.

For example, in a social network scenario, User A updates her photo access permissions to prevent user B from viewing her photos, and then posts a photo mocking user B. When B checks A's social feed, his request to access A's permissions cache may not reflect A's recent update (preventing B from viewing A's photos), but the cache accessed to retrieve the feed might already be updated, so the feed will include the photo A posted. The authentication component incorrectly allows B to view A's photo, causing a data consistency anomaly. User A's operations have a causal relationship: the update of access permissions is the cause of the update of the album content. The operations of updating access permissions and updating album content are likely not executed in the same process. If this system uses PRAM consistency, it will not satisfy this causal relationship, ultimately allowing B to see the photo posted by A, resulting in a very poor user experience.

When we introduced linearizability, we talked about the happens-before relationship [73]. The core idea of causal consistency is the happens-before relationship. However, Leslie Lamport did not propose a corresponding consistency model based on the happens-before relationship, only explaining how to obtain a partial order set based on potential causal relationships. Causal consistency first appeared in the 1995 paper "Causal memory: Definitions, implementation, and programming" [3]. In this paper, the authors first reviewed the sequential consistency model and the PRAM consistency model, and then proposed the causal consistency model between the two. The definition of the causal consistency model is as follows:

- L_i. The local execution history of process P_i.
- $O_1 > O_2$. Operation O_1 on process P_i is executed before O_2 in the partial order relation of the corresponding consistency model.
- H. The set of local execution histories of all processes $L_1, L_2, \ldots, L_n$.

For linearizability and sequential consistency, since they have the SingleOrder attribute, all processes can see a unified execution order history, which is the global order. This order satisfies the requirements of linearizability or sequential consistency (e.g., linearizability requires RealTime, while sequential consistency requires PRAM order).

For the PRAM consistency model, each process has its own execution order history, which will not be unified among all processes, and **in PRAM, operations with potential causal relationships in different processes will not have a global order** but only a global order within a single process.

For the causal consistency model, like the PRAM consistency model, each process has its own execution order history, which will not be unified among all processes. However, all operations with potential causal relationships must have a global order, and the execution order of these operations is the same from the perspective of all processes. Potential causal relationships have been mentioned many times in this section, but in practical systems, the definition of potential causal relationships will be adjusted according to different business systems. In the abstract introduction of causal consistency in this section, apart from specific cases, potential causal relationships can generally be determined as there is a causal relationship between read and write operations of the same data item by different processes, and there is also a causal relationship between read and write operations performed by the same process on the same data item. This does not cover all situations, such as in some business systems, where an operation modifying data item Y may change due to a previous operation reading data item X. This is also a causal relationship. Here, we are only giving an example to illustrate causal consistency and will not integrate it into actual business system scenarios.

Next, we illustrate the causal consistency model with read and write operations of different processes (nodes) as shown in Fig. 5.9, where the dashed lines represent the causal relationships before operations.

First, suppose the execution follows PRAM consistency but not causal consistency, then the execution order within each process (node) may be inconsistent, as follows:

- S_{PA}. $W1 \quad W2 \quad W4 \quad W5 \quad W3 \quad W6$
- S_{PB}. $W3 \quad W6 \quad W1 \quad W2 \quad W4 \quad W5$

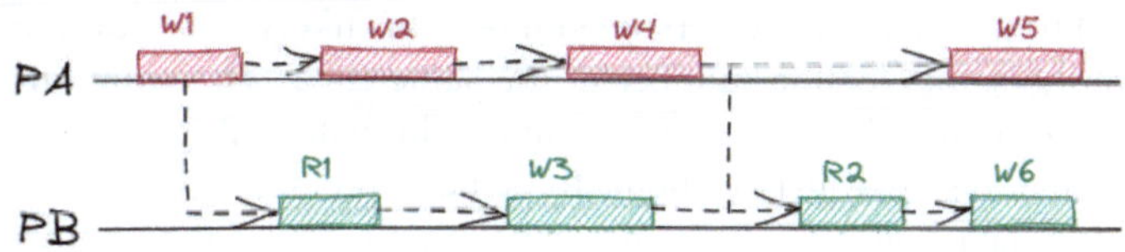

Fig. 5.9 The process of different processes performing read and write operations on shared data items

It can be seen that under PRAM consistency, the execution order of operations is not only inconsistent within each process but also does not follow causal relationships. For example, in S_{PB}, $W3$ is executed before $W1$, which obviously violates the actual causal relationship, ultimately breaking the system constraints.

What about the execution order under causal consistency? If the system follows causal consistency (causal consistency inherently includes PRAM consistency), we might get the following execution order:

- S_{PA}. $W1$ $W3$ $W2$ $W4$ $W5$ $W6$
- S_{PB}. $W1$ $W2$ $W3$ $W4$ $W5$ $W6$

At the level of causal consistency, the operation order seen within different processes is not consistent. Why is this so? Because causal consistency only ensures that operations with causal relationships have a globally unified order on top of PRAM consistency, and it cannot achieve a unified order for operations without causal relationships in different processes. For example, in the above example, the following operations have causal relationships: $W1 \rightarrow W3 \rightarrow W6$, $W1 \rightarrow W2 \rightarrow W4 \rightarrow W5$, and $W1 \rightarrow W2 \rightarrow W4 \rightarrow W6$. We can see that the execution orders S_{PA} and S_{PB} obtained above satisfy these causal relationships. The reason for different execution orders is the lack of a convergent way to unify the execution order. For example, $W2$ and $W3$ are concurrent operations without a causal relationship, so the order of these operations seen by different processes is disordered. We need a globally unified arbitration method to unify these execution orders.

Causal + Consistency (CC+), a stricter version of causal consistency, also known as convergent causal consistency, was proposed in the 2011 paper "Don't settle for eventual: Scalable causal consistency for wide-area storage with COPS" [84]. CC+ consistency includes two properties: causal consistency and a convergence strategy. For concurrent operations, there is no causal relationship between the two operations, so these operations may be disordered under the execution of different nodes. Therefore, CC+ consistency adds a convergence strategy on top of causal consistency. The convergence strategy can avoid data inconsistency. Conflict convergence strategies can be customized according to different business systems, and there are also some simple conflict resolution strategies, such as last-write-win: for concurrent writes to the same data item, both carry the information to be retained, check the operation version number, and discard the one with the lower version number. Version numbers generally face two issues:

1. If physical time is used as the version number, it is likely that multiple operations will have the same version number at the same time. The solution is to add a unique value such as ID to the timestamp to eliminate ambiguity: <time, ID >.
2. In distributed systems, it is difficult to achieve precise synchronization of the time in each data center, so the common solution is to use the Lamport Clock:
   - Each server maintains a $T_{\max}$: the highest version number seen by each server from all places.
   - When it needs to use the version number, it calculates the version number based on $\max(T_{\max} + 1, \text{real time})$.

Each new version number will be higher than the highest version number seen by the node before.

In COPS [84], causal consistency is achieved using dependency sets and version numbers, where an operation is only executed after its dependent operations have completed. The COPS context maintains a (key, version, deps) table. During a read, the client adds the read key and its dependencies to the context; during a write, the client first retrieves the dependencies from the most recent item of the corresponding key, computes new dependencies, and then sends them to the server. Upon success, it returns the version number and adds the new version number and new dependencies, along with the key, to the context.

The dependencies corresponding to Fig. 5.10 are shown in Fig. 5.11, which includes Nearest Deps and All Deps.

Nearest Deps only saves the most recent dependent operations and does not need to store dependencies because the obtained values are closer than all other

Fig. 5.10 Dependency relationships between operations

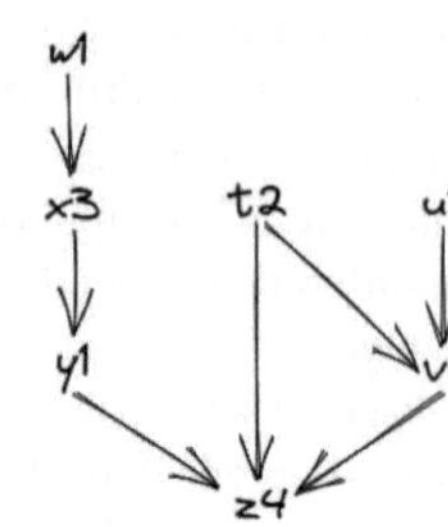

Value	Recent Dependencies	All Dependencies
t2	-	-
u1	-	-
v6	t2,u1	t2,u1
w1	-	-
x3	w1	w1
y1	x3	x3
z4	y1,v6	t2,u1,v6,w1,x3,y1

Fig. 5.11 Dependency sets

Listing 5.1 User operation pseudocode

```
C1: get_trans(ACL, list)
C1: get(ACL) -> v1, no deps
    // Intermediate insert operation
    C2: put(ACL, v2)
    C2: put(list, v2, deps=ACL/v2)
C1: get(list) -> v2, deps: ACL/v2
(C1 checks dependencies against value versions)
C1: get(ACL) -> v2
(now C1 has a causally consistent pair of get() results)
```

dependencies. Therefore, the context only stores a (key, version) table, which represents the nearest dependencies: during a read, the obtained keys and version numbers are added to the context; during a write, the current context is used as the nearest dependencies; then the context is cleared and filled with the results of the current write. Nearest Deps is generally used to handle causal consistency that does not involve transactions.

Previously, we explained causal consistency using album access permissions. The actual operation is generally to first execute get(ACL) to obtain the album permission list, and then execute get(album) to obtain the data in the album. Conflicts may occur because, during the execution of these two steps by client C1, the album owner client C2 inserts operations, updating album permissions or photos, causing the data obtained by client C1 to conflict. In COPS transactions, the client obtains a complete dependency set. Each get operation has a dependency list, and C2's intermediate insert operation will modify the dependencies of the get, so C1's second get operation will not satisfy the dependencies and will re-execute the previous operations to avoid data conflicts. The pseudocode is shown in Listing 5.1.As can be seen, after C2 inserts an operation, C1 knows that its dependencies have been updated (ACL/v2), but it does not have the dependency, so it will re-get(ACL). Tracking causal relationships incurs significant overhead, so the method of building causal consistency based on dependency sets is not widely used in the actual industry.

5.2.4 Eventual Consistency and Weak Consistency

We have introduced several levels of data consistency, but many modern distributed systems do not necessarily require strict data consistency levels. Definitions for these near non-consistency levels have been established: **eventual consistency** and **weak consistency**. The earliest definition of weak consistency is any consistency model weaker than sequential consistency. A weak consistency system cannot guarantee that a read returns the most recent written value, and certain conditions (usually unspecified) must be met to return a value. In fact, weak consistency does not provide order guarantees—thus, no synchronization protocol is required.

Although the applicability of this consistency model seems limited, it was proposed when the cost of synchronization protocols was too high, and occasional data exchanges between replicas were sufficient without complex synchronization protocols. For example, typical use cases of weak consistency are lenient caching strategies that can be applied across various layers of web applications, or even caches implemented in web browsers.

Eventual consistency means that if no new updates are made to a given data item, **eventually** all accesses to that item will return the last updated value [41]. This consistency level does not guarantee that the same data on any node is the same at any moment, but over time, the same data item on different nodes will converge to a consistent state. Eventual consistency was first proposed in a paper published in 1994 [125], where the authors introduced four session guarantees, some of which we have already covered:

- "read your writes." If a process performs a write operation w, then a subsequent read operation r by that process must observe the effect of w.
- "monotonic writes." If a process performs write operation w1 followed by write operation w2, then all processes will observe w1 before w2.
- "monotonic reads." If a process performs read operation r1 followed by read operation r2, then r2 cannot observe a state that precedes the write operations reflected in r1; intuitively, read operations cannot go backward.
- "writes follow reads." Write operations are propagated after the read operations they depend on.

Many people argue that eventual consistency, which provides no strong guarantees on data consistency, cannot be considered true consistency. However, from a narrow perspective of consistency, apart from linear consistency, which truly guarantees data consistency, other consistency levels, such as causal consistency, do fit the definition of eventual consistency. Therefore, in a sense, causal consistency can be seen as a stricter version of eventual consistency. Eventual consistency has numerous implementations in the industry, such as Cassandra [21] and DynamoDB [38], and it is more of an engineering implementation pattern that aligns with the AP system in the CAP theorem.

Marc Shapiro summarized three important properties for achieving eventual consistency in his 2011 paper [118]:

- **Eventual Delivery**. An update to a correct replica will eventually be delivered to all replicas.
- **Convergence**. Correct replicas that have applied the same updates will **eventually** reach an equivalent state.
- **Termination**. All methods terminate.

An eventually consistent system performs updates, but in case of conflicts with other replicas, it may perform rollbacks to resolve conflicts, which is wasteful of resources. Typically, consensus must be reached to ensure all replicas arbitrate conflicts in the same way. Therefore, the authors proposed **strong eventual consistency**, which will not be elaborated here. In short, strong eventual consistency

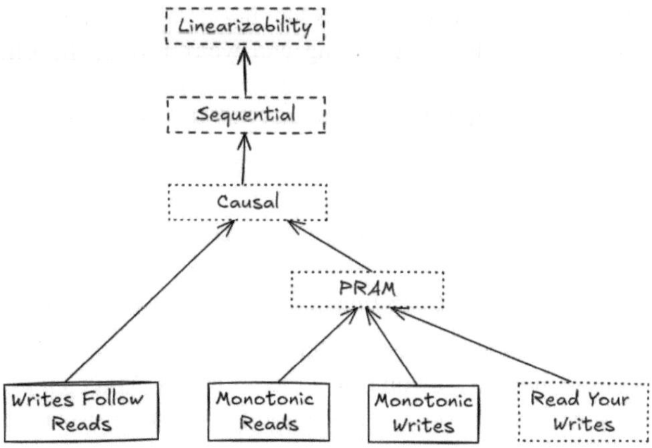

Fig. 5.12 Consistency models

adds an extra constraint compared to eventual consistency: correct replicas that receive the same list of updates will have consistent states, thus avoiding data conflict rollbacks.

Figure 5.12 is a summary of consistency models by Jepsen [60]. The dashed line part indicates that this consistency level is unavailable during certain types of network failures; to ensure safety, some or all nodes must pause operations. The dotted line part indicates that this consistency level is available on each non-faulty node, as long as clients communicate with the same server and do not switch to a new one. The solid line part indicates that it is available on each non-faulty node, even during total network failure. The consistency levels are ordered from low to high from bottom to top.

5.3 Distributed Data Consistency/Consensus Algorithms

Many articles introducing distributed systems often confuse the terms consistency and consensus. Consensus is a broader concept than data consistency. Consensus not only focuses on the data layer but also refers to the process by which multiple nodes in a distributed system reach an agreement on each other's state. For example:

- Consensus on time among nodes to ensure the order of events. This can be achieved using logical clocks or methods that tolerate physical clock drift.
- Consensus on mutual exclusion to determine which node has the right to use the data.
- Consensus on the coordinator to decide which node is the leader.

Consensus algorithms are more like universal algorithms for solving coordination problems among distributed system nodes. If consensus among multiple distributed nodes can be achieved, then achieving data consistency is even easier. Unlike data consistency, which has strong and weak forms, the ultimate goal of consensus—and its only goal—is for all nodes to reach agreement.

Any form of consensus algorithm must satisfy all the following characteristics:

- Termination. All correct processes (non-faulty) will eventually decide on some value.
- Integrity. If all correct processes propose the same value, then any correct process must decide on that value.
- Agreement. All correct processes must decide on the same value.

These three properties seem very ordinary, which also shows that consensus algorithms are never methods to solve specific problems. You can see the presence of consensus in any scenario: consensus among nodes in a distributed system on physical or logical clocks, allocation of file access permissions in a shared file system, and so on.

A Special Case of the Consensus Problem: The Byzantine Generals Problem
The Byzantine Generals Problem is an unavoidable topic when discussing distributed consensus and is a particularly notable issue within the field. The problem originates from Leslie Lamport's 1982 paper "The Byzantine Generals Problem" [74]. In this paper, Leslie Lamport abstractly represents the failure problem caused by conflicting information being sent to different components of the system as the Byzantine Generals Problem

Suppose several armies of Byzantium are preparing to attack an enemy city. Each army is commanded by a general who directs the attack, but since the armies are far apart, the generals can only communicate with each other through messengers. To simplify the problem, Leslie Lamport limits the generals' strategies to either attack or retreat, and they must make a unified decision; any disagreement would result in a catastrophic outcome. Traitors exist among the generals, whose ultimate goal is to prevent the generals from reaching a unified strategy. The generals need an algorithm to ensure:

1. Condition A. All loyal generals will decide on the same plan of action. The loyal generals will follow the algorithm, but the traitors can do anything they want. No matter what the traitors do, the algorithm must guarantee Condition A. The honest generals should not only reach an agreement but also come to a reasonable plan of action. This means that, in reality, we want to guarantee Condition B.
2. Condition B. A few traitors cannot cause the loyal generals to adopt a bad plan. A bad plan is difficult to define. Consider how the generals make their decisions. Each general observes the enemy and communicates the observation to others. Let v(i) be the information transmitted by the i-th general. Each general combines the values v(1)... v(n) into a plan of action, where n is the number of generals. Condition A is achieved by having all generals use the same method to combine information, and Condition B is achieved by using a more robust method. For example, if the decision is simply to attack or retreat, then v(i) could be General

i's opinion on which choice is better, and the final decision could be based on the majority vote. A few traitors can influence the decision only when the loyal generals are almost evenly divided between the two possibilities, in which case neither attack nor retreat can be considered a bad plan.

Condition A requires Condition C: Every loyal general must receive the same information v(1),v(2),. . . ,v(n). Condition C means that a general does not necessarily use the v(i) value obtained directly from General i because General i may be a traitor and may send different values to different generals. For each i, there is Condition D: If General i is loyal, then every loyal general must use the value he sends as v(i). Condition C is modified to Condition C': Any two loyal generals Cuse the same value for v(i). Summarizing the conditions:

- Every loyal general must receive the same information v(1),v(2),. . . ,v(n).
- Any two loyal generals use the same value for v(i).
- If General i is loyal, then every loyal general must use the value he sends as v(i).

Now the problem is restricted to how one general transmits commands to others, so the Byzantine Generals Problem can be restated: The commander must send orders to his n–1 lieutenants, with the following conditions:

- Condition 1. All loyal lieutenants obey the same order.
- Condition 2. If the commander is loyal, then each loyal subordinate must obey the order he sends.

These two conditions are called Interactive Consistency (IC). When the commander is loyal, Condition 1 can be deduced from Condition 2, but the commander may not be loyal. Leslie Lamport concluded through proof that the Byzantine Generals Problem is unsolvable when the number of loyal members does not exceed two-thirds. Specifically, when the total number of generals is 3, no method can solve this problem in the presence of a traitor.

When the commander is disloyal, the execution scenario is as shown in Fig. 5.13. The commander is a traitor and sends the "attack" command to Lieutenant 1 and the "retreat" command to Lieutenant 2. Lieutenant 1 does not know who the traitor is, nor does he know what command the commander actually sent to Lieutenant 2. Therefore, when Lieutenant 1 receives the commander's "attack" command, he must obey. Similarly, Lieutenant 2, upon receiving the commander's "retreat" command, must obey it due to IC condition 2, even if Lieutenant 1 might tell him that the commander's order was "attack." Lieutenant 1 executes the attack command, and Lieutenant 2 executes the retreat command, violating IC condition 1. When a lieutenant is disloyal, assume one of the lieutenants is a traitor, as shown in Fig. 5.14. The commander is loyal and issues the "attack" order to both lieutenants, but Lieutenant 2 is a traitor and reports to Lieutenant 1 that he received the "retreat" order. To satisfy IC condition 2, Lieutenant 1 must obey the attack order. Lieutenant 1 executes the attack command, and Lieutenant 2 executes the retreat command, violating IC condition 1.

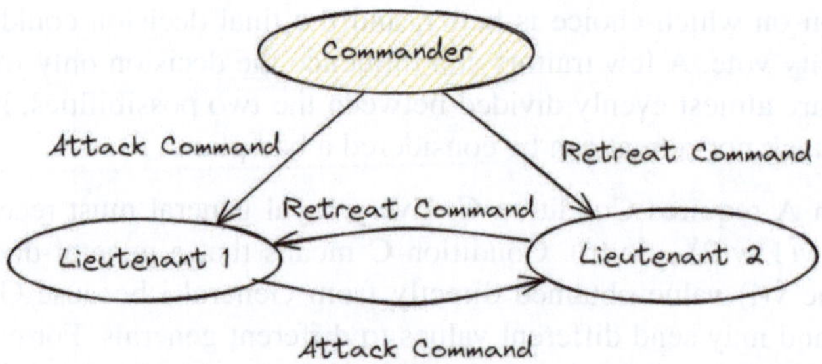

Fig. 5.13 Scenario 1: The commander is a traitor

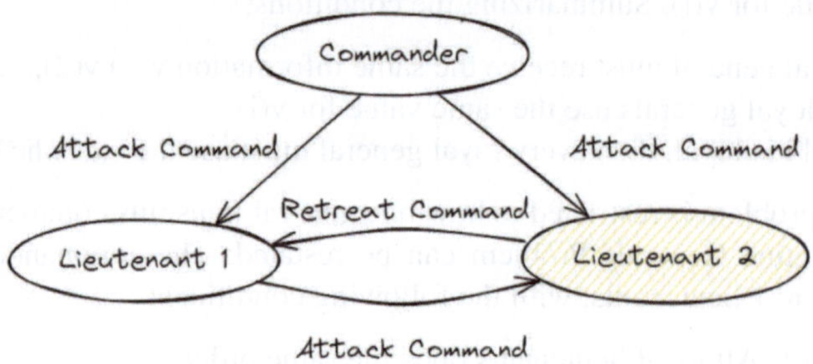

Fig. 5.14 Scenario 2: The lieutenant is a traitor

Generalizing this conclusion, if the number of generals is $3n+1$, then the problem cannot handle n traitors. Applying the conclusion of the Byzantine Generals Problem to the field of distributed systems: in a synchronous network with $3n + 1$ nodes, if the number of faulty nodes does not exceed n (where the faults referred to are "traitor" faults—faulty nodes do not stop functioning but propagate erroneous messages, also known as Byzantine faults), then the problem is solvable. However, in reality, synchronous network situations in distributed systems are rare, and attempting to achieve consensus in asynchronous systems with unreliable channels is impossible. This conclusion comes from the FLP Impossibility Theorem [44]: in asynchronous systems, if nodes have a risk of failure, there is no stable algorithm that can achieve consensus among distributed nodes. The node failures referred to in the FLP Impossibility Theorem are fail-stop failures, where nodes stop functioning upon failure, unlike the "traitor" situation in the Byzantine Generals Problem. Other nodes do not know the faulty node's status and only assume message delay.

So, is it still meaningful to continue studying consensus despite the FLP Impossibility Theorem? Yes, because the FLP Impossibility Theorem has preconditions. The theorem is based on an asynchronous system model that is very restrictive, requiring all algorithms to use no clocks or timeout mechanisms to ensure the safety and liveness properties of the distributed system. However, in practical distributed systems, timeout mechanisms can be used, which relaxes the liveness requirement, allowing for stable consensus algorithms.

The safety and liveness properties were proposed by Leslie Lamport in his paper "Proving the Correctness of Multiprocess Programs" [68], initially used to verify the correctness of multiprocess programs and later extended to the field of distributed systems:

- Safety. Some "bad situations" will never occur. For example, multiple processes accessing a critical resource simultaneously is strictly forbidden in a distributed system because if this happens, the distributed system becomes meaningless, and high concurrency will lead to completely unpredictable node states.
- Liveness. Some "good situations" will eventually occur, even if we do not know when they will happen. The most obvious example is "eventual consistency," where the states of multiple nodes will eventually synchronize, although we do not know when the synchronization will be completed.

We mentioned two types of failures above: one is a failure that causes a crash, and the other is a failure that continues to run and send erroneous messages. Consensus algorithms can be classified into two types based on these failures:

- CFT Consensus Algorithm: CFT stands for "Crash Fault Tolerance." It can handle non-Byzantine faults in a distributed system, such as network, disk failures, and server crashes. However, if a node behaves contrary to the consensus principle, such as tampering with data, the algorithm cannot ensure the system's safety. Well-known consensus algorithms such as Paxos [70] and Raft [98] belong to CFT consensus algorithms.
- BFT Consensus Algorithm: BFT stands for "Byzantine Fault Tolerance." It can handle not only non-Byzantine faults but also Byzantine faults where some nodes send erroneous messages. BFT algorithms are often used in blockchain systems. PBFT [23] and PoW [94] are examples of BFT consensus algorithms.

This section mainly introduces algorithms from the perspective of data consistency in distributed databases, including consensus algorithms and consistency algorithms. Generally, the consensus algorithms commonly used in distributed databases are CFT (Crash Fault Tolerant) consensus algorithms, while BFT (Byzantine Fault Tolerant) algorithms are often used in the blockchain field. For the sake of completeness in introducing consensus algorithms, we will present a significant BFT algorithm to the readers. The remaining algorithms are CFT consensus algorithms or consistency algorithms specifically used to achieve consistency among multiple data replicas.

5.3.1 ViewStamped Replication Algorithm

The ViewStamped Replication algorithm, abbreviated as the VR algorithm, was initially proposed in the 1988 paper [97]. One of the authors of this paper is Barbara Liskov, who also proposed the PBFT algorithm, which we will introduce later. In 2012, the authors updated the VR algorithm and published it in the paper [82].

The authors state that the VR algorithm does not fall within the scope of consensus algorithms but rather uses a consensus mechanism similar to the Paxos algorithm to handle replica replication. Thus, it is more accurate to refer to the VR algorithm as a consistency algorithm. The VR algorithm operates in distributed asynchronous networks and handles node failures due to crashes. It supports a replicated service running on multiple replica nodes. This replicated service maintains a state and allows a set of client machines to access that state. The VR algorithm provides a replicated state: clients can run regular operations to observe and modify the service state. Therefore, this method is suitable for implementing replicated services, such as lock managers or file systems. This section introduces the VR algorithm primarily based on the 2012 version, which, compared to the 1988 version, does not require any I/O persistence operations and introduces a reconfiguration protocol to allow changes in the identity of replica nodes.

5.3.1.1 Assumptions of the VR Algorithm

The VR algorithm deals with non-Byzantine faults, where the only result of a node failure is a crash. Therefore, machines either run normally or completely stop. The VR algorithm operates in an asynchronous network where the non-arrival of a message does not imply the sender's status. Messages may be lost, delayed, or fail to be sent and may be sent multiple times; it is assumed that if a message is resent, it will eventually be delivered to the target node.

The VR algorithm can handle up to f replica failures in a distributed system with $2f + 1$ replicas. The VR algorithm architecture is shown in Fig. 5.15. The figure shows some clients using the VR algorithm, running on three replicas; thus, in this

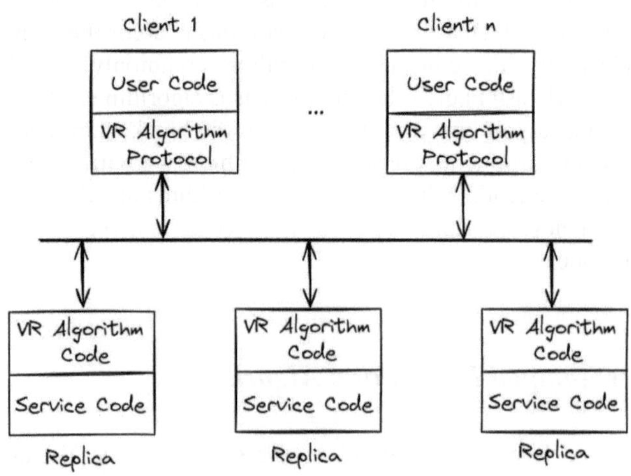

Fig. 5.15 VR algorithm architecture

example, $f = 1$. Client machines run user code on the VR protocol. The user code communicates with VR algorithm nodes through protocol operation calls.

The replicated state machine requires replica nodes to start with the same initial state and for operations to be deterministic. If replicas execute the same sequence of operations, they will end up in the same state. The challenge for the replication protocol is how to ensure that operations are executed in the same order on all replica nodes while tolerating high concurrent client requests and node crashes. The VR algorithm is based on the master-slave synchronization method introduced in the data synchronization chapter, where the master node specifies the order and the slave nodes passively accept these orders. What happens if the master node fails? The VR algorithm addresses this issue by allowing different replicas to take on the role of the master node over time. In the VR algorithm, the period during which a particular replica acts as the master node (similar to the "Term" in the Raft algorithm, which we will discuss later) is called a view. If the master node is updated, the view number is incremented by 1. In each view, a replica is chosen as the master node. The slave nodes monitor the master node, and if the master node fails, a view change protocol is executed to select a new master node.

If a view change protocol is executed, the new master node must execute client operations in the order determined by the previous master node. The VR algorithm accounts for this by having the master node wait for at least $f+1$ replicas (including itself) to recover before executing client requests and initialize the state of the new view by querying at least $f + 1$ replicas. Therefore, every request is known to the majority of nodes, and the new view starts from the majority. The VR algorithm provides a method for resuming incomplete operations and continuing execution after node recovery. The VR algorithm can be divided into the following three subprotocols to handle various situations:

1. Handling user operations under no-fault conditions.
2. View change to select a new master node.
3. Recovery of failed replicas to allow them to rejoin the VR algorithm cluster normally.

The state in the VR algorithm is as follows:

- configuration. The set of IP addresses of all replicas.
- replica number. The index of a particular replica address.
- current view number. The view number, initially 0.
- current status. Divided into normal, view-change, and recovering.
- op-number. The most recently received operation number, initially 0.
- log. An array of op-number entries containing the requests received so far in the specified order.
- commit-number. The op-number of the most recently committed operation.
- client table. Records the most recent request number for each client and the result if the request has been executed.

User Operation Handling in the Absence of Faults When the primary node is operating normally, replicas communicate through messages containing the view number information. Only when the view number meets the requirements will the message be processed. If the received view number is too small, the replica is outdated and will be discarded; if the view number is too large, the replica will change state: it requests the missing information from other replicas and updates its state with this information before processing the message. Under normal circumstances, the execution flow of the VR algorithm is as follows:

1. The client sends a <REQUEST *op*, *c*, *s*> message to the primary node, where *op* is the operation the client wants to run (and its parameters), *c* is the client ID, and *s* is the request number assigned to the request. When the primary node receives the request, it compares the request number *s* with the information in the client table. If request number *s* is not greater than the information in the table, it will discard the request. However, if the request is the most recent request from this client and has already been executed, it will resend the response.
2. The primary node determines the operation number op-number for the request, adds the request to the end of the log, and updates the client's information in the client table to include the new operation number *s*. It then sends a <PREPARE *v*, *m*, *n*, *k*> message to the other replicas, where *v* is the current view number, m is the message received from the client, *n* is the operation number op-number assigned to the request, and *k* is the commit number commit-number.
3. The backup nodes process the PREPARE messages in order: a backup node can only process this message after it has processed all earlier messages with operation numbers less than op-number *n*. When backup node *i* receives a PREPARE message, it waits until its log has entries for all earlier requests (if necessary, it performs state transfer to obtain the missing information). It increments the op-number, adds the request to the end of its log, updates the information in the client table, and sends a <PREPAREOK *v*, *n*, *i*> message to the primary node to indicate that this operation and previous operations have been prepared locally.
4. The primary node waits for at least $f + 1$ PREPAREOK messages from different backup nodes; at this point, it considers the operation (and all earlier operations) committed. It then executes all earlier operations (those assigned smaller operation numbers) by calling the service code, increments its commit number commit-number, and sends a <REPLY *v*, *s*, x> message to the client, where *v* is the view number, *s* is the number provided by the client in the request, and x is the result of the up-call. The primary node also updates the entry in the client table to include the result.

Typically, the primary node notifies the backup nodes of the commit when it sends the next PREPARE message; this is the purpose of carrying the commit number in the PREPARE message. However, if the primary node does not receive new client requests in time, it will notify the backup nodes of the most recent commit by sending a <COMMIT *v*, *k*> message to the backup nodes, where *k* is the commit number (note that in this case, commit-number = op-number).

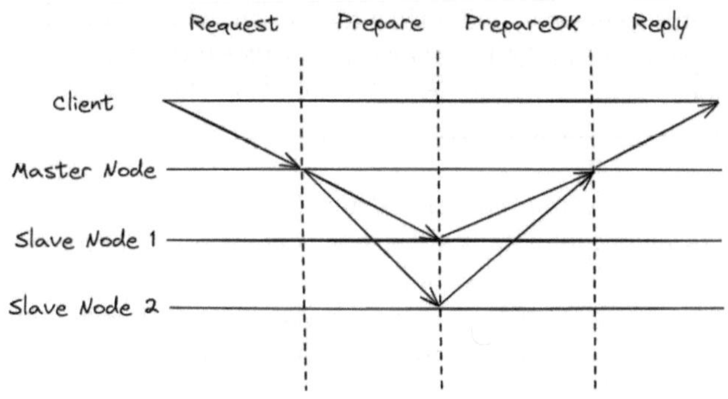

Fig. 5.16 Normal execution flow of the VR algorithm

When the primary node commits, backup node waits until it has the request in its log (which may require state transfer) and has executed all earlier operations. It then executes the operation by performing an up-call to the service code, increments its commit number, updates the client table, but does not send a reply to the client.

The normal execution flow of the VR algorithm is shown in Fig. 5.16.

If a primary node failure occurs, the VR algorithm will execute the **view change phase**. The process is as follows.

A backup node i that notices the need for a view change increments its view number view–number by 1 and sets its status to view change. It sends a <STARTVIEWCHANGE v, i> message to all other backup nodes, where v represents the new view. Backup nodes will enter the view change state based on their own timers or if they receive a STARTVIEWCHANGE or DOVIEWCHANGE message with a number greater than their current view number.

When replica i receives STARTVIEWCHANGE messages from f other replicas carrying their view numbers, it sends a <DOVIEWCHANGE v, l, v', n, k, i> message to the new primary node in the new view. Here, v is its view number, l is its log, v' is the view number of its latest stable state, n is the operation number, and k is the commit number.

When the new primary node receives $f + 1$ DOVIEWCHANGE messages from different replicas (including itself), it sets its view–number to the view number in the messages and selects the log with the largest v' from the messages as the new log. If multiple messages have the same v', it selects the one with the largest n. It sets its operation number to the operation number of the latest entry in the new log, changes its status to normal, and sends a <STARTVIEW v, l, n, k> message to the other replicas, where l is the new log, n is the operation number, and k is the commit number.

The new primary node begins accepting client requests. It also sequentially executes any unexecuted committed operations, updates its client table, and sends replies to the clients. When other replicas receive the STARTVIEW message, they replace their logs with the log in the message, set their operation number to the latest entry in the log, set their view number to the view number in the message, change their status to normal, and update their client table information. If there are uncommitted operations in the log, they send a <PREPAREOK v, n, i> message to the primary node, where n is the operation number. They then execute all known committed operations that were previously unexecuted and update their client table information.

When a replica recovers from a crash, it cannot participate in request processing or view changes unless its state is at least as good as it was before the failure. If it reenters without considering the state difference, the system may fail. For example, if it forgets that it prepared an operation, even if the operation was committed, fewer than a quorum of replicas may know about the operation, potentially causing the operation to be discarded during a view change. A replica that recovers after a crash and rejoins the cluster sets its state to recovering and executes the recovery protocol. During this time, it does not participate in processing requests or the view change protocol. The **recovery protocol** process is as follows.

The recovering replica i sends a <RECOVERY i, x> message to all other replicas, where x is a nonce. Replica j only replies to the RECOVERY message if its status is normal. In this case, it sends a <RECOVERYRESPONSE v, x, l, n, k, j> message to the recovering replica, where v is its view number, x is the nonce from the RECOVERY message. If j is the primary node of its view, l is its log, n is its operation number, and k is its commit number; otherwise, these values are nil.

The recovering replica waits to receive at least $f + 1$ RECOVERYRESPONSE messages from different replicas, all containing the nonce it sent in the RECOVERY message, and including the nonce in the latest view's primary node's information. It then updates its state using the information from the primary node, changes its status to normal, and completes the recovery protocol.

The first version of the VR algorithm was the primary copy method, where the primary node executed the request locally and then synchronized the other nodes. The second version of the algorithm was based on a replicated state machine, confirming the order of each operation, making it more like a consensus algorithm. However, the authors explicitly stated in the paper [82] that this is only a replication protocol, so it is more application-oriented rather than a general-purpose consensus protocol.

5.3.2 *Paxos Algorithm*

The history of the Paxos algorithm dates back to Leslie Lamport's 1990 paper "The Part-Time Parliament" [71]. However, due to the paper's complexity and Lamport's storytelling approach—similar to his earlier work on the Byzantine Generals Problem—the paper was not rigorously proven and thus was not widely accepted. For a long time, the Paxos algorithm did not receive much attention. It wasn't until 1998 that Nancy Lynch, one of the authors of the FLP impossibility theorem, revisited and rewrote the paper, eventually getting it published.

The Part-Time Parliament In this paper, Leslie Lamport fictionalized an island named Paxos. The island is a commercial trade center where the inhabitants use a parliamentary system to establish laws to ensure the orderly operation of the city. However, for the people of Paxos, making money is paramount, and serving as a parliamentarian is only a part-time job. Thus, this parliamentary system is referred to as the part-time parliament, which is the origin of the paper's title. The part-time parliamentary system allows parliamentarians to maintain consistent meeting records despite frequently entering and leaving the meeting room and the messengers being quite forgetful. This system corresponds to the problems faced by today's high-availability distributed data service systems: parliamentarians can be seen as processes or nodes, and their absence can be seen as process or node failures. Therefore, the solutions of the Paxos part-time parliament system may be of interest to computer science.

After this paper was published, it did not gain much traction due to the difficulty in understanding the story. In 2001, Leslie Lamport abandoned the storytelling approach and explained the Paxos algorithm using a method more suitable for computer science, publishing the paper "Paxos Made Simple" [70].

Paxos Made Simple The abstract of this paper is as follows: "The Paxos algorithm, when presented in plain English, is very simple," but in reality, the Paxos algorithm is still very complex for most people. In this paper, Leslie Lamport directly illustrates the practical use of Paxos in distributed systems and how the algorithm is applied. Paxos can be used to determine the value of an immutable variable, which can be any binary data. Once determined, it will not change and can be retrieved. These immutable variables can be a sequence of operations in distributed storage systems, so the Paxos algorithm can directly apply consensus to construct state machines in distributed systems. This allows each node replica to execute the operation sequence in the same order, constructing the same state machine, solving a significant portion of distributed system problems.

The essence of applying consensus algorithms in state machines is to have multiple distributed nodes reach a consensus on a sequence of data updates: [op1, op2, ..., opi], determining what the i-th operation opi is. This update sequence in distributed storage systems can be viewed as a Write Ahead Log (WAL). Below is the description of the Paxos algorithm from the original paper.

Consensus Algorithm Problem Description Assume a set of processes that can propose values. The consensus algorithm ensures that one of the proposed values is chosen. If no value is proposed, no value should be chosen. If a value is chosen, processes should be able to learn the chosen value. The safety requirements for consensus are:

- Only a proposed value can be chosen.
- Only one value is chosen.
- A process never learns that a value has been chosen unless it has been chosen.

The three roles in the consensus algorithm are executed by three types of agents: proposers, acceptors, and learners. In practical implementations, a single process can act as multiple agents, but the mapping of agents to processes is not considered. Agents communicate with each other by sending messages using an asynchronous non-Byzantine model, where:

- Agents run at arbitrary speeds and may fail by crashing and subsequently restarting. Since all agents might fail after choosing a value and then restart, no solution is possible unless agents that crash and restart remember some information.
- Message delivery can take arbitrarily long, messages can be duplicated or lost, but they are not corrupted (non-Byzantine failure).

How to Choose a Value The simplest way to choose a value is to have a single acceptor agent and have one proposer send a proposal to that acceptor, who then chooses the first value it receives. However, this method is flawed because if the acceptor crashes, consensus cannot be achieved. Therefore, let's consider a different method of choosing a value. Instead of having a single acceptor, let's use multiple acceptors (multiple acceptor agents). Proposers send their proposed values to a group of acceptors. Acceptors may accept these proposals. When a sufficient number of acceptors have accepted a value, that value is chosen. But what constitutes a sufficient number? To ensure that only one value is chosen, we need any sufficiently large set to include a majority of agents. Because any two majority sets intersect in at least one acceptor, if an acceptor accepts at most one value, this method works. In the absence of failures or lost messages, we want a value to be chosen even if only one value is proposed by a single proposer, which requires the following conditions:

- **P1. An acceptor must accept the first proposal it receives**. It is not guaranteed that there will always be a first proposal because multiple proposals may be proposed simultaneously, leading to a tie. To avoid this conflict, different proposals can have different numbers, a unique globally ordered ID. Each proposal can be represented as an (ID, Value) pair. Reaching consensus means reaching consensus on the value, so multiple different proposals may be selected, but all chosen proposals must have the same value.
- **P2. If a proposal with value v is chosen, then any higher-numbered proposal that is chosen must have value v.** Since a chosen proposal is selected from accepted proposals, the following conditions must be met to satisfy P2:
  - **P2a. If a proposal with value v is chosen, any higher-numbered proposal accepted by any acceptor must have value v.** Even if a proposal with value v has been chosen, an acceptor c may not have accepted any proposal. If it then accepts a higher-numbered proposal with a different value, due to condition P1, acceptor c must accept this proposal, violating P2a. To ensure compatibility between P1 and P2a, we need to satisfy a stronger condition P2b.

- **P2b. If a proposal with value v is chosen, every higher-numbered proposal issued by any proposer must have value** v. If this condition is satisfied, then P2a and P2 are naturally satisfied. But how do we satisfy P2b? Let's assume a proposal with ID m and value v has been chosen, which means any proposal with ID n and $n > m$ also has value v. To simplify the proof, we use induction on n. Therefore, we can prove this as follows. Suppose every proposal issued with ID $m \ldots (n-1)$ has value v, where $i \ldots j$ denotes a range of numbers from i to j. Then, the proposal with ID n must have value v. For a proposal with ID m to be chosen, there must be a set of acceptors C, consisting of a majority of acceptors, each of whom has accepted the proposal. Combining this with the induction hypothesis, we find that the assumption that m is chosen implies that each acceptor in C accepted a proposal with an ID in $m \ldots (n-1)$, and every proposal accepted by any acceptor with an ID in $m \ldots (n-1)$ has value v. Because any majority set S includes at least one acceptor from C, we deduce that to ensure proposal ID n has value v, the following condition must hold.
- **P2c. For any proposal (n,v), there must be a majority set S such that either (a) no acceptor in S has accepted any proposal with an ID less than n, or (b) v is the value of the highest-numbered proposal among all proposals with IDs less than n accepted by acceptors in S**. To satisfy P2c, when constructing a proposal, it is necessary to obtain the IDs and corresponding values of proposals with IDs less than the current proposal that have been or will be accepted (due to asynchronous conditions, some proposals with smaller IDs might not have been accepted when the new proposal is being constructed). While obtaining accepted proposals is straightforward, predicting which proposals will be accepted is challenging. Thus, a clever way is constructed here: When constructing a proposal, require acceptors to make a promise not to accept proposals with IDs less than n. The entire proposal process is as follows:

  1. A proposer selects a new proposal with ID n and sends a request to each member of a set of acceptors, asking them to respond as follows: to promise never to accept proposals with IDs less than n; and to promise never to accept proposals with IDs less than n among the proposals they have already accepted (if any). We call such a request a prepare request with ID n.
  2. If the proposer receives responses to its request from a majority of acceptors, it can propose a proposal with ID n and value v, where v is the value of the highest ID proposal among the responses, or any value chosen by the proposer if no proposals were reported. The proposer makes the proposal by sending a request to accept this proposal to a set of acceptors (which does not need to be the same set that responded to its initial request). We call this request an accept request.

This describes the algorithm of the proposer. What about the algorithm of the acceptor? The acceptor may receive two types of requests from the proposer: prepare requests and accept requests. The acceptor may accept or ignore any request without compromising safety. So, we only need to describe when it may respond to a request. An acceptor can always respond to a prepare request. An acceptor can respond to an accept request and accept its proposal only if it has not promised not to accept it. In other words: **P1a: An acceptor can accept a proposal with ID n if and only if it has not responded to a prepare request with a proposal ID greater than n.** P1a effectively implements P1.

Suppose an acceptor receives a prepare request with proposal ID n, but it has already responded to a prepare request with a proposal ID greater than n (therefore, it has promised not to accept a new proposal with ID n). In this case, there is no reason for the acceptor to respond to this new prepare request, as it will not accept the proposal with ID n that the proposer wants to make. Thus, we let the acceptor ignore such prepare requests. We also let the acceptor ignore prepare requests for proposals it has already accepted.

With this optimization, an acceptor only needs to remember the highest proposal ID it has accepted and the highest proposal ID from the prepare requests it has responded to. Since P2c must be guaranteed regardless of failures, even if the acceptor fails and subsequently restarts, it must be able to remember this information. Note that a proposer can always discard a proposal and forget everything about it, as long as it does not try to make another proposal with the same proposal ID. The algorithm operations include the following two phases:

- Phase 1. The proposer selects a proposal ID n and sends prepare requests with proposal ID n to a majority of acceptors. If an acceptor receives a prepare request and its proposal ID is greater than any proposal ID it has responded to, it promises not to accept any proposals with IDs less than n and (if any) the highest ID proposal it has accepted.
- Phase 2. If the proposer receives responses to its prepare requests (with proposal ID n) from a majority of acceptors, it will send an accept request for proposal ID n and value v to each of these acceptors, where v is the value of the highest ID proposal among the responses, or any value if no proposals were reported in the responses. If an acceptor receives an accept request for proposal ID n, it will accept the proposal unless it has already responded to a prepare request with a proposal ID greater than n.

To know the chosen value, a learner needs to discover the proposal accepted by a majority of acceptors. A straightforward method is for each acceptor to notify all learners whenever it accepts a proposal, but this results in high overhead due to the total number of responses being the product of the number of acceptors and learners. A more general method can be used, as our context is based on non-Byzantine failures, making the transmission of proposals among learners relatively reliable. Acceptors can respond to some higher-ranked learners, who will then notify all learners when a value is chosen. Using a larger set of learners provides better reliability but at the cost of more complex communication.

From a practical perspective, let's summarize the roles of each component in the Paxos algorithm:

- Proposer. The proposer is responsible for generating proposals based on client requests. There can be multiple proposers, each proposing a unique proposal ID and value. Each proposal is numbered, with a globally unique ID.
- Acceptor. Acceptors receive and confirm proposals from proposers. There are multiple acceptors working together. Suppose the number of acceptors is N, then a proposal must be confirmed by at least $N/2 + 1$ (a majority) of the acceptors to be accepted.
- Learner. Learners can be seen as the executors of proposals. For proposals accepted by acceptors, learners execute the proposal content and return the result to the client. There are multiple learners.

The Paxos algorithm process can be divided into two phases, the Prepare phase and the Accept phase, forming a typical two-phase protocol. In acceptor nodes, several values are maintained to record: the maximum proposal ID received in the Prepare phase (maxPrepareID), the maximum proposal ID received in the Accept phase (maxAcceptID), and the value of the proposal corresponding to the maximum proposal ID in the Accept phase (maxAcceptValue). Below is the complete operation process of the Paxos algorithm:

1. **Prepare phase**. In the Prepare phase, a proposer selects a globally unique proposal ID x and sends prepare requests to all acceptors in the system. The proposal does not include a value at this point. If the acceptor has not received any proposals or the proposal ID is greater than any previously received proposal ID, the acceptor will return Success and promise not to accept proposals with IDs less than x, setting maxPrepareID to x. Additionally, if maxAcceptID is not NULL, the acceptor will also return maxAcceptID and maxAcceptValue. If the proposer receives successful responses from a majority of acceptors, it proceeds to the Accept phase; otherwise, it retries with a larger proposal ID.
2. **Accept phase**. In the Accept phase, the proposer chooses a value for proposal ID x. If the Prepare phase responses do not include maxAcceptValue, it means the acceptors have not accepted any proposals, and the proposer can choose any value. After determining the proposal ID and value, the proposer sends accept requests to all acceptors. If proposal ID x is greater than or equal to maxPrepareID, the acceptor will return Success. If the proposer receives successful responses from a majority of acceptors, it can confirm that the current proposal value is the final chosen value.

After the Prepare and Accept phases, the system completes the confirmation of the value. In the first phase, the Prepare phase, the proposal ID is confirmed, and then in the second phase, the Accept phase, the proposal value is confirmed. The learner will execute the operation with this proposal value. The execution sequence of the Paxos algorithm is shown in Fig. 5.17. The Paxos algorithm completes the decision for a single value through two rounds of message passing, Prepare and Accept. However, it does not specify how to decide on multiple values. Therefore, the Paxos

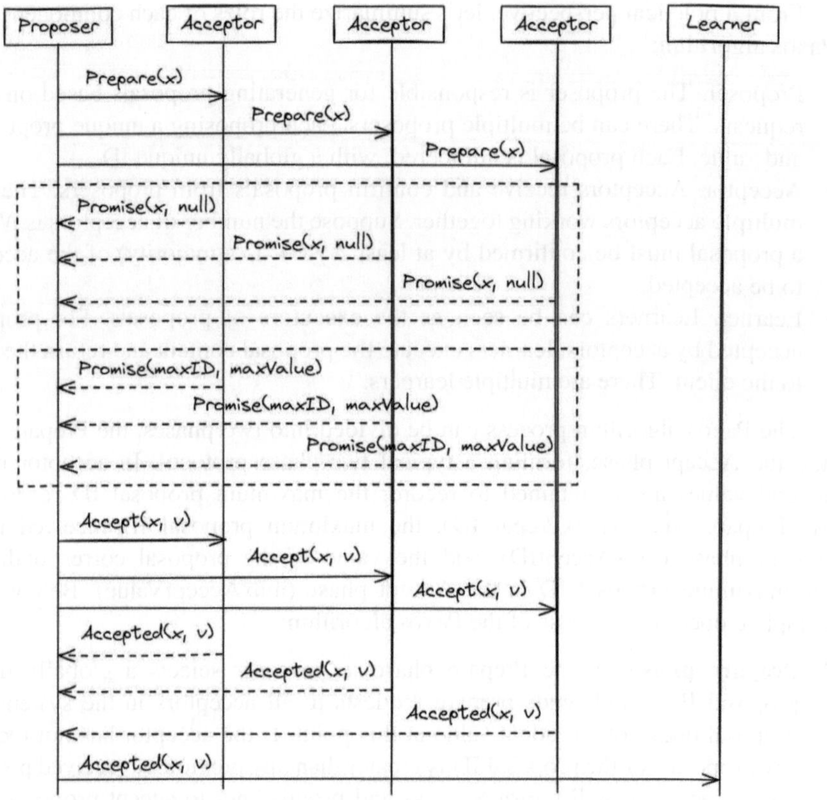

Fig. 5.17 Paxos algorithm sequence diagram

algorithm introduced above is also known as the Basic Paxos algorithm. To solve the problem of deciding on multiple values, Leslie Lamport proposed an improved version of the Basic Paxos algorithm called Multi-Paxos. In this algorithm, a leader is elected from the proposers, which itself is a decision process. Basic Paxos can use one round of decision to elect a leader. In subsequent executions, only the leader is allowed to submit proposals to the acceptors. Since only the leader can submit proposals, the Prepare phase is no longer needed, thus improving the algorithm's efficiency. The consecutive multiple values are identified by incrementally generated IDs by the leader, allowing a series of values to be decided. The idea of the Multi-Paxos algorithm paved the way for many later consensus algorithms, such as Raft [98] and ZAB [57]. These consensus algorithms are also known as Multi-Paxos-like consensus algorithms.

5.3.3 *Practical Byzantine Fault Tolerance Algorithm*

Practical Byzantine Fault Tolerance algorithm, abbreviated as PBFT, is a practical Byzantine fault-tolerant algorithm. Previously, we introduced the Paxos algorithm, which assumes the consensus problem is based on non-Byzantine faults. However, the authors of the PBFT algorithm believe that malicious attacks and software errors will become more frequent, causing failed nodes to exhibit arbitrary behavior, and faults will no longer simply result in nodes crashing. Early Byzantine fault-tolerant algorithms were mostly based on synchronous networks, but the performance of synchronous networks is too low to be practical, so an algorithm is needed that can solve Byzantine fault tolerance in asynchronous networks.

The PBFT algorithm comes from a paper of the same name published in 1999 [23]. One of the authors is Barbara Liskov, the proposer of the Liskov substitution principle, who won the Turing Award in 2008. PBFT is a state machine replication algorithm that can tolerate Byzantine faults. We also mentioned state machine replication in the Paxos algorithm, which is necessary in distributed systems to avoid stopping operation due to a single failure. The intuitive method to achieve fault tolerance is through multiple replicas, allowing multiple nodes with the same state to work together. If a node fails, another node with the same state can take over, solving the problem of stopping operation due to a single failure in distributed systems.

System Model An asynchronous network distributed system where nodes are connected through a network. The network may fail to deliver messages, delay messages, duplicate messages, or deliver messages out of order. Using the Byzantine fault model, where faulty nodes can behave arbitrarily, it is assumed that **node failures occur independently**. To ensure this assumption holds even in the presence of malicious attacks, some steps need to be taken, such as running different service code and operating systems on each node, and having different root passwords and administrators. **Cryptographic techniques are used to prevent spoofing and replay attacks and to detect corrupted messages**. Messages propagated in the system include public key signatures (e.g., RSA algorithm), message authentication codes (MAC), and message digests generated by collision-resistant hash functions. We use m to denote a message, m_i to denote a message signed by node i, and $D(m)$ to denote the digest of message m. Conventionally, only the digest of the message is signed, and the signature is appended to the message text. It is also assumed that all nodes know the public keys of other nodes for signature verification.

State Replication Service Properties A deterministic replication service with state and certain operations. The operations are not limited to simply reading or writing parts of the service state; they can perform any deterministic computation using the state and operation parameters. Clients send requests to the replicated service to invoke operations and block until they receive a response. The replicated service is implemented by replicas consisting of n nodes.

The algorithm ensures safety and liveness in the presence of up to $(n-1)/3$ faulty replicas. Safety means that the replicated service satisfies linearizability: clients' operations on multiple nodes in the distributed system behave as if they were executed atomically, one operation at a time. Safety requires limiting the number of faulty replicas to within the system's fault tolerance limit, as faulty replicas can exhibit arbitrary behavior. The system can limit malfunctioning clients, audit clients, and prevent them from initiating operations beyond their authorized permissions. The algorithm relies on synchronous operations to ensure liveness, meaning all clients can receive responses to their requests as long as the number of faulty nodes does not exceed $(n-1)/3$.

The PBFT algorithm uses a three-phase voting design, known as a three-phase protocol, which includes Pre-Prepare, Prepare, and Commit phases. We will introduce the PBFT algorithm using the Byzantine Generals Problem, with the following background.

- There are 4 generals ($n \geq 3f + 1$), so the PBFT algorithm can tolerate up to 1 traitor ($f = 1$).
- Each general has a unique identifier, labeled 0, 1, 2, 3. The generals can recognize each other's signatures.
- Each action has a **Sequence Number**. The attack/retreat actions form a sequence ordered by the sequence number.
- One general is the **Primary**, and the others are **Validators**.
- The primary may change, and each change increments an epoch number. This epoch number is called the **View**.
- Generals can propose to rotate the primary, known as a View-Change mechanism.

The operational flow of each phase of the PBFT algorithm is as follows:

1. **Pre-Prepare**. The primary general receives the attack or retreat request from the Byzantine Emperor (i.e., the Client of the distributed system). The primary then initiates a proposal, including the message of attack or retreat, the epoch number (View), and the sequence number. The primary sends its signed Pre-Prepare message to the other validators through messengers, as shown in Fig. 5.18.
2. **Prepare**. Upon receiving the Pre-Prepare message from the primary node, all validator nodes independently evaluate the proposal to determine whether they concur with it. If they agree, they send a signed Prepare message to all generals; if they disagree, they do not send any message. If a general receives more than 3 Prepare messages, it enters the "Prepared" state. The collection of these Prepare messages is called a "Prepared Certificate", as shown in Fig. 5.19.
3. **Commit**. A general in the Prepared state, if it decides to execute, will send a signed Commit message to all generals; if it decides not to execute, it will not send any message. Generals that send the Commit message enter the "Commit" phase. If a general receives more than three Commit messages, it executes the message content, indicating that multiple generals have reached a consensus on the proposal. Generals that execute the message content enter the "Committed"

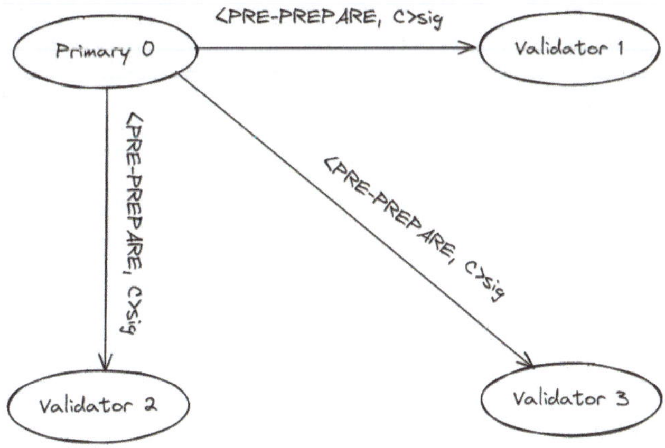

Fig. 5.18 Pre-prepare phase

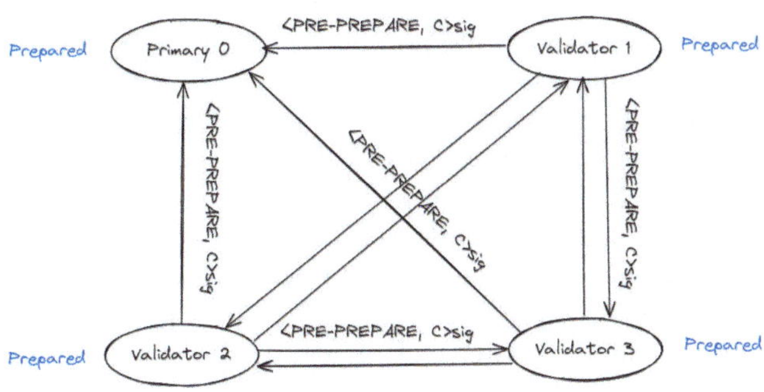

Fig. 5.19 Prepare phase

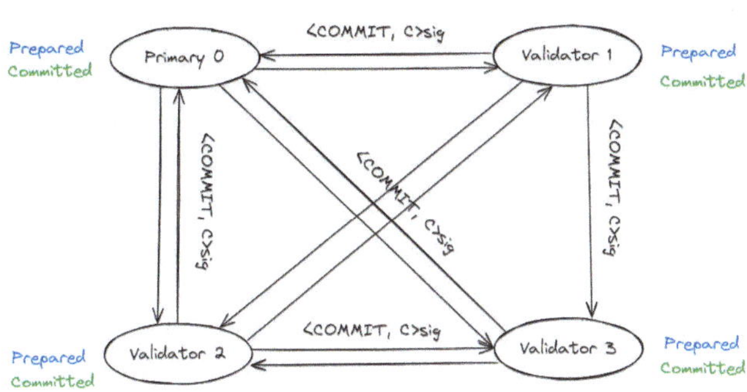

Fig. 5.20 Commit phase

phase and then reply with the execution result to the Byzantine Emperor (Client), as shown in Fig. 5.20.

The execution sequence of PBFT is shown in Fig. 5.21. In the Pre-Prepare phase, the primary node p receives the client's request m, assigns a sequence number n to the request, and broadcasts the message <Pre-Prepare, v, n, $D(m)$> to all backup

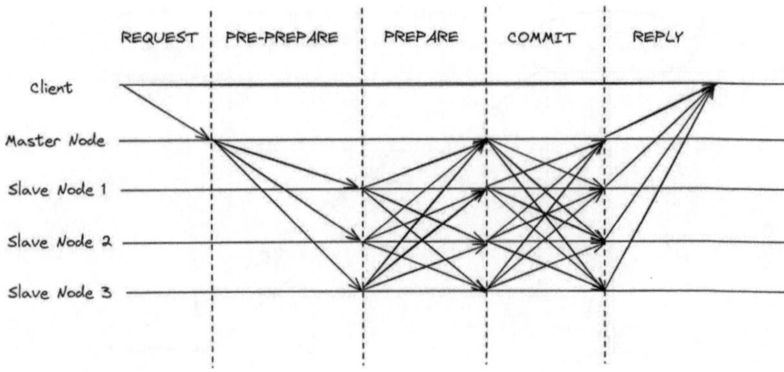

Fig. 5.21 Timing diagram of the PBFT algorithm

nodes. As long as a backup node approves the sequence number n assigned by the primary node and accepts the <Pre-Prepare> message, the backup node enters the Prepare phase. Nodes that enter the Prepare phase send <Prepare, v, n, $D(m)$, i> messages to all other nodes except themselves. Each node accepts Prepare messages from other nodes. When a node receives $2f$ Prepare messages consistent with the <Pre-Prepare> message, the request m is considered to be in the Prepared state. At this point, the backup node broadcasts a <Commit, v, n, $D(m)$, i> message to the entire network. When a node receives $2f + 1$ (including itself) validated <Commit> messages, the request m is considered to be in the Committed state and is executed. The execution result is returned to the client and saved in the local state database.

One very special case of Byzantine error is the case where the Dominant is a traitor. In PBFT, if the Primary node is a traitor, it may give the same serial number to different requests, or not assign serial numbers, or make neighboring serial numbers non-contiguous. It should be the duty of the backup node to proactively check the legitimacy of these serial numbers. If the master node drops or does not broadcast the client's request, the client sets up a timeout mechanism and broadcasts a request message to all replica nodes if it times out. The replica node detects that the master node is a renegade or is offline and initiates a **View Change Protocol**.

1. Each general starts timing after receiving a Prepare message, and stops timing if it switches to the Committed state. If a general fails to execute a message after the timeout expires, then that general sends a View-change message with the new generation (old generation + 1) and other information. If the dominant traitor fails to propose, then each loyal verifier sends a View-change message because of the timeout, as shown in Fig. 5.22.
2. If the new dominator receives more than three View-change messages, the new dominator can send the new generation message New-view, which contains the new generation, all Pre-prepare messages with prepared proofs but against execution, and other messages. After receiving the New-view message, the other verifiers vote one by one for the Pre-prepare messages that have not yet been

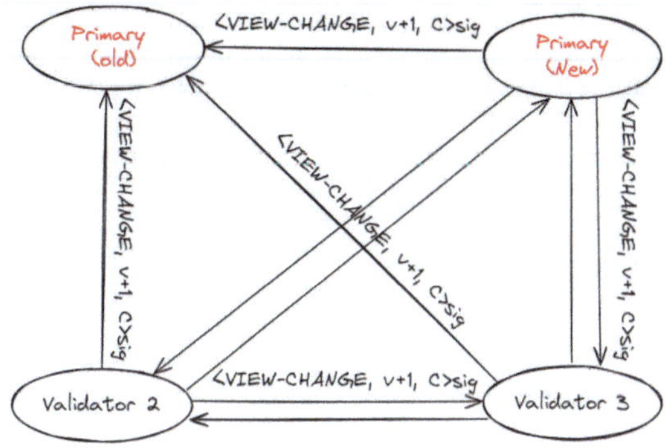

Fig. 5.22 View change-1

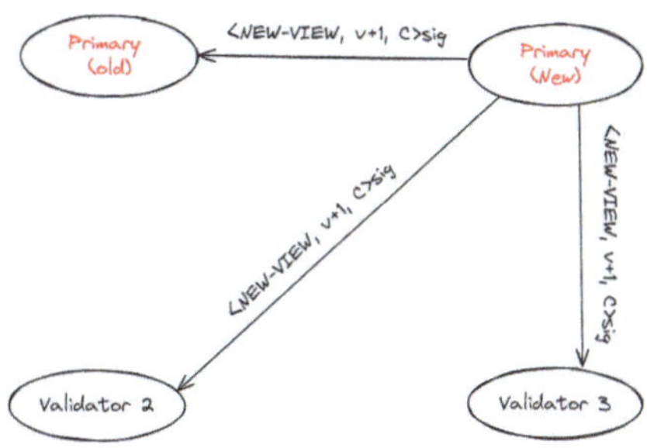

Fig. 5.23 View change-2

executed. The new dominator is responsible for receiving Byzantine Monarch (Client) commands, as shown in Fig. 5.23.

PBFT serves as the foundation for numerous subsequent consensus algorithms implementing Byzantine Fault Tolerance. Following the advent of blockchain technology, PBFT garnered renewed interest, with many researchers reinvesting their efforts in studying the algorithm. Although certain aspects of PBFT may not be directly applicable to blockchain systems, they continue to provide valuable insights and inspiration for consensus algorithms in the blockchain domain.

5.3.4 Raft Algorithm

In this section, we will introduce the Raft algorithm. Afterward, we will implement a distributed storage system based on the Raft algorithm, providing an overview of its application in a real-world system.

Figure 5.24 illustrates a distributed KV system implemented using the Raft algorithm. The design goal of our system is to ensure consistency across all nodes in the cluster, meaning that the state of the KV table (a simplified concept of "table", where the actual data is stored in a storage engine) in each node will eventually be consistent.

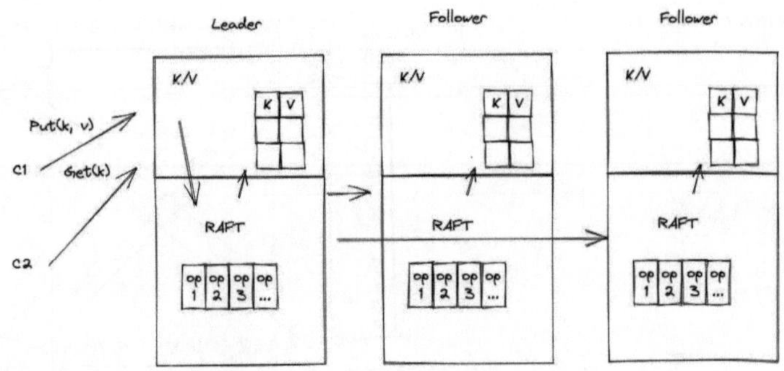

Fig. 5.24 Raft algorithm overview

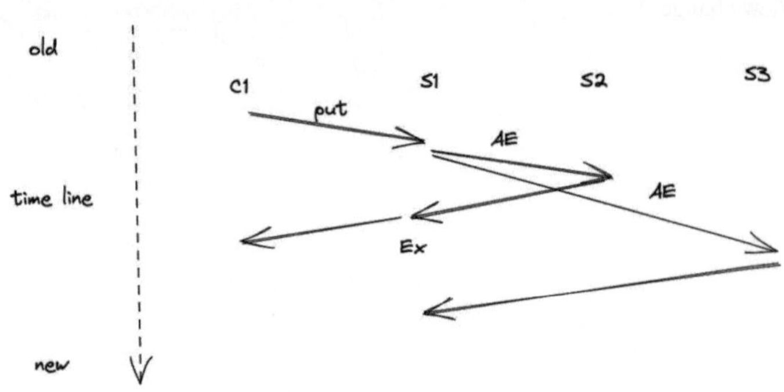

Fig. 5.25 Raft KV operation sequence

Let's first examine how the system operates under normal conditions without considering fault scenarios. We will analyze the process of a Put operation through this system.

1. First, the client sends a Put request to the KV application layer of the current Leader node in the Raft cluster.
2. The Leader encapsulates the Put operation as an entry and submits it to the Raft layer. The Raft layer generates a log entry for this Put request and stores it in its log sequence.
3. Concurrently, the Raft layer replicates this log entry to the Follower nodes in the cluster. Once a majority of the nodes have replicated the log entry and responded, the Leader commits the log entry, applies it by writing the data to the KV table, and notifies the application layer that the operation was successfully executed.
4. At this point, the KV layer responds to the client. Simultaneously, the Leader includes the commit information in the next replication request sent to the Followers. The Followers then apply the log entry, writing the data to their KV tables. Ultimately, the state across all nodes in the cluster becomes consistent. The operation sequence of the entire system is illustrated in Fig. 5.25.

This is an example of how Raft ensures consistency in a system. At first glance, it seems simple. However, when delving into the algorithm's details, the system becomes more complex. For instance, during log replication, many constraints are in place to ensure the consistency of committed logs. Additionally, how do we correctly elect the next leader during faults? How do we ensure log consistency after multiple failures? These issues will be the focus of our subsequent analysis.

There are numerous distributed key-value storage systems based on consensus algorithms similar to the one described above. Examples include Etcd [40] and TiKV [127], which utilize the Raft algorithm to ensure consistency, and FlexSync [136], which allows for dynamic switching between consistency models based on user requirements. In the following sections, we will combine specific code examples and the Raft algorithm paper [98] to make understanding Raft's solutions to these problems as straightforward as possible.

5.3.4.1 Split-Brain in Distributed Systems

Before introducing the Raft algorithm, let's examine the split-brain problem in distributed systems. Literally, "split-brain" means a brain that is split apart, which would cause chaos as the brain is the control center of the body.

In the context of distributed systems, this typically means that the nodes in the cluster are divided into different partitions due to network failures or other issues. During this time, since different partitions cannot communicate, the system may become inconsistent. If the system does not account for handling such scenarios, it will fail to ensure correctness when the network recovers.

Figure 5.26 shows the situation of a network partition in a distributed system. There are five nodes in the system, labeled A–E. Due to a fault, nodes A and B are separated into one partition, while nodes C, D, and E are in another partition. The green circles represent two clients. If they write data to nodes in different partitions, can the system ensure consistency once the partitions recover? The Raft algorithm addresses this issue. The Majority Vote, also known as **majority agreement**, mentioned in Raft's paper, is the key to solving the split-brain problem. First, let's explain how the majority vote works. Assume a distributed system has $2f+1$ servers. To make a decision, the system requires a majority of nodes to agree,

Fig. 5.26 Split-brain

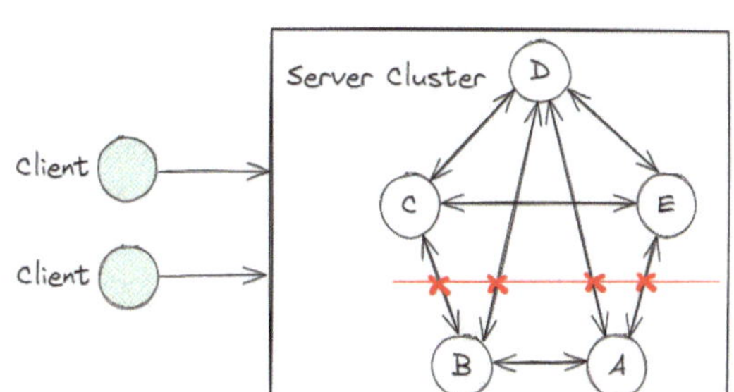

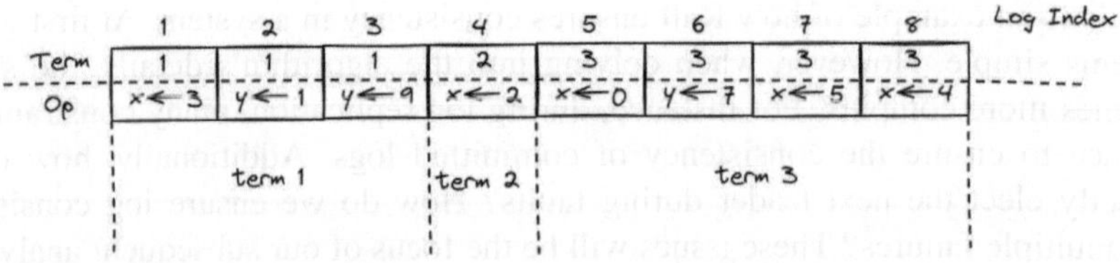

Fig. 5.27 Structure of Raft node logs

meaning at least $f+1$ servers must be alive for the system to operate normally. Thus, the system can tolerate up to f server failures.

Raft uses the majority vote to solve the split-brain problem. Suppose our distributed system consists of an odd number of nodes (3, 5,..., $2n+1$). In the event of a network partition, there will inevitably be a partition with a majority of nodes, allowing the majority vote strategy to function correctly. This prevents the system from becoming unavailable, making the majority vote crucial for resolving the split-brain issue.

5.3.4.2 Raft's Log Structure

Earlier, we provided an overview of the Raft algorithm's process, where requests are initially written to the Raft layer's log. What does this log structure look like? Let's explore Raft's log structure next. Figure 5.27 depicts the log structure of a Raft node. The log is primarily used to record user operations, which are sequentially numbered. Each log entry (1~8) has a unique log index. Additionally, each log entry has a term number, such as entries 1~3 being from term 1. The term number indicates the election state of the log, which we will explain later. Each log entry contains an operation. This operation modifies the state machine, such as entry 1 setting x to 3.

5.3.4.3 Raft's State Transitions

The Raft protocol operates in a Leader and multiple Followers model. In the Raft protocol, each node maintains a state machine with three states: Leader, Follower, and Candidate. At any point during system operation, each node in the cluster is in one of these three states. Figure 5.28 shows the state transition process.

Each node starts in the Follower state. Upon reaching the election timeout, it transitions to the Candidate state and initiates an election. If the node receives a majority of votes, it transitions from Candidate to Leader. Alternatively, if the Candidate node discovers a new Leader or receives a message from a new term, it reverts to Follower. Similarly, a Leader who discovers a higher term will also become a Follower. During normal system operation, nodes continually transition between these three states.

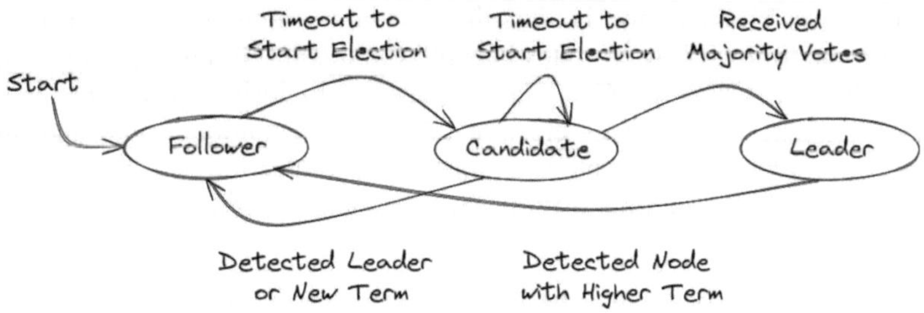

Fig. 5.28 Raft state transitions

5.3.4.4 Leader Election

Before introducing the Leader election process, let's explain two timeouts related to elections in the Raft protocol: election timeout and heartbeat timeout. If a Follower node does not receive a heartbeat message from the Leader within the election timeout, it transitions to the Candidate state to start a new election. The election timeout is typically set to a random value between 150 ms and 300 ms. The randomness prevents nodes from simultaneously initiating elections, which could result in a tie and necessitate a re-election. Increasing the randomness of the election timeout facilitates quicker Leader election. The heartbeat timeout is the interval at which the Leader sends heartbeat messages to Follower nodes.

Let's outline the election process.

1. When the cluster is initialized, all nodes become Follower nodes.
2. After some time (election timeout), if a Follower has not received a heartbeat message from the Leader, it transitions to the Candidate state and initiates an election.
3. Upon becoming a Candidate, the node's term number increases. The Candidate votes for itself and sends RequestVoteRPC messages to all other nodes in the cluster in parallel.
4. If the Candidate node wins the majority of votes, it becomes the Leader. The new Leader then sends heartbeat messages to the other nodes in the cluster, announcing its leadership.

This is the general process for Raft Leader election, but there are two important details to note.

- While waiting for votes, a Candidate might receive a heartbeat message from another node that has become the Leader. If the term number in this message is greater than the Candidate's current term, the Candidate will recognize this Leader as legitimate and transition to a Follower. If the term number is less than the Candidate's current term, the Candidate will reject the message and remain in its current state.

- Another possible outcome is that the Candidate neither wins nor loses the election: multiple Follower nodes become Candidates simultaneously, resulting in a split vote where no Candidate receives a majority. When this happens, each Candidate will reset its election timeout to a random value and attempt the election again. Due to the randomness of the election timeouts, it is likely that one node will win a majority in the next election round, becoming the new Leader.

Log Replication Let's assume there are three nodes in the cluster: A, B, and C, with node A as the Leader and nodes B and C as Followers. When a client sends an operation to the cluster.

1. The client request is first sent to the cluster's Leader node. Upon receiving the request, Leader node A updates the operation record in its local log.
2. Leader node A sends AppendEntries messages to the other nodes in the cluster. These messages contain the log information of the latest client request received by Leader node A (entries that have not yet been synchronized with the Followers).
3. When Followers B and C receive the AppendEntries messages from Leader node A, they record the operations in their local logs and notify the Leader of the successful log append.
4. When Leader node A receives successful append responses from a majority of nodes, it updates the committed log index.
5. Leader node A responds to the client and, in the next AppendEntries message, includes the commit index for Followers B and C.
6. Upon receiving the message, Followers B and C update their local commit index.

Note: The above process describes the normal operation. If a Follower node crashes or the Leader's messages are lost, the append request might fail. In such cases, the Leader will keep retrying until all Followers have stored all log entries.

5.3.4.5 Log Compaction and Snapshotting

Chapter 7 of the Raft paper introduces log compaction. According to our previous discussion on Raft replication, as long as new client operations come in, Raft will write to our log files. After the Leader synchronizes the log with the Followers, the log size will continuously grow as operations increase. In eraft, LevelDB is used to store Raft log entries. If the log size grows indefinitely, accessing logs will take longer. If a node crashes and rejoins the cluster, appending a large amount of log data will consume significant I/O resources, impacting system performance.

How can we solve this problem? First, let's analyze the log structure.

Figure 5.29 illustrates operations on variables x and y. Each log entry, once committed, is applied to the state machine. However, we recognize that storing every log entry is redundant. Our primary concern is maintaining a consistent state, which is defined as the state of the state machine after all committed logs have been

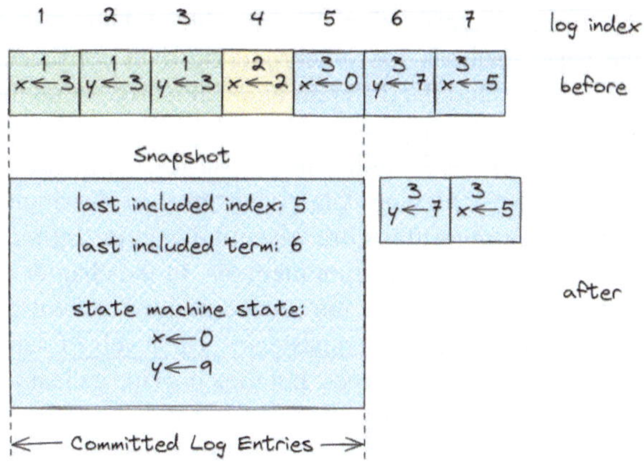

Fig. 5.29 Raft log snapshot

applied. In the figure, we can record the state of the state machine after applying logs 1 through 4, resulting in $x \leftarrow 0$ and $y \leftarrow 9$. By recording this state and the information of the first log entry thereafter, we can safely delete logs 1~4.

The entire operation described above is called a snapshot in Raft. Raft periodically takes snapshots to record the historical state. When a node in the cluster crashes and its log cannot be recovered, the Leader node in the cluster will first send the snapshot to the newly joined node using an InstallSnapshot RPC to transmit the snapshot data. After the node installs the snapshot data, it continues to synchronize the incremental logs, allowing the new node to restore its state quickly.

The Raft algorithm is similar to the Viewstamped Replication (VR) algorithm we introduced earlier. In the VR algorithm, each node needs to maintain state information such as view-number, status, op-number, etc. Similarly, the Raft algorithm has corresponding properties to maintain these states. At a high level, VR and Raft algorithms are indeed similar. However, at a lower level, Raft's AppendEntry encapsulates VR's Prepare, StartView, PrepareOK, Commit, GetState, and NewState messages. Raft introduces a timer mechanism for leader election, providing a simpler and faster solution to election conflicts. Therefore, the Raft algorithm can be considered an optimized variant of the VR algorithm.

5.4 Summary

In this chapter, we introduced data consistency in distributed data services. Consistency refers to maintaining uniform data across multiple data sources (usually at the database level) in a distributed environment. Previously, in the chapter on distributed transactions, we discussed the ACID properties of transactions, mentioning consistency (the C in ACID, Consistency). However, the data consistency in data services differs from transaction consistency. The consistency property within a transaction refers to maintaining data state constraints before and after the transaction execution. We introduced data consistency from three perspectives: the basic methods for synchronizing data consistency, the various levels of data consistency, and the different algorithms for achieving data consistency.

Data consistency synchronization methods include master-slave replication, multi-master replication, and masterless replication. These are the foundational methods for achieving data consistency in distributed systems, and all distributed consistency algorithms are essentially optimizations and improvements based on these three synchronization methods. In the chapter on levels of data consistency in distributed systems, we introduced major levels of data consistency, such as linear consistency and causal consistency. The levels of data consistency we introduced do not cover all possible types, but they provide a clear and comprehensive understanding of data consistency. Besides introducing data consistency, we also expanded on the concept of consensus in distributed systems. Consensus encompasses the concept of data consistency, and in this chapter, we introduced several consensus algorithms, such as Paxos and Raft. As the cornerstone of distributed systems, consensus algorithms are continually being updated and optimized. We just provide a preliminary understanding of consensus algorithms, and for deeper knowledge, readers can seek additional learning resources.

Part II
Distributed Systems Classic Case Studies and Practice

Chapter 6
Distributed System Case Analysis: GFS

In the second part of this book, we will start with the classic case of distributed systems, GFS (Google File System), and gradually guide readers through building the distributed consensus algorithm Raft, based on Raft to construct the distributed key-value storage system eraftkv, and addressing the scalability problem of Raft single-group by implementing Multi-Raft. GFS is designed by Google for large-scale distributed data-intensive applications, featuring a scalable distributed file system. In this chapter, we will begin with GFS to provide readers with a basic understanding of the architecture, reading and writing, consistency, and more in distributed file systems.

6.1 GFS Design Goals

GFS runs on a large number of inexpensive commercial servers, offering strong performance, excellent scalability, high availability, and fault tolerance. Its inception was to meet Google's rapidly growing data processing needs. GFS shares the design goals of early distributed file systems: performance, scalability, reliability, and availability. However, Google found that the design principles of early distributed file systems could not meet current and anticipated application workloads and technological environments. The designers of GFS reexamined the design principles of early distributed file systems and proposed the following six design goals:

- GFS consists of many inexpensive servers that frequently experience failures. GPS requires periodic self-checks and has the ability to tolerate component failures and recover promptly.
- GFS stores a moderate amount of large files, expecting to number in the millions. Each file has a size of 100 MB or larger. GFS is capable of efficiently managing a vast amount of gigabyte-sized files. While GFS supports small files, it does not optimize specifically for them.

Z. Xu, *Distributed Data Services*, https://doi.org/10.1007/978-981-96-9392-4_6

- The workload of GFS primarily consists of two types of reading methods: large-scale streaming reads and small-scale random reads. A single operation of large-scale streaming reads typically reads several hundred KB or 1 MB. Sequential operations from the same client typically read continuous regions of the file. Small-scale random reads usually read several KB at any position in the file. Performance-oriented applications typically batch and sort these small reads to steadily progress through the file rather than moving back and forth.
- GFS involves numerous large sequential append writes. The typical size of these write operations is similar to read operations. Once data is written into a file, the file is seldom modified. While GFS supports small writes at any position in the file, it may not always be efficient.
- GFS supports record append operations, allowing multiple clients to append data to the same file simultaneously while ensuring the atomicity of each client's append.
- GFS supports sustained high-bandwidth read and write operations. Sustained high bandwidth is prioritized over low latency. Most Google applications require high-speed bulk data processing, with few having strict response time requirements for individual reads or writes.

6.2 GFS Master Node

The GFS cluster, as illustrated in Fig. 6.1, consists of one master node and multiple chunkserver nodes. In GFS, files are divided into several fixed-size blocks, known as chunks. When a file block is created, the master node assigns it an immutable, globally unique 64-bit file block handle, referred to as a chunk handle. Chunkservers store chunks as Linux files on local disks. Chunkservers can read from or write to block data specified by file blocks and byte ranges. To achieve high availability, each file block is replicated on different chunkservers, with a default replication factor of three.

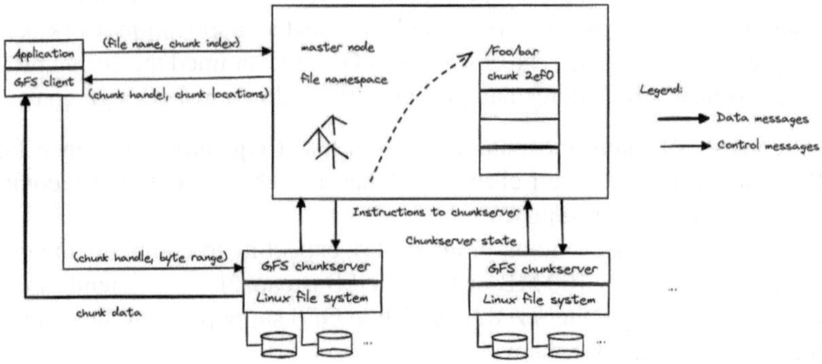

Fig. 6.1 GFS architecture diagram

The master node is responsible for maintaining the metadata of files. File metadata includes namespace, access control information, mapping information between files and file blocks, and the location information of file blocks. Additionally, the master node is responsible for block lease management, garbage collection, block migration, namespace management, and replica management. The master node communicates periodically with chunkservers via heartbeat messages, issuing instructions to chunkservers and collecting their status. Having a single-master node greatly simplifies the design of GFS. It allows the master node to make complex decisions regarding block placement and replication using global information. However, a single-master node can become a performance bottleneck. To prevent this situation, GFS minimizes the involvement of the master node in the read/write process as much as possible. Clients do not directly read from or write data to the master node. Next, we will introduce the specific read and write file processes in GFS.

6.3 GFS Read Operations

Combining with Fig. 6.1, let's introduce how the GFS client reads files from the GFS cluster:

1. According to a fixed file block size, the client converts the file name and byte offset specified by the application into block index within the file.
2. The client sends a request to the master node that includes the file name and the block index.
3. The master node returns the corresponding file block handle and the location of the block replica.
4. The client uses the file name and block index as keys to cache this information.
5. The client sends a request to one of the chunkservers that holds a replica of the block (likely the most recent one), specifying the block handle and the byte range to be accessed within the block.
6. The chunkserver returns the corresponding block data to the client.

Before cached information expires or a file is reopened, the client does not need to communicate with the master node for reading the same file block data. In fact, the client often requests information for multiple file blocks at once, and the master node also returns information for multiple file blocks at the same time. Returning multiple file block information at once can avoid frequent interactions between the client and the master node.

6.4 GFS Write Operations

As illustrated in Fig. 6.2, next we will introduce the process of how the client writes files to GFS:

1. The client queries the master node about which chunkserver holds the current lease of the file block and the location of the block replicas. If no chunkserver holds the lease, the master node will grant the lease to a selected replica.
2. The master node returns to the client the locations of the Primary Replica and the Secondary Replica of the file block stored on chunkservers. When the Primary Replica becomes inaccessible or no longer holds the lease, the client caches this location information.
3. The client pushes data to all replicas in any order. Each chunkserver stores the data in an internal LRU cache buffer until the data is used or expires.
4. Once all replicas confirm receiving the data, the client sends a write request to the master node. This write request identifies the data previously pushed to all replicas. The master node serializes write requests from multiple clients, assigning consecutive sequence numbers. The master node applies the write requests to its local data in sequence order, modifying its local state.
5. The Primary Replica forwards write requests to all Secondary Replicas. Each Secondary Replica applies the changes to its local data in the same sequence number order assigned by the Primary Replica, modifying its local state..
6. All Secondary Replicas reply to the Primary Replica indicating that they have completed the write operation.
7. The Primary Replica replies to the client that the write operation has been completed. During the process of writing data to replicas, any errors encountered will be reported to the client. In the event of a write error, the data may have already been written to the Primary Replica and some of the Secondary Replicas. If the write operation fails on the Primary Replica, no sequence number will be

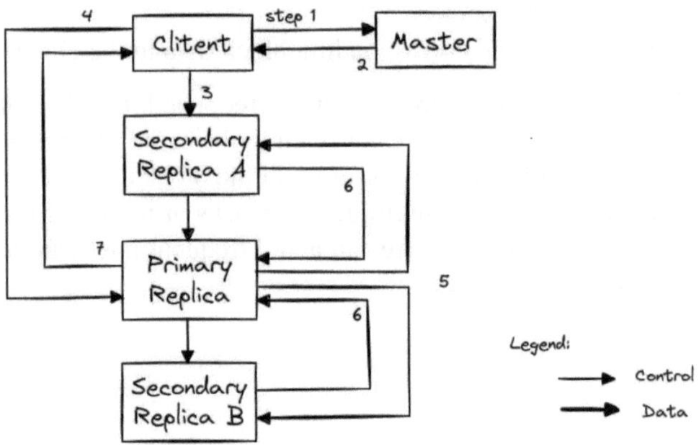

Fig. 6.2 GFS file writing process

assigned and forwarded. When a client's write request fails, the data modification areas of the replicas are in an inconsistent state. GFS handles such errors by retrying failed write operations. GFS first retries several times from step 3 to step 7, and if the retries fail, it will start the write process from the beginning.

GFS breaks down a large write operation into multiple smaller write operations. These smaller write operations follow the aforementioned control flow but may be interleaved and overlap with concurrent operations from other clients. As a result, the shared file region may ultimately contain fragments from different clients, but the replicas of the file remain identical because each operation is successfully completed in the same order across all replicas. This involves aspects of GFS consistency, which we will discuss in the subsequent section.

6.5 GFS Consistency Guarantees

In distributed systems, whenever there are multiple replicas of data, consistency issues must be addressed. In GFS, file blocks have multiple replicas, and due to the instability of network communication, it takes time to synchronize the content of file blocks to multiple replicas. During this period, if the client or chunkserver encounters problems, it will cause inconsistencies between the replica data. There are three types of operations in GFS that affect consistency:

1. **Modifying Metadata**. In the absence of concurrency, the modification operation either succeeds or fails, and the metadata remains consistent at all times. However, in a concurrent scenario, GFS employs a locking mechanism to execute multiple modification operations sequentially, ultimately ensuring the consistency of the metadata. Nonetheless, due to the uncertain execution order of multiple modification operations, the modified value after concurrency is also indeterminate.
2. **Writing Data**. When writing data to a file block, the client needs to specify the offset within the file and the content of the data. In the absence of concurrency, the Primary Replica directly performs the write operation, and after all Secondary Replicas have successfully written the data, the Primary Replica returns a message indicating successful writing. This operation guarantees strong consistency of the data, meaning that all replicas have the same value, corresponding to the defined state of the file block in GFS. In a concurrent scenario, after the Primary Replica receives multiple write requests from clients, it will determine an execution order. Due to the uncertain write order, after concurrent writes, the data read by the client may not necessarily be the one they wrote, but the data is identical, similar to eventual consistency. After concurrent writes, the corresponding file block state is consistent but undefined.
3. **Appending Data**. In the absence of concurrency, when a client appends a data, if the first attempt fails halfway through, the status of all replicas of the file block is as follows:

- Primary Replica: original data, offset1 : data.
- Secondary Replica1: original data, offset1 : data.
- Secondary Replica2: original data.

The client resends the append write request. Since the Primary Replica executes the operation first and then sends the request to all Secondary Replicas, so the Primary Replica's current file is the longest. The Primary Replica continues to append data at offset2 (the current file tail) and notifies all Secondary Replicas to append data at offset2 as well. However, because the end of Secondary Replica2's file is not at offset2, Secondary Replica2 needs to pad zeros. If this append operation is successful, the final data will be like this:

- Primary Replica: original data, offset1: data, offset2: data.
- Secondary Replica1: original data, offset1: data, offset2: data.
- Secondary Replica2: original data, offset1: *, offset2: data.

And return offset2 to the client. The data in the middle part of the file block at offset1 is inconsistent, but for the appended data, it is defined. The client reads offset2 and can determine that it reads the data. This is the defined interspersed with inconsistent state of the append operation.

In the case of concurrency, two clients respectively append data1 and data2 to the same file:

- Client1: append data1.
- Client2: append data2.

After receiving the append operation, the Primary Replica performs operation serialization, assuming the following:

- Primary Replica: data2, data1.

Then perform the append operation. Assuming the append operation of client2 fails, client2 sends the append operation again sequentially. The Primary Replica reappends it, resulting in the following data situation:

- Primary Replica: original data, offset1: data2, offset2: data1, offset2: data2.
- Secondary Replica1: original data, offset1: data2, offset2: data1, offset2: data2.
- Secondary Replica2: original data, offset1: *, offset2: data1, offset2: data2.

The client1 receives offset2 from GFS, indicating that data1 is appended to the position of offset2 in the file. And the client2 receives offset3. The states of offset2 and offset3 are defined, while offset1 is inconsistent. The overall state of the file is defined interspersed with inconsistent. We can see that, regardless of concurrency, append operations cannot guarantee that all data is defined, only that there is defined data, but there may be inconsistent data mixed in.

In GFS, there are three states for file blocks, with consistency levels ranging from high to low: defined, consistent but undefined, and inconsistent. Defined and undefined refer to the ordering of concurrent writes by multiple clients to the same file block offset. Consistent and inconsistent refer to whether the contents of multiple replicas at the same offset are the same. Now let's explain the three states of file blocks:

1. **Defined**. After a client writes data to a certain offset in a file, when reading the file again, it will always read the data it wrote.
2. **Consistent but undefined**. Multiple clients concurrently writing to the same offset of the same file, with the writing order determined by the Primary Replica. After writing, when reading again, the read data may be what the client wrote or data written by other clients. However, after all clients have finished concurrent writing, the data at the same offset in all replicas is consistent, ensuring eventual consistency.
3. **Inconsistent**. After a client modifies a file, the data at the same offset in all replicas is not entirely identical.

6.6 Summary

In this chapter, we discussed GFS, which leverages a large number of inexpensive commercial servers to implement a highly available file system. GFS treats machine failures as normal occurrences, focusing on fault tolerance, which is crucial in distributed system design. GFS systems are typically deployed on hundreds or even thousands of inexpensive commercial servers, with a significant number of GFS clients accessing GFS services. Therefore, application failures, operating system bugs, connectivity issues, network failures, and even machine power failures are common occurrences. GFS systems can support system monitoring, fault detection, fault tolerance, and automatic recovery, providing very high reliability. Furthermore, files in GFS systems are generally large, and file operations are mostly appends rather than overwrites. Once a file is written, most operations involve reading the file sequentially. We reviewed the design goals of GFS, the role of the master node, file read and write operations in GFS, and GFS consistency in this chapter. GFS consistency serves as a typical example of distributed data consistency, and we will continue to expand upon in subsequent chapters, guiding readers through the manual implementation of a distributed key-value storage system.

Chapter 7
Go Language Basics for Distributed Systems Design

In the previous chapter, through the study of GFS, we have gained a basic understanding of distributed file systems. In the following chapters, we will teach you how to implement a distributed key-value storage system, eraftkv, built on the strongly consistent algorithm Raft. The eraftkv is a distributed key-value storage system implemented in the Go language. The Go language benefits from the design of goroutines and channels, making it more convenient for network programming in distributed systems. Programming in Go under high-concurrency environments ensures that the logic is simple and clear, while performance can also be guaranteed to a certain extent. This is the reason for using the Go language to build distributed systems. Before delving into the implementation details of eraftkv, we will first introduce the basic knowledge of the Go language.

7.1 Advantages of Go

The inspiration for the birth of the Golang came from the compilation and build process of a C++ project within Google. At that time, Google predominantly used C++ to build various systems, but C++ was noted for its high complexity, slow compilation and build speeds, and lack of convenient support for concurrency in writing server programs. Based on such pain points, Rob Pike, Robert Griesemer, and Ken Thompson, three Google programmers, conceived the idea of designing a new programming language. Ken Thompson, notably, is the inventor of the UNIX operating system. The main design concept was to start with the foundation of the C language, remove some criticized features, and add missing functionalities.

Based on the initial ideas of the three programmers, the iterative design and implementation process of the Golang began. On November 10, 2009, the Golang project was officially open-sourced, with its main code repository located at go.googlesource.com/go. After the open-sourcing of the Golang, it attracted attention from developers worldwide. With the influence of the three authors in the

Z. Xu, *Distributed Data Services*, https://doi.org/10.1007/978-981-96-9392-4_7

industry and the support of Google, more and more talented programmers joined the Go development team. As a result, an increasingly diverse group of contributors contributed to the Golang project, forming a vibrant Golang community.

After the open-sourcing of Golang, some major players and startups in the cloud computing domain astutely turned their attention to the language, giving rise to exemplary projects such as the container engine Docker, the cloud-native de facto standard platform Kubernetes, as well as products like etcd, Consul, Cloudflare CDN, and Qiniu Cloud Storage. Consequently, Golang earned the reputation of being the "cloud computing infrastructure programming language." Not only is Golang well-suited for the cloud computing infrastructure domain, but it has also shone brightly in areas such as basic backend device software, microservices, and Internet infrastructure. Examples include distributed relational databases like TiDB and CockroachDB, Go-kit, micro, Monzo Bank's Typhon, Bilibili, blockchain public chains like Ethereum, and consortium chains like Hyperledger Fabric. The widespread adoption of Golang in the cloud-native domain in recent years has positioned it as a leading programming language of the cloud-native era.

The widespread application of the Golang is attributed to the following advantages.

7.1.1 Simple and Easy to Use

Unlike mainstream programming languages that continuously add new features through mutual borrowing, the designers of Golang rejected the path of language feature fusion from the very beginning of its design. They chose to leave complexity to the design and implementation of the language itself, to the core development of Golang, while keeping simplicity, ease of use, and clarity for the vast majority of Golang users. This design significantly reduces the cognitive burden for developers in choosing paths and understanding others' chosen paths. For example, complex concurrency can be achieved through a simple keyword "go," backed by the meticulous design and continuous efforts of the Golang development team. It is precisely because the Golang development team adheres to the principle of "pursuing simplicity, less is more" that Golang is very friendly to beginners and easy to pick up quickly. Golang has only 25 keywords, inheriting the concise genes of the C language, with embedded support for C syntax; it features object-oriented characteristics including inheritance, polymorphism, and encapsulation; and it is a cross-platform language. We can quickly get started with Golang through the code examples provided on this website: https://gobyexample.com.

7.1.2 Advocating Composition over Inheritance

To increase code reusability, there are two main approaches in language design patterns: composition and inheritance. Simply put, composition is generally understood as a "has-a" relationship, while inheritance is an "is-a" relationship. Both approaches can couple various parts of the program together.

Many mainstream object-oriented languages (such as C + + and Java) couple various parts of the program through extensive top-down type systems, inheritance, and explicit interface implementation mechanisms. However, inheritance is intrusive; once a subclass inherits, it must possess all methods and properties of the superclass. This reduces the flexibility of the code. Moreover, the inheritance approach amplifies coupling between modules. When the properties and methods of the superclass are modified, it often necessitates consideration of modifications in subclasses, and in an unregulated environment, such modifications may lead to extensive code refactoring.

In the design of Golang, parts are seamlessly coupled together using composition. Each type and package is orthogonally independent, without the concept of subclasses.

In the Golang, **we can embed already implemented functionalities into new types through type embedding, to quickly meet the functional requirements of the new types**. The syntax of type embedding replaces inheritance to achieve vertical expansion of functionalities. Listing 7.1 is an example of type embedding. In the admin struct type, we embed the type Mutex. The method set of the embedded Mutex type will be promoted to the admin type. At this point, the admin type will have the Lock and Unlock methods of Mutex. When calling the Lock and Unlock methods, the method calls will still be passed to the Mutex type within admin.

We can also **horizontally compose various parts of the program through interfaces**. Listing 7.2 is an example of composition using a regular function that accepts parameters of interface type. The function Copy is implemented through the

Listing 7.1 Type embedding

```
//The source code package src/sync/mutex.go defines the data structure
    for a mutex lock
//The Mutex provides two methods to the outside: the locking method Lock
    () and the unlocking method Unlock()
type Mutex struct {
  state int32
  sema  uint32
}

type admin struct{
  Mutex //Embedding Mutex
  level int
}
```

Listing 7.2 Interface horizontal composition

```
//$GOROOT/src/io/io.go
func Copy(dst Writer, src Reader)(written int64, err error)
```

Listing 7.3 Interface aggregation

```
//$GOROOT/src/io/io.go
type ReaderAndWriter interface{
    Reader
    Writer
}
```

io.Reader and io.Writer interfaces, and is horizontally composed with the package that contains Copy in a loosely coupled manner.

We can also **achieve aggregation of interface behaviors and compose large interfaces by embedding interface types within the definition of an interface**, as shown in Listing 7.3.

Golang uses composition to genuinely reduce the coupling between various modules, making the binding of methods more flexible.

7.1.3 Native Concurrency

After reaching the limits of improving performance solely through increasing CPU clock speeds, hardware development shifted toward multicore processing. This means placing multiple execution cores in a single processor, allowing each core to operate in parallel at lower frequencies to enhance performance. At this point, efficiently utilizing multiple cores and improving parallel processing capabilities became essential for program development. However, programming languages created during the single-core era (such as C + + and Java) did not take hardware scalability into consideration. They typically rely on the operating system for concurrent operations, where the program creates threads and the operating system schedules them. This traditional method of supporting concurrency is very complex and difficult to scale.

The design of the Golang took into account the trend toward multicore processors and natively incorporated concurrent support. Golang utilizes user-level lightweight threads, known as goroutines, for concurrent operations. A Golang program can create tens of thousands of concurrent, lightweight goroutines, and pieces of code in Golang execute within these goroutines. The goroutines are not perceived by the operating system, and their scheduling is handled by Golang itself. Moreover, developers can easily create, destroy, and control goroutines concurrently using the built-in concurrent syntax elements and mechanisms provided by the Golang. Golang's native support for concurrency enables it to fully utilize multicore processors, switch between cores with low overhead, and significantly reduces the burden of writing highly concurrent applications.

7.1.4 Possessing an Open Ecosystem

In the powerful Golang community, skilled technicians from around the world have developed a rich ecosystem of tools, many of which effectively address engineering scalability issues. For example, Golang has a runtime system scheduling mechanism that assists programmers with tasks such as garbage collection, load balancing Go routines, and other optimization mechanisms. It features an efficient GC (garbage collection) mechanism (since Golang version 1.8, GC has included a tricolor marking and hybrid write barrier, making it highly efficient). Additionally, Golang offers a wealth of standard libraries, such as the io library for read and write operations, the net library for network programming, and the database/sql library for database operations. The extensive toolkit can largely meet the everyday engineering needs of developers.

7.2 Slices

Slice, in the Golang, is an abstraction of arrays. In Go, the length of an array is fixed once it is defined, which makes arrays unsuitable for many specific scenarios. To address this issue, Golang has designed the slice data structure. Slices are very similar to arrays and can be understood as dynamic arrays. We will first introduce arrays and then discuss the operations related to slices.

7.2.1 Arrays in Go

The length of arrays in Golang is fixed, and when passing arrays as arguments, they strictly match the array type. In Listing 7.4, we define three arrays and attempt to print and modify them using functions.

We can observe that when a function's formal parameter is an array of length n, the passed actual parameter must also be an array of the same type (with length n). Furthermore, the array parameter is passed by value; thus, it cannot be assigned a new value within the function, which significantly complicates the usage of arrays. Therefore, Golang provides a flexible and powerful built-in type called slices (dynamic arrays). In contrast to arrays, slices have variable lengths and can have elements appended to them, potentially increasing the slices' capacity upon appending. Next, we will familiarize ourselves with how to create and use the slice, and delve into its underlying implementation.

Listing 7.4 Array operations

```
//Print the array
func printArray(Arrayn [4]int){
    for index, value := range Arrayn{
        fmt.Println("index=",index,",value=",value)
    }
    //Change the value of the array
    Arrayn[0]=100
}

func main(){
    //Defining fixed-length arrays
    var Array1 [10]int
    Array2 := [10]int{1,2,3,4}
    Array3 := [4]int{11,22,33,44}

    //Attempt to print Array2
    //A type mismatch error is reported, where the formal parameter is
     of type int[4], but the actual parameter is of type int[10].
    printArray(Array2)

    //Attempt to print Array3
    printArray(Array3)//Print index=0,value=11...

    //Verify if the array values have been modified
    for index, value := range Array3{
        fmt.Println("index=",index,",value=",value)
        //Output: index=0, value=11... Failed to change the value of
     Array[0]
    }
}
```

7.2.2 Slice Declaration

Listing 7.5 are several common ways to create slices, all of which ultimately return a reference to the slice.

7.2.3 Appending to Slices

In Golang, the len() function is used to retrieve the length of an array, while cap() is used to retrieve the capacity of an array. As shown in Listings 7.6–7.9, let's observe how the length and capacity properties change, when continuously appending elements to a slice with a length of 3 and a capacity of 5. The length increases by one with each execution of the append method until it reaches the maximum capacity. The capacity automatically expands based on the cap base. For

Listing 7.5 Slice declaration

```
//Declare slice1 as a slice, and initialize it with default values of 1,
    2, 3, with a length len=3
slice1 := []int{1,2,3}

//Declare slice as a slice, allocate 3 spaces for slice, and initialize
    the values as 0
//Method 1
var slice1 [] int
slice1=make([]int,3)

//Method 2
var slice1 []int = make([]int,3)

//Method 3
slice1 := make([]int,3)
```

Listing 7.6 Slice append step1

```
//Step1.Declare a slice with a length of 3 and a capacity of 5
var slice = make([]int,3,5)//len=3, cap=5, slice=[0,0,0]
fmt.Printf("len=%d,cap=%d,slice=%v\n",len(slice),cap(slice),slice)
```

Listing 7.7 Slice append step2

```
//Step2. Append element 1
slice=append(slice,1)//len=4,cap=5,slice=[0,0,0,1]
fmt.Printf("len=%d,cap=%d,slice=%v\n",len(slice),cap(slice),slice)
```

Listing 7.8 Slice append step3

```
//Step3. Append element 2
slice=append(slice,2)//len=5,cap=5,slice=[0,0,0,1,2]
fmt.Printf("len=%d,cap=%d,slice=%v\n",len(slice),cap(slice),slice)
```

example, initially with a capacity cap $=5$, when it reaches the maximum capacity for the first time, the cap automatically expands to 10; when it reaches the maximum capacity for the second time, the cap automatically expands to 15, and so on.

The underlying structure of the slice in memory at this step is shown in the Fig. 7.1.

The underlying structure of the slice in memory at this step is shown in Fig. 7.2.

The underlying structure of the slice in memory at this step is shown in Fig. 7.3.

The underlying structure of the slice in memory at this step is shown in Fig. 7.4.

Listing 7.9 Slice append step4

```
//Step 4. Append element 3. When appending an element to a slice that
    has already reached its full capacity, the cap capacity doubles (
    expansion mechanism).
slice=append(slice,3)//len=6,cap=10,slice=[0,0,0,1,2,3]
fmt.Printf("len=%d,cap=%d,slice=%v\n",len(slice),cap(slice),slice)
```

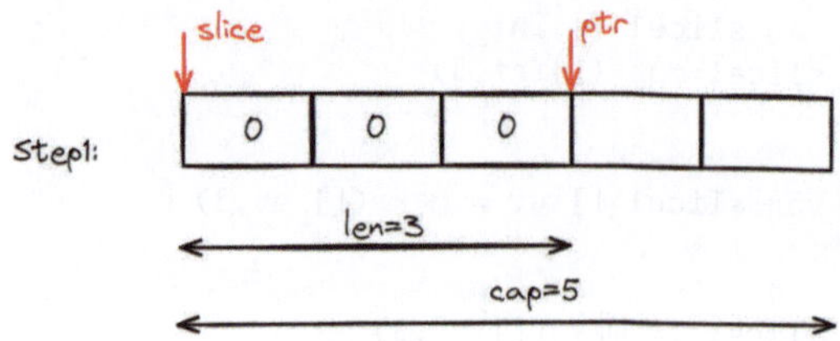

Fig. 7.1 Slice append step1

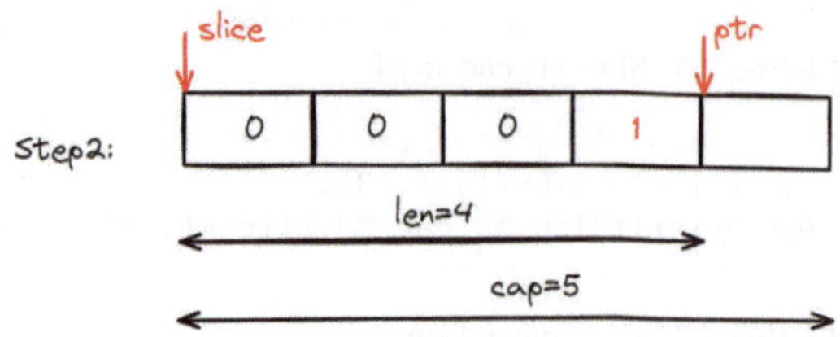

Fig. 7.2 Slice append step2

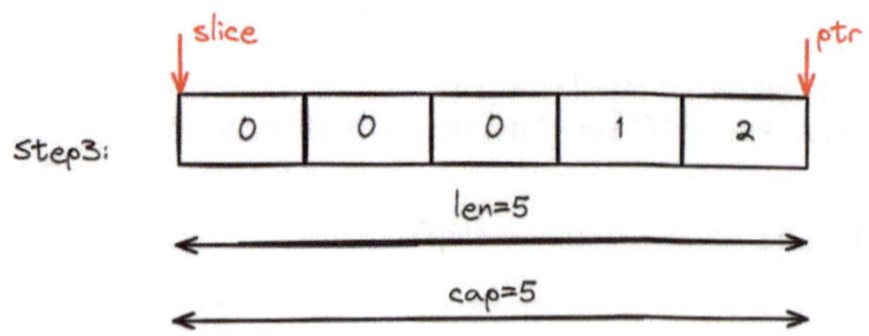

Fig. 7.3 Slice append step3

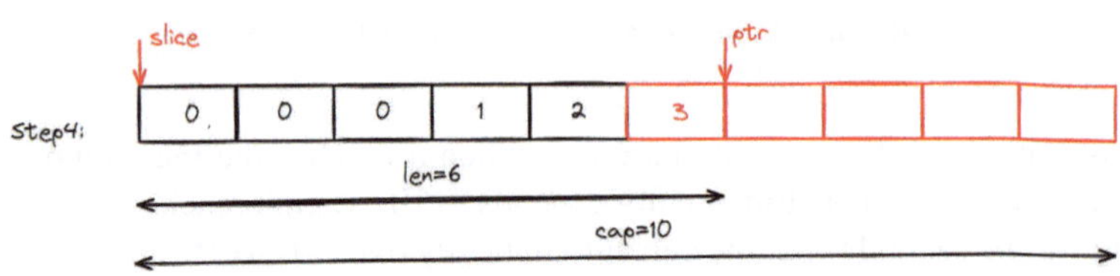

Fig. 7.4 Slice append step4

7.2.4 Slicing Slices

Slicing slices in Golang is similar to slicing operations in Python. We can run Listing 7.10 to deepen our understanding of slice slicing operations. In this example, we slice the array a to obtain a slice. This operation results in a slice containing the elements with indices [low, high), including the low index but excluding the high index.

Listing 7.10 Slicing slices

```
package main
import "fmt"

func main() {
  var a = [5]string{"Alpha", "Beta", "Gamma", "Delta", "Epsilon"}

  //Creating a slice from the array
  var s []string = a[1:4]
  var s1 []string = a[1:]
  var s1 []string = a[:2]

  fmt.Println("Array a = ", a)
  fmt.Println("Slice s = ", s)
  fmt.Println("Slice s1 = ", s1)
  fmt.Println("Slice s2 = ", s2)
}

//Output
Array a =  [Alpha Beta Gamma Delta Epsilon]
Slice s =  [Beta Gamma Delta]
Slice s1 =  [Beta Gamma Delta Epsilon]
Slice s2 =  [Alpha Beta]
```

Listing 7.11 Modifying slice elements

```
s := []int{1,2,3} //len=3,cap=3

//[0,2) left-closed, right-open slicing
s1 := s[0:2] //s1=[1,2]
fmt.Println(s1)

//Attempt to modify a sliced portion of a slice
s1[0] = 100
fmt.Println(s)//Output: s=[100,2,3]
fmt.Println(s1)//Output: s1=[100,2]

//Copy copies the contents of a slice from the underlying array to a new slice
s2 := make([]int,3) //s2=[0,0,0]
copy(s2,s)
fmt.Println(s2) //s2=[100,2,3]
```

7.2.5 Modifying Slice Elements

Slices are reference types that point to the underlying array. When we use a slice reference to modify the corresponding element in the array, other slice objects that reference the same array will also see the result of this modification. It is illustrated in Listing 7.11.

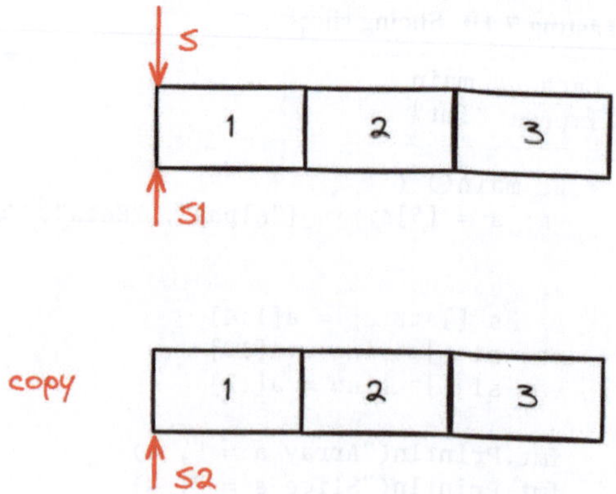

Fig. 7.5 Copy function copy array schematic

Listing 7.12 Using slices to print and modify a slice

```
//Print Array
func printArray(Arrayn []int){
    for index, value := range Arrayn{
        fmt.Println("index=",index,",value=",value)
    }
    //Modify array values
    Arrayn[0]=100
}

func main(){
    //Define a dynamic array and use slicing on it
    Array1 := []int{1,2,3,4}

    //Attempt to print Array1
    printArray(Array1)//Output: index=0,value=1...

    //Verify if the array values have been modified
    for index, value := range Array1{
        fmt.Println("index=",index,",value=",value)
        //Output: index=0, value=100... can change the value of Array[0].
    }
```

In the example above, modifications made to s are visible in s1, and modifications made to s1 are also visible in s. The reason for this is that s and s1 point to the same memory space. To modify the array data independently, you can allocate a new memory space and use the copy function to duplicate the array before making modifications. The schematic diagram is shown in Fig. 7.5.

Finally, using Listing 7.12, we try to print and modify the array using slices.

In the above example, we defined a slice named Array1 and passed it to the printArray function. We found that it was possible to change the value of Array1[0]. This indicates that the passing of function parameters is done by **reference passing**, where a reference pointer to the underlying array is transmitted. Therefore, the function is able to modify the value of the array within the function.

7.3 Goroutine and Channels

Generally speaking, the applications we write execute sequentially, completing functional tasks independently. In this programming mode, the execution steps of the program are clear and easy to debug and maintain. However, when dealing with certain complex tasks that have long waiting times, a single-threaded program will remain idle for extended periods, thereby reducing processor resource utilization. In such cases, we can divide a task into multiple subtasks and use concurrent programming to allow the program to handle multiple different tasks simultaneously, improving the program's execution efficiency.

To further illustrate the difference between sequential execution and concurrent execution, let's assume we need to scrape the content of three webpages and calculate the size of each page based on the returned HTML content. As shown in Figs. 7.6 and 7.7, programs implement their functions in two modes: sequentially and concurrently. From a functional perspective, we only need to utilize the methods defined in the **net/http** package in the Golang to send GET requests based on the input URL and retrieve the content of the corresponding webpage. After that, we convert the Body content of the returned object into a byte type. During the process of web crawling, the program calculates the size of the webpage. The program sends requests to the corresponding webpages, but due to factors such as webpage size, network latency, and the location of the website, the response times vary. While waiting for the crawl to complete, the application needs to wait for the content to be returned before proceeding to the next step of calculation and processing. In Fig. 7.6, the program adopts a sequential execution mode to crawl the content and calculate the size of three webpages. This means that the crawling and calculation of one webpage have to wait until the previous webpage's crawling and calculation are completely finished, which will lead to the program being idle due to waiting for request responses. In contrast, as shown in Fig. 7.7, the program adopts a concurrent execution mode to crawl the content of three webpages. After sending a request to one webpage, it immediately sends requests to the remaining webpages instead of idling while waiting for the content to return. As soon as any of the requests returns the crawled content, the program can proceed to calculate the size of that page. Finally, after calculating the sizes of all three webpages, the program outputs the results and terminates. The concurrent mode allows the program to crawl the content of three pages simultaneously, reducing idle wait time and improving the utilization of computing resources.

Golang provides **Goroutine** support for concurrent execution of different tasks. By creating functions or code segments as Goroutines, the Golang runtime schedules them as independent work units for execution. In the process of handling complex tasks concurrently, it is sometimes necessary for multiple Goroutines to

Fig. 7.6 Sequential execution of web content crawling

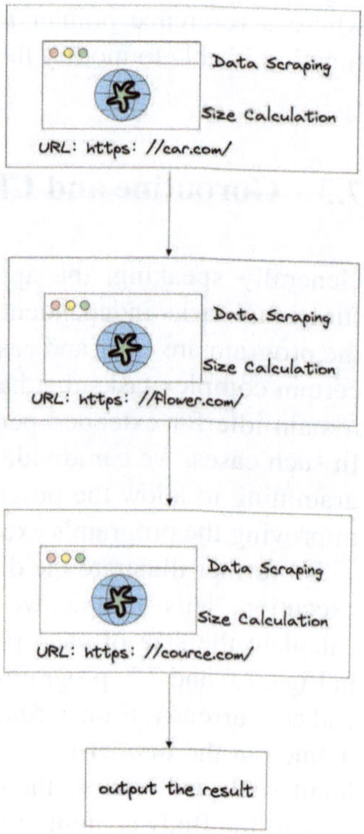

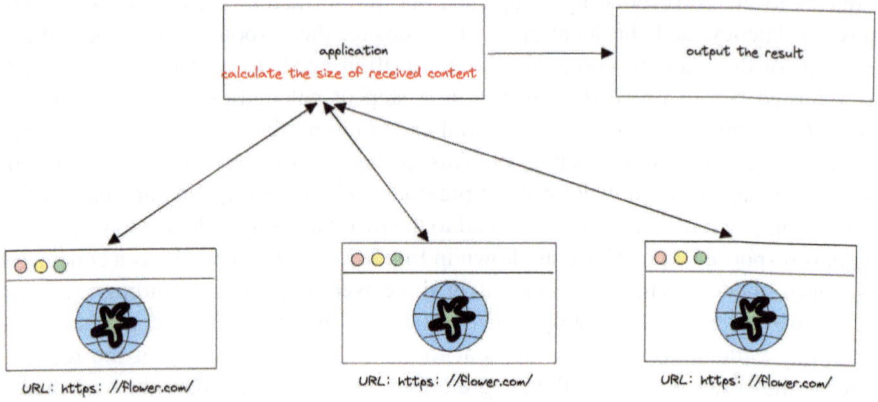

Fig. 7.7 Concurrent execution of web content crawling

follow a certain sequence and cooperate to complete functional business tasks. This brings up the issue of synchronized access among the Goroutines. Golang adopts the **Communicating Sequential Processes (CSP)** model as an example for its concurrent synchronization model. CSP is a message-passing model that achieves synchronized access by introducing a channel structure that allows data transmission between Goroutines.

In the following subsections, we will introduce how to use Goroutines and Channels in the Golang to implement concurrent programming.

7.3.1 Goroutine Concepts

Goroutine is a lightweight thread managed by the Go runtime. As shown in Table 7.1, compared with threads implemented in other programming languages, Goroutine has advantages such as low usage cost and low communication latency. However, all Goroutines in a Golang program run in the same address space. Therefore, for access to shared memory, Goroutines need to perform synchronized access processing. Moreover, as a thread adhering to the pure function concept, Goroutine does not have a unique identifier to implement local storage space, in order to prevent the abuse of thread-local storage, compared to traditional threads.

In addition, the Golang runtime includes an independent scheduler that utilizes the M:N technique to map m Goroutines to n operating system threads for execution. Compared to traditional operating system threads, which rely on hardware clock interrupts to trigger the CPU to call kernel functions for context switching and thread scheduling, Golang's scheduler is triggered by specific Golang constructs and

Table 7.1 Comparison between Goroutine and traditional threads

Comparison	Thread type	
	Goroutine	Thread
Scheduler	Go runtime scheduler	Operating system kernel
Communication method	Channel	Locking and semaphore mechanism
Communication delay	Based on channel communication, the communication latency is low	Communication latency between threads is high
Local storage	Null	Threads have unique identifiers, enabling local storage
Stack memory	Scalable size	Fixed size
Startup latency	Low	High
Thread switching cost	Low	High

does not require switching to the system kernel context for scheduling. Therefore, scheduling a Goroutine is significantly less costly than scheduling a thread.

7.3.2 Goroutine Practice

Every program contains at least one Goroutine, which is called the main Goroutine. The simple example we run below (Listing 7.13) is a main Goroutine. In this example, the say function prints the input string three times, and the main function calls the say function with the strings "world" and "hello" as inputs, printing them respectively.

Next, we will run the two calls to the say function concurrently in different Goroutines. To create a new Goroutine to execute a function, simply use the "go" keyword as a prefix to the function definition or function call. Therefore, we modify the above code to Listing 7.14 and run it.

Listing 7.13 Main Goroutine example

```
package main

import (
  "fmt"
  "time"
)

func say(s string) {
  for i := 0; i < 3; i++ {
    fmt.Println(s)
  }
}

func main() {
  say("world")
  say("hello")
  fmt.Println("\n terminating program")
}

// output
world
world
world
hello
hello
hello
terminating program
```

Listing 7.14 Different Goroutine example

```
package main

import (
  "fmt"
  "time"
)

func say(s string) {
  for i := 0; i < 3; i++ {
    fmt.Println(s)
  }
}

func main() {
  go say("world")
  go say("hello")
  fmt.Println("\n terminating program")
}

// Output
terminating program
```

Since the two say functions run in separate new Goroutines, concurrently with the main Goroutine, when a new Goroutine is called, the main program does not wait for the function to finish and return, as it would with a regular function call. Instead, after the function call is executed, it immediately proceeds to the next line of code, ignoring the return value of the Goroutine. Therefore, in the program example, after the say("world") and say("hello") functions are created as independent Goroutine calls, the main Goroutine continues executing directly. After printing the program termination prompt, the program finishes and terminates, and all the created Goroutines also terminate accordingly. Thus, in the output, we only see the program termination prompt's print output, without the print outputs generated by the two function calls.

To address this issue, we need to ensure that the main Goroutine remains running until the other Goroutines have finished executing. This is the synchronization issue between Goroutines that we mentioned in Sect. 7.3.1. We can introduce another feature of Golang, Channels, to solve this, but for now, we will simply use the time.Sleep(time.Second) method from Go's time package to pause the main Goroutine for a certain duration. The modified code example, Listing 7.15, is as follows.

We rerun the modified code and can see that the functions say("world") and say("hello") successfully print their respective characters. This is because we use the time. Sleep method in the main Goroutine to make it idle and wait, allowing enough time for the two function call Goroutines to complete their printing before the main Goroutine terminates.

Listing 7.15 Main Goroutine sleeps and waits

```
package main

import (
  "fmt"
  "time"
)

func say(s string) {
  for i := 0; i < 3; i++ {

    fmt.Println(s)
  }
}

func main() {
  go say("world")
  go say("hello")
  time.Sleep(time.Second)
}

// Output

hello
hello
hello
world
world
world
```

Apart from using the time. Sleep method to make the main Goroutine pause for a while and wait for other Goroutines to complete, we can also use synchronization primitives provided in Golang's sync package, such as Mutex, Once, and WaitGroup types, to achieve synchronized access. Suppose we create two Goroutines in the main Goroutine to call the say("hello") and say("world") functions, respectively. To prevent the main Goroutine from ending before the two functions finish executing, we use a WaitGroup to keep track of the number of Goroutines that need to complete and block the main Goroutine until all the child Goroutines have finished their tasks. The WaitGroup structure primarily provides the following three methods to address the synchronization issue:

1. **Add()**. Before creating and using a new Goroutine each time, call Add() to set or increment the waiting counter, indicating the number of Goroutines that need to complete.
2. **Done()**. Before each Goroutine that needs to be waited upon completes, it should call that method to indicate its completion, which will decrement the waiting counter by 1.
3. **Wait()**. Used in a Goroutine that needs to be blocked, it will block that Goroutine until the waiting counter reaches 0.

Listing 7.16 WaitGroup usage example,

```
package main

import (
  "fmt"
  "sync"
)

func say(s string, wg *sync.WaitGroup) {
  for i := 0; i < 3; i++ {
    fmt.Println(s)
  }
  // Call the Done() method when the function exits to notify the main
     Goroutine that the work is completed.
  wg.Done()
}

func main() {
  // Create WaitGroup
  // Increment the waiting counter by 2, indicating that two goroutines
     need to complete their execution
  var wg sync.WaitGroup
  wg.Add(2)
  go say("world", &wg)
  go say("hello", &wg)
  // Block the main Goroutine until the WaitGroup counter returns to 0
  wg.Wait()
}

// Output

hello
hello
hello
world
world
world
```

The following Listing 7.16 is a simple example of using WaitGroup.

7.3.3 *Channel Concepts*

When a function call is modified with go to run as an independent Goroutine, the function cannot return a result that is received by a variable in the main Goroutine. Refer to Listing 7.17, where square() is used to calculate the square of an input number. Since the function is modified with go, it runs in an independent Goroutine. Therefore, the main Goroutine will not wait but will execute the next line of code directly, while the square() function takes some time to compute. This leads to the

Listing 7.17 Function return values not received by the main Goroutine

```
func main() {
  result = go square(5) // Compilation error
  fmt.println(result)
}
```

Listing 7.18 Build channel

```
package main

import (
  "fmt"
)

func square(result chan int, x int) {
  result <- x * x
}

func main() {
  result := make(chan int)
  go square(result, 5)
  fmt.Println(<-result)
}
```

potential risk that the variable in the main Goroutine's subsequent print statement cannot receive the return value. Hence, the Go compiler will generate a compile-time error to prevent the use of the return result of a function modified with go.

To address the issue of data communication between different Goroutines, the Golang provides the channel structure. Channels allow Goroutines to pass data while ensuring that the receiving Goroutine has access to the transmitted value only after the sending Goroutine has finished processing and sent the result, thus achieving synchronized access.

Listing 7.18 briefly introduces how to create a channel and solve the issue mentioned above, enabling the transmission of numerical results between the main Goroutine and the spawned child Goroutines. To create and use a channel, we need to complete the following steps:

1. Declare a channel parameter in the function you want to run in a separate Goroutine to pass data to other Goroutines.
2. Create a channel of the same type in the main Goroutine.
3. Use go to decorate the function call and pass the channel variable as a parameter to receive the return value.

Analyze each step of creating a channel one by one.

Function Declaration with Channel Variables and Result Passing As shown in Listing 7.19, channels of the same type are declared as function parameters to receive the created channels. Afterward, the computed results are passed into the channels through the operator (<-).

Listing 7.19 Data transfer via channels

```
func square(result chan int, x int){
    result <- x * x
}
```

Listing 7.20 Unbuffered channel and buffered channel

```
// Unbuffered int channel
unbuffered := make(chan int)

// Buffered int channel
buffered := make(chan string, 10)
```

Create Channel] Golang provides the built-in function make to create channels, specifying the type of data to be transmitted through the channel when declaring it. Channels can share values or pointers of built-in types, named types, structure types, and reference types, but a channel can only transmit values of its element type.

During creation, we need to pass the chan keyword to the make function to indicate the creation of a channel and specify the data type at the same time, as shown in Listing 7.20.

In the creation of the second channel mentioned above, apart from specifying the chan keyword and the data type we are submitting, we also pass an additional parameter to indicate the buffer size of the channel, thus creating a buffered channel. The key difference between an unbuffered channel and a buffered channel is that an unbuffered channel will cause the Goroutine that performs the receive or send operation first to block, waiting for the other Goroutine to complete its receive or send operation before continuing. In contrast, a buffered channel allows Goroutines to communicate with each other to send and receive data without both completing their operations simultaneously. Blocking of Goroutines only occurs in the following two scenarios: when the receiver Goroutine executes the receive operation first but there is no value in the channel to receive, the receiver Goroutine will be blocked. On the other hand, when the buffer in the channel is full, the sender Goroutine will be blocked, waiting for a free slot in the buffer to receive the sent value.

Receiver Invokes and Receives the Result After creating a channel in the receiving Goroutine, pass the channel to a function and run the function in a new Goroutine. Once the function computation is completed, the result will be passed into the channel, from which the receiving side retrieves the transmitted value. As shown in Listing 7.21, when receiving a value from a channel, the "<-" operator should be on the left side of the channel variable being operated on.

Listing 7.21 Transmission of values via receiving channels

```
func main() {
  result := make(chan int)
  go square(result, 5)
  fmt.println(<-result)
}
```

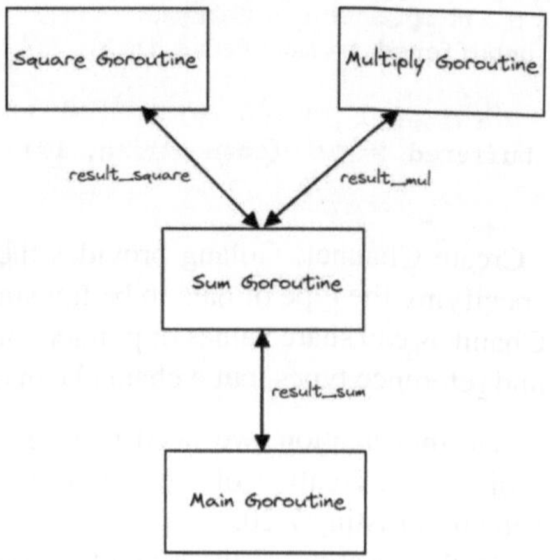

Fig. 7.8 Synchronous calculation order

7.3.4 Synchronization via Channels

In Sect. 7.3.3, we mentioned that channels can ensure that a Goroutine receiver will have access to the transmitted data only after the sender has passed the data into the channel for retrieval. This is achieved through blocking operations, facilitating synchronized data transmission between multiple concurrent Goroutines. Taking an unbuffered channel as an example, the send operation blocks the sender until another Goroutine performs the receive operation. Conversely, the receive operation blocks the receiver until another Goroutine performs the send operation on the channel. By blocking and pausing the execution of the Goroutines involved, synchronization of operations between the two is achieved.

Below, we use an example to illustrate how to utilize channels to help multiple Goroutines collaborate to accomplish a business function. We define three functions, square(), multiply(), and sum(), each running in a different Goroutine. Through channels, we synchronize the computation of Eq. 7.1's value. As depicted in Fig. 7.8, the square() function performs square calculations, and the multiply() function performs multiplication calculations concurrently. Subsequently, the results are passed to the sum Goroutine for summation calculation. Finally, the main Goroutine waits for the summation result to be returned and prints it:

$$\text{result} = 5^2 + 6 * 7 \tag{7.1}$$

Listing 7.22 Multi-Goroutine cooperation example

```
package main

import (
  "fmt"
)

func square(result chan int, x int) {
  result <- x * x
}

func multiply(result chan int, x int, y int) {
  result <- x * y
}

func sum(result_square chan int, result_mul chan int, result chan int) {
  x := <-result_square
  y := <-result_mul
  result <- x + y
}

func main() {
  result_square := make(chan int)
  result_mul := make(chan int)
  result_sum := make(chan int)
  go square(result_square, 5)
  go multiply(result_mul, 6, 7)
  go sum(result_square, result_mul, result_sum)
  result := <-result_sum
  fmt.Println(result)
  fmt.Println("Terminating program.")
}
```

Based on the characteristics of channels, blocking the receiving or sending Goroutine allows the execution of the next step only when the other party is ready to complete the data transmission operation. In Listing 7.22, we create three channels to facilitate data transmission and collaboration between Goroutines. The **result_square** and **result_mul** channels restrict the Sum Goroutine from executing until the square and multiplication functions have completed their calculations, respectively. The **result_sum** channel transmits the final calculation result between the Sum Goroutine and the main Goroutine. The main Goroutine waits for the completion of the sum function calculation, prints the result, and terminates the program. In summary, multiple Goroutines collaborate by utilizing channels to accomplish a simple calculation function.

7.4 Scheduler

In Sect. 7.3.1, it was mentioned that the Go language runtime includes an independent scheduler to manage Goroutines, thereby achieving low-cost Goroutine thread switching.

7.4.1 Scheduler Design Decisions

We introduce the functionality and role of the scheduler when managing multiple Goroutines through Listing 7.23.

The above program cyclically creates 11 Goroutines in the main Goroutine to perform printing, and each printed value can be regarded as the identifying ID of that Goroutine. Compared to the orderly output of printing in a loop using ordinary functions, we can see that the printing execution of Goroutines in the above program is in disorder, and it does not print out in the order of invocation. Regarding this output result, we have the following questions:

Listing 7.23 Multi-Goroutine execution

```
func main() {
    var wg sync.WaitGroup
    wg.Add(11)
    for i := 0; i <= 10; i++ {
        go func(i int) {
          defer wg.Done()
          fmt.Printf("loop i is - %d\n", i)
        }(i)
    }
    wg.Wait()
    fmt.Println("Hello, Welcome to Go")
}

// Output

loop i is - 0
loop i is - 4
loop i is - 1
loop i is - 2
loop i is - 3
loop i is - 8
loop i is - 7
loop i is - 9
loop i is - 5
loop i is - 10
loop i is - 6
Hello, Welcome to Go
```

- How are these 11 Goroutines executed concurrently?
- Is there a specific execution order among the Goroutines, or how can we determine the execution sequence of each Goroutine?

To answer these two questions, we need to consider:

- How do we distribute multiple Goroutines to run on a machine with limited CPU cores?
- To ensure fair access to CPU resources among these Goroutines, in what order should these Goroutines be scheduled to run on multiple CPU cores?

These issues are precisely the factors that the scheduler needs to consider during the scheduling process. In the following sections, we will address the above questions.

7.4.2 *Go Runtime Scheduler Model*

To address the scheduling issue of concurrent execution of multiple goroutines, the Golang has designed a scheduler model depicted in Fig. 7.9.

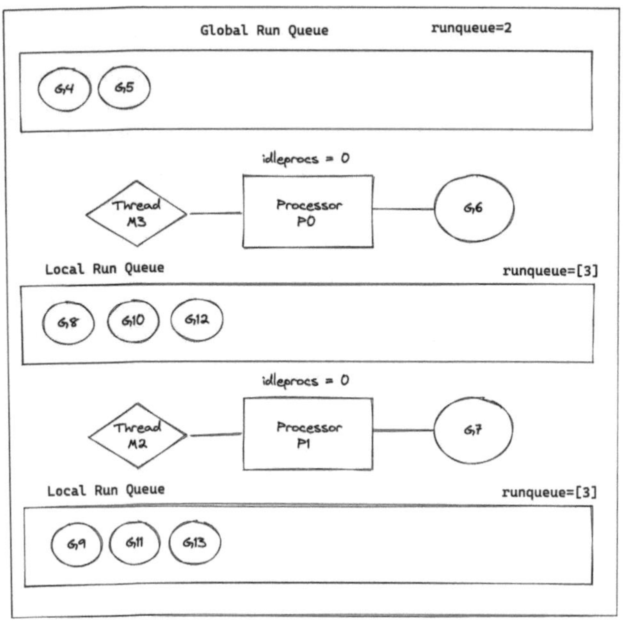

Fig. 7.9 Go runtime scheduler model

Golang employs the GMP scheduling model, in which:

- **G**: represents a Goroutine object, which is a concurrently executable code block created using the go keyword. Every time the go keyword is invoked, a G object is generated.
- **M**: represents an operating system kernel thread. Each time an M is created, a corresponding underlying kernel thread is also created. All G run on M.
- **P**: represents a logical processor. Each running M owns a corresponding P, which is responsible for scheduling the execution of G on M. Each P maintains a local queue of G tasks.

In Fig. 7.9, two P are responsible for scheduling and managing eight Goroutines. As seen in the figure, there are two types of queues managing the Goroutines, and the roles of these two types of queues are:

- **Local Run Queue**. The Local Run Queue resides in the P logical processor and holds G waiting to run. The capacity of this queue is limited and typically cannot exceed 256 Goroutines. When a user creates a new Goroutine and this queue is full, the Go runtime will move half of the G in the Local Run Queue to the Global Run Queue.
- **Global Run Queue**. It holds G that is waiting to run. When other Local Run Queues become full, they will move G over to the Global Run Queue.

The Role of the Local Queue Compared to the GM model scheduler initially used in the Golang, the GMP model scheduler introduces the concept of P logical processors, and each P has its own local queue. This significantly reduces the direct dependency on the global queue during Goroutine scheduling, decreasing the number of lock contentions when accessing the global queue, and thus lowering the performance overhead of scheduling. As explained in the definition of the queues mentioned above, the GMP model implements work stealing functionality. When a P's local queue is empty, it will steal runnable Goroutines from the global queue or other P's local queues, improving the utilization of computing resources. Conversely, if the local queue of a P reaches its maximum capacity for storing Goroutines, the Go runtime will transfer half of the Goroutines to the global queue for management.

The Workflow of the Go Scheduler As shown in Fig. 7.10, the general steps in the process of a Go function being created as a Goroutine and scheduled for execution by the scheduler are as follows:

1. By using go func() or prefixing a function call with go to create a Goroutine, denoted as G.
2. The newly created G is managed by placing it into a queue. It will first be saved in the local queue of P. Once the local queue of P is full, G will be moved to the global queue.
3. M is the entity that runs G, and each M must be bound to a P. M will look for an executable G in the local queue of P to execute. If P's local queue is empty, M will utilize work-stealing functionality to steal an executable G from another P's

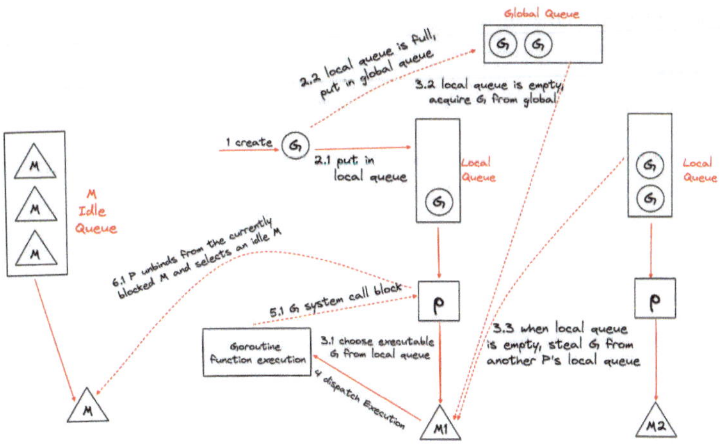

Fig. 7.10 Workflow of GMP model scheduler

local queue. M will search the global G queue if there is no executable G in other local queues.

4. During the execution of G, if a channel blockage or a user-level blockage occurs, M will not be blocked. Instead, M will continue to search for an executable G to run. Once the blocked G recovers, it will be re-enqueued to wait for execution. In another scenario, when G makes a system call, M will be blocked, and P will be detached from M. A new idle thread M will be found or created to bind with P, continuing to schedule and execute Goroutines in the queue.
5. After G finishes its system call, it will attempt to acquire an unbound and idle P to execute. If no idle P is available, thread M will enter a dormant state and be added to the idle thread queue, while G will be placed in the global queue to wait.

Tracking the Workflow of the Go Scheduler Golang provides the GODEBUG tool built-in, allowing programmers to generate Go runtime debugging information for reference and analysis without implementing additional functionalities. For the usage of the Go scheduler, we can request summary information and debugging details from the scheduler. If we want the program to print corresponding tracing debug information during runtime, we need to set the GODEBUG environment variable before compiling and running the program. Below, we will use a Listing 7.24 as example to demonstrate how to use the GODEBUG tool and scheduler tracing information.

In the above code, we create ten Goroutines using a for loop. We utilize the WaitGroup structure as introduced in Sect. 7.3.2, setting the number of Goroutines that need to be waited for completion. Each Goroutine first sleeps for 1 second using the time.Sleep() function, then accumulates the counter variable through a for loop

Listing 7.24 GODEBUG tracing example

```
package main

import (
  "sync"
  "time"
)

func main() {
  var wg sync.WaitGroup
  wg.Add(10)
  for i := 0; i < 10; i++ {
    go work(&wg)
  }

  wg.Wait()

  // Wait to see the global run queue deplete.
  time.Sleep(3 * time.Second)
}

func work(wg *sync.WaitGroup) {
  time.Sleep(time.Second)

  var counter int
  for i := 0; i < 1e10; i++ {
    counter++
  }

  wg.Done()
}
```

Listing 7.25 GODEBUG tracing instructions

```
GOMAXPROCS=2 GODEBUG=schedtrace=1000 ./go_demo
```

for one billion times. After completing the counting, it calls the wg.Done() method to notify the main Goroutine of its completion.

Compile the Listing 7.24.

```
go build go_demo.go
```

To use the GODEBUG tool to analyze and observe the scheduling of Goroutines by the scheduler in the Go runtime, we need to execute the following command, as shown in Listing 7.25.

In the execution command mentioned above, we need to set the value of the GODEBUG environment variable to output scheduling-related debugging information. The value of the GODEBUG environment variable can be composed of several key-value pairs, with the key and value separated by an equal sign "=" and multiple key-value pairs separated by commas ",". Different key-value pairs correspond to

Listing 7.26 Scheduling output

```
colin@book % GOMAXPROCS=2 GODEBUG=schedtrace=1000 ./go_demo
SCHED 0ms: gomaxprocs=2 idleprocs=1 threads=4 spinningthreads=0 idlethreads=1 runqueue=0 [0 0]
SCHED 1009ms: gomaxprocs=2 idleprocs=0 threads=4 spinningthreads=0 idlethreads=1 runqueue=0 [8 0]
SCHED 2009ms: gomaxprocs=2 idleprocs=0 threads=4 spinningthreads=0 idlethreads=1 runqueue=2 [3 3]
SCHED 3016ms: gomaxprocs=2 idleprocs=0 threads=4 spinningthreads=0 idlethreads=1 runqueue=2 [3 3]
SCHED 4017ms: gomaxprocs=2 idleprocs=0 threads=4 spinningthreads=0 idlethreads=1 runqueue=7 [0 1]
SCHED 5027ms: gomaxprocs=2 idleprocs=0 threads=4 spinningthreads=0 idlethreads=1 runqueue=5 [0 3]
SCHED 6031ms: gomaxprocs=2 idleprocs=0 threads=4 spinningthreads=0 idlethreads=1 runqueue=3 [2 3]
SCHED 7037ms: gomaxprocs=2 idleprocs=0 threads=4 spinningthreads=0 idlethreads=1 runqueue=5 [1 2]
SCHED 8045ms: gomaxprocs=2 idleprocs=0 threads=4 spinningthreads=0 idlethreads=1 runqueue=4 [2 2]
SCHED 9052ms: gomaxprocs=2 idleprocs=0 threads=4 spinningthreads=0 idlethreads=1 runqueue=8 [0 0]
SCHED 10065ms: gomaxprocs=2 idleprocs=0 threads=4 spinningthreads=0 idlethreads=1 runqueue=4 [0 4]
SCHED 11069ms: gomaxprocs=2 idleprocs=0 threads=4 spinningthreads=0 idlethreads=1 runqueue=4 [1 3]
```

different debugging information outputs. Here, we focus on two values related to scheduler tracing information, schedtrace and scheddetail:

1. **Schedtrace**. When setting the value of schedtrace to X, the Go runtime monitor will print a summary of the scheduler's status to the error output every X milliseconds.
2. **Scheddetail**. Based on the use of schedtrace to print out the scheduler's summary information, scheddetail additionally provides relevant detailed scheduling information. If it is set to 1, the monitor will also print multiple lines of information to the error output every X milliseconds, including the status information of the scheduler as well as all existing M, P, and G.

After executing the above command, we can obtain and see the following output Listing 7.26. Naturally, the output may vary depending on the machine being used. For this experiment, the laptop used has four CPU cores, and we have specified two logical processing cores through the instruction **GOMAXPROCS**.

We have selected the second line of output information and provided an analysis and explanation of the various terms in Table 7.2.

```
SCHED 1009ms: gomaxprocs=2 idleprocs=0 threads=4 spinningthreads=0 idlethreads=1 runqueue=0 [8 0]
```

More Tracing Information Details Among the key-value pairs in the GODEBUG environment variable mentioned above, schedtrace generates summary information

Table 7.2 Analysis of scheduled trace information

Terminology	Meaning
1009 ms	The time from when the program was launched to when the trace captured the information
gomaxprocs = 2	The number of Configured logical processing cores
idleprocs = 2	The number of currently idle logical cores
thread = 4	The number of threads being managed by the scheduler
spinningthread = 0	The number of spinning threads
idelthreads = 1	The current number of idle threads, indicating 1 thread idle and 3 threads running
runqueue = 0	The number of Goroutines in the global queue
[8 0]	The number of Goroutines queued in the local queue on a logical core. As we have configured two logical processing cores, one has 8 Goroutines queued, and the other has 0. However, in subsequent scheduling, Goroutines are also queued in the local queue

Listing 7.27 GODEBUG detailed tracing command

```
GOMAXPROCS=2 GODEBUG=schedtrace=1000,scheddetail=1 ./go_demo
```

of the scheduler's status at set time intervals. In addition to this, we can set the scheddetail key-value pair to obtain status information about existing G, M, and P.

We executed the following command for the above successfully compiled program example to obtain more scheduling trace information, as shown in Listing 7.27.

As the amount of information generated is excessive, we have extracted the scheduler state information from the second and presented it below, as shown in Listing 7.28.

In the above information output, the scheduler's summary section remains largely the same, but it provides more detailed output regarding the state information of Processors P, Threads M, and Goroutines G. We will analyze the contents one by one. First, Table 7.3 represents the detailed meanings of each parameter in the scheduler information for Processors P. We have extracted the relevant scheduler information related to Processors P from Listing 7.28 and presented it below as Listing 7.29. According to the parameter meanings indicated in the table, we know that there are currently two Processors P, and *status* = 1 indicates that a Thread M is currently bound to the Processor. The *m* parameter specifically represents the number of the Thread M bound to the Processor P. At the same time, *runqsize* indicates the number of Goroutines in the local run queue of the current Processor P.

Based on Listing 7.28, there are four threads present in the current program. The detailed M thread scheduling information output Listing 7.30 shows the relevant status information for the four M threads. According to the meanings of the various parameters in Table 7.4 for thread M, we can see that threads M3 and M0 have processors bound to them. Furthermore, the parameter *curg* indicates the Goroutine

Listing 7.28 Detailed scheduling information output

```
SCHED 2009ms: gomaxprocs=2 idleprocs=0 threads=4 spinningthreads=0
    idlethreads=1 runqueue=2 gcwaiting=0 nmidlelocked=0 stopwait=0
    sysmonwait=0
  P0: status=1 schedtick=52 syscalltick=0 m=0 runqsize=3 gfreecnt=0
    timerslen=0
  P1: status=1 schedtick=48 syscalltick=0 m=3 runqsize=3 gfreecnt=0
    timerslen=0
  M3: p=1 curg=7 mallocing=0 throwing=0 preemptoff= locks=0 dying=0
    spinning=false blocked=false lockedg=-1
  M2: p=-1 curg=-1 mallocing=0 throwing=0 preemptoff= locks=0 dying=0
    spinning=false blocked=true lockedg=-1
  M1: p=-1 curg=-1 mallocing=0 throwing=0 preemptoff= locks=2 dying=0
    spinning=false blocked=false lockedg=-1
  M0: p=0 curg=-1 mallocing=0 throwing=0 preemptoff= locks=1 dying=0
    spinning=false blocked=false lockedg=-1
  G1: status=4(semacquire) m=-1 lockedm=-1
  G2: status=4(force gc (idle)) m=-1 lockedm=-1
  G3: status=4(GC sweep wait) m=-1 lockedm=-1
  G4: status=4(GC scavenge wait) m=-1 lockedm=-1
  G5: status=1(sleep) m=-1 lockedm=-1
  G6: status=1(sleep) m=-1 lockedm=-1
  G7: status=2(sleep) m=3 lockedm=-1
  G8: status=1(sleep) m=-1 lockedm=-1
  G9: status=1(sleep) m=-1 lockedm=-1
  G10: status=1(sleep) m=-1 lockedm=-1
  G11: status=1(sleep) m=-1 lockedm=-1
  G12: status=1(sleep) m=-1 lockedm=-1
  G13: status=1(sleep) m=-1 lockedm=-1
  G14: status=1(sleep) m=-1 lockedm=-1
```

Table 7.3 Meaning of processor P parameter information

Terminology	Meaning
status	The operational status of the P logical processing core, When it is 1, it indicates that the current P is bound to M
schdtick	The number of scheduling events for the P logical processor
syscalltick	The number of system calls for the P logical processor
m	Binding situation with Thread M
runqsize	The number of G in the local run queue
gfreecnt	The number of available G
runqueue $=0$	The number of Go in the global queue
timerslen	The number of the timers

currently running on the thread, with Goroutine number 7 running on M3, while M0 does not have a Goroutine running on it currently. Among the two unbound threads, the parameter $blocked = true$ for M2 indicates that the current thread is blocked.

Listing 7.29 P processor scheduling information output

```
P0: status=1 schedtick=52 syscalltick=0 m=0 runqsize=3 gfreecnt=0
  timerslen=0
P1: status=1 schedtick=48 syscalltick=0 m=3 runqsize=3 gfreecnt=0
  timerslen=0
```

Listing 7.30 M thread scheduling information output

```
M3: p=1 curg=7 mallocing=0 throwing=0 preemptoff= locks=0 dying=0
  spinning=false blocked=false lockedg=-1
M2: p=-1 curg=-1 mallocing=0 throwing=0 preemptoff= locks=0 dying=0
  spinning=false blocked=true lockedg=-1
M1: p=-1 curg=-1 mallocing=0 throwing=0 preemptoff= locks=2 dying=0
  spinning=false blocked=false lockedg=-1
M0: p=0 curg=-1 mallocing=0 throwing=0 preemptoff= locks=1 dying=0
  spinning=false blocked=false lockedg=-1
```

Table 7.4 The meanings of the parameters for thread M

Terminology	Meaning
p	The bound P logical processor
curg	Which Goroutine is currently executing on this thread
mallocing	Is memory being allocated
throwing	Is there an exception
preemptoff	When it is not an empty string, keep the current G running on the thread
lock	The number of locks held on the current thread
spinning	Is it spinning thread
blocked	Is it blocked

Table 7.5 Goroutine parameter information meanings

Terminology	Meaning
status	The status of G
m	On which thread M it runs
lockedm	If it has locked M

As shown in the goroutine scheduling information output Listing 7.31, there were 14 Goroutines existing in the current program at the second. The G status information parameters are relatively simple, consisting of three parameters as indicated in Table 7.5. The current status of G is mainly reflected by the status parameter, and the explanation of each status value is shown in Table 7.6.

Based on the meanings of the Goroutine status values represented in Tables 7.5 and 7.6, we analyze the status information of some Goroutines from the second-second scheduling trace. As shown in the following output Listing 7.32, the running status of each Goroutine is indicated by the comments.

By studying the basic model concepts and fundamental scheduling processes of the scheduler, and using the GODEBUG tool provided by Golang to obtain detailed debugging information during the scheduling process, we can gain a deep

Listing 7.31 Goroutine scheduling information output

```
G1: status=4(semacquire) m=-1 lockedm=-1
G2: status=4(force gc (idle)) m=-1 lockedm=-1
G3: status=4(GC sweep wait) m=-1 lockedm=-1
G4: status=4(GC scavenge wait) m=-1 lockedm=-1
G5: status=1(sleep) m=-1 lockedm=-1
G6: status=1(sleep) m=-1 lockedm=-1
G7: status=2(sleep) m=3 lockedm=-1
G8: status=1(sleep) m=-1 lockedm=-1
G9: status=1(sleep) m=-1 lockedm=-1
G10: status=1(sleep) m=-1 lockedm=-1
G11: status=1(sleep) m=-1 lockedm=-1
G12: status=1(sleep) m=-1 lockedm=-1
G13: status=1(sleep) m=-1 lockedm=-1
G14: status=1(sleep) m=-1 lockedm=-1
```

Table 7.6 Meanings of Goroutine status.

Status	Meaning
0 idle	Just allocated, uninitialized
1 runnable	In run queue, not executed yet
2 running	Assigned M and P, executable, not in queue
3 syscall	Executing a system call
4 waiting	Blocked, waiting for execution
5 moribund_unused	Not yet used, but hardcoded in gdb
6 dead	Not yet used, recently terminated or just initialized but not yet executed
7 enqueue_unused	Not yet used
8 copystack	Copying stack, not executing code

understanding of how the Go scheduler works under the GMP model, as well as the real-time status information of each P, M, and G.

7.5 Summary

In this chapter, we introduced the basics of the Go programming language. Go's features such as goroutines and channels have made it popular among programmers working in distributed systems and high-concurrency environments. In this chapter, we covered the background of Golang, slice operations, fundamentals of goroutines and channels, and knowledge related to the scheduler associated with goroutines. In the subsequent chapters, we will implement a distributed key-value storage system using Golang.

Listing 7.32 Classification of Goroutine status information

```
// The following Goroutine is waiting to run on a logical processor P
G5: status=1(sleep) m=-1 lockedm=-1
G6: status=1(sleep) m=-1 lockedm=-1
G8: status=1(sleep) m=-1 lockedm=-1
G9: status=1(sleep) m=-1 lockedm=-1
G10: status=1(sleep) m=-1 lockedm=-1
G11: status=1(sleep) m=-1 lockedm=-1
G12: status=1(sleep) m=-1 lockedm=-1
G13: status=1(sleep) m=-1 lockedm=-1
G14: status=1(sleep) m=-1 lockedm=-1

// Running on logical processor P1's thread M3
G7: status=2(sleep) m=3 lockedm=-1

// The main thread is blocked by the WaitGroup, waiting
G1: status=4(semacquire) m=-1 lockedm=-1

// Blocked waiting to be cleared and reclaimed
G2: status=4(force gc (idle)) m=-1 lockedm=-1
G3: status=4(GC sweep wait) m=-1 lockedm=-1
G4: status=4(GC scavenge wait) m=-1 lockedm=-1
```

Chapter 8
Building a Strong Consistency Algorithm Library

In the previous chapter, we learned the basics of the Go programming language. In this chapter, we will build a strongly consistent algorithm Raft using Go. We will name our custom-built Raft library eraft.

8.1 Core Data Structure Design

In the consensus algorithm chapter, we introduced the main content of the Raft algorithm. In this chapter, we will discuss its specific implementation. First, we need to abstract the data structures we require. Let's start by outlining the possible data structures. Since nodes need to access each other, we need to define a network client for accessing other nodes. This should include the node's ID, address, and an RPC client. We abstract this entire structure as RaftClientEnd, and the main data content is shown in Listing 8.1.

We previously described three node states, which are defined as shown in Listing 8.2.

To complete the election process, we need two timeout durations. Here, we use the Timer from the Golang time package, which can periodically send messages to a channel. We can utilize it to implement election timeouts and heartbeat timeouts, as shown in Listing 8.3.

Additionally, we have constructed a Raft struct. The Raft struct keeps track of the current node's ID, the current term number, the candidate it voted for, the vote count statistics, the committed log entry index, and the last log entry index that was applied to the state machine, and if the node is a Leader, it also needs to record the latest matched log entry index for replication to other nodes. The definition of the Raft struct is shown in Listing 8.4.

Z. Xu, *Distributed Data Services*, https://doi.org/10.1007/978-981-96-9392-4_8

Listing 8.1 Client data structure

```
type RaftClientEnd struct {
  id             uint64
  addr           string
  raftServiceCli *raftpb.RaftServiceClient  // GRPC Client
}
```

Listing 8.2 Raft node states

```
const (
  NodeRoleFollower NodeRole = iota
  NodeRoleCandidate
  NodeRoleLeader
)
```

Listing 8.3 Timer

```
electionTimer    *time.Timer
heartbeatTimer   *time.Timer
```

When the system starts, it needs to construct the Raft structure. This process is implemented in MakeRaft, which mainly initializes variables and two timers, and launches related goroutines. Here, we have a Replicator goroutine for replication with each peer node, a Tick goroutine to trigger two timeout periods, and an Applier goroutine to apply the committed logs.

8.2 Coroutine Model

Figure 8.1 illustrates the coroutine model within our raftcore. When a proposal from the application layer arrives, at the Leader node, the main coroutine appends the log locally, then sends a BroadcastAppend message. This triggers the wakeup of the coroutines responsible for replicating the log to the Follower nodes, initiating a round of replication. Once replicateOneRound successfully replicates the log to a majority of nodes, a commit is triggered, and rf.applyCond.Signal() wakes up the Applier coroutine that was waiting to perform the Apply operation.

In addition, the Tick coroutine listens to signals from two timeout channels of the Timer. Once a heartbeat timeout occurs and the current node's state is Leader, the Tick coroutine will invoke BroadcastHeartbeat to send the heartbeat, and during the heartbeat process, it will also replicateOneRound. Similar to the aforementioned process, if a quorum of nodes successfully replicates the log, a commit will be triggered, and rf.applyCond.Signal() will wake up the Applier coroutine that is waiting to perform the Apply operation.

Listing 8.4 Raft struct

```
type Raft struct {
  mu              sync.RWMutex
  peers           []*RaftClientEnd // RPC Client.
  me_             int  // Own id
  dead            int32 // Node status.
  applyCh         chan *pb.ApplyMsg // Apply coroutine channels, which
    will be discussed in the coroutine model.
  applyCond       *sync.Cond // Apply process control semaphore
  replicatorCond []*sync.Cond // Replication control semaphore
  role            NodeRole  // The current state of the node
  curTerm         int64   // The current term
  votedFor        int64   // Who it voted for
  grantedVotes    int     // The number of votes obtained
  logs            *RaftLog  // Info log
  commitIdx       int64     // The ID of the largest committed log entry
  lastApplied     int64     // The last log ID applied to the state
    machine
  nextIdx         []int     // The information of the next matching log
    ID for other nodes
  matchIdx        []int     // The information of the currently matching
    log ID for other nodes

  leaderId           int64   // The ID of the current Leader node in the
    cluster
  electionTimer      *time.Timer   // Election timeout timer
  heartbeatTimer     *time.Timer   // Heartbeat timeout timer
  heartBeatTimeout uint64         // Heartbeat timeout duration
  baseElecTimeout  uint64         // Election timeout duration
}
```

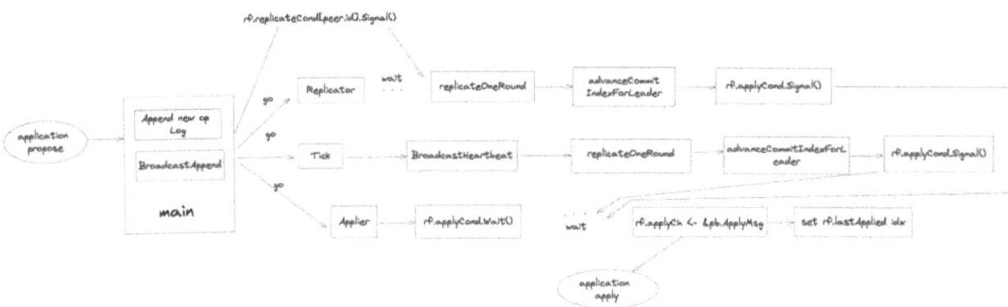

Fig. 8.1 Goroutine model

After the Applier coroutine applies a message, it will write the ApplyMsg to the rf.applyCh channel to notify the application layer. The application layer's coroutine can monitor this channel and, if an ApplyMsg arrives, it will apply it to the state machine.

8.3 RPC Definition

The RPC definition file for eraft is located in the pbs directory and named raftbasic.proto. The main messages defined in this file are log entry Entry, vote request RequestVote, and log append request AppendEntries. We will provide a detailed introduction to these messages in the following sections.

8.3.1 Entry

Listing 8.5 represents a log entry information structure. As described previously, it consists of a term number(term), an index number(index), and serialized data (data) for the operation. We store it in a byte stream. There are two types of log entries: one is a normal log entry (EntryNormal), and the other is a log entry for configuration changes (EntryConfChange).

8.3.2 RequestVote

Listing 8.6 is the definition of the request vote RPC, which remains largely consistent with the description in the paper: The request vote contains the candidate's term number, its ID, and information about the index and term number of its last log entry. The response includes a term number, which is used to update the candidate's term when the election fails, and a vote_granted flag indicating whether the request vote operation has been accepted by the peer node.

Listing 8.5 RPC definition of log entry

```
enum EntryType {
    EntryNormal = 0;
    EntryConfChange = 1;
}

message Entry {
    EntryType entry_type = 1;
    uint64    term = 2;
    int64     index = 3;
    bytes     data = 4;
}
```

Listing 8.6 Definition of RequestVote RPC

```
message RequestVoteRequest {
    int64 term = 1;
    int64 candidate_id = 2;
    int64 last_log_index = 3;
    int64 last_log_term = 4;
}

message RequestVoteResponse {
    int64 term = 1;
    bool  vote_granted = 2;
}

service RaftService {
    //...
    rpc RequestVote (RequestVoteRequest) returns (RequestVoteResponse)
     {}
}
```

8.3.3 *AppendEntries*

The definition of the log append operation is as follows, which is essentially consistent with the paper:

The request contains the Leader's term and ID (used to inform the Follower; this way, after a Client accesses a Follower, it can be informed which node is the Leader). Prev_log_index indicates the index of the log entry immediately preceding the first log entry to be synchronized in the message, prev_log_term is its term information, leader_commit is the commit number of the Leader (which can be used to inform Follower nodes of the current commit progress), and entries represent the log entry information.

In the response message body, term is used to inform the Leader if there is a new term, which can be used to update the Leader's term number. Success indicates whether the log append operation was successful. Conflict_index is used to record the index of the conflicting log, and conflict_term is used to record the term of the conflicting log, as shown in Listing 8.7.

8.4 Leader Election Implementation Analysis

The Raft visualization website [108] provides an animated demonstration of the Raft algorithm. First, let's intuitively understand the Leader election process through this visualization, and then we will explain the process in conjunction with the code.

In the Raft algorithm, there are two timeout periods that control the Leader election process. The first is the election timeout, which is the duration a Candidate

Listing 8.7 The definition of the AppendEntries RPC

```
message AppendEntriesRequest {
    int64     term = 1;
    int64     leader_id = 2;
    int64     prev_log_index = 3;
    int64     prev_log_term = 4;
    int64     leader_commit = 5;
    repeated Entry entries = 6;
}

message AppendEntriesResponse {
    int64  term = 1;
    bool   success = 2;
    int64 conflict_index = 3;
    int64 conflict_term = 4;
}

service RaftService {
    //...
    rpc AppendEntries (AppendEntriesRequest) returns (
     AppendEntriesResponse) {}
}
```

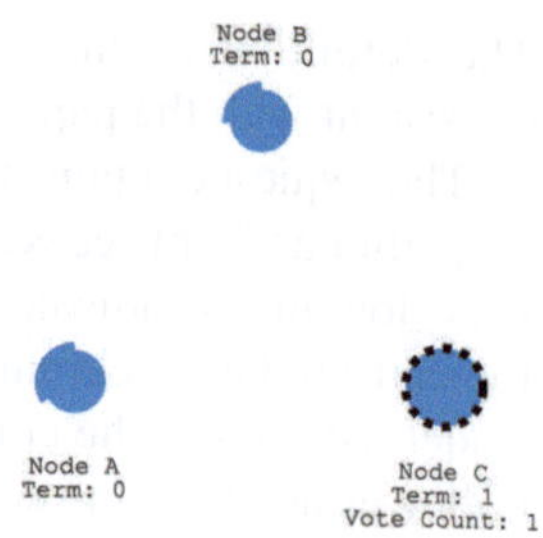

Fig. 8.2 Election process 1

waits to become a Leader. If the Candidate is not elected as Leader within this time, the timeout timer will be reset. As previously mentioned, this timeout period is generally set between 150 ms and 300 ms.

At startup, the election timeout for all nodes is set to a random value between 150 ms and 300 ms. Most likely, one node will reach the timeout first. As shown in Fig. 8.2, Node C reaches the timeout before Nodes A and B (the black dashed line indicates that the node has become a Candidate). Node C transitions from Follower to Candidate and initiates a new term's election. It votes for itself and then sends RequestVoteRequest RPC to the other nodes in the cluster, requesting their votes:

1. In the eraft code, at startup, which is when the application layer calls the MakeRaft function, it passes in baseElectionTimeOutMs and heartbeatTimeOutMs. Here, the heartbeat timeout is fixed. We use the MakeAnRandomElectionTimeout construct to generate a random election timeout, as shown in Listing 8.8.

Listing 8.8 Random election timeout

```
func MakeRaft(peers []*RaftClientEnd, me int, newdbEng storage_eng.KvStore, applyCh chan *pb.ApplyMsg, heartbeatTimeOutMs uint64, baseElectionTimeOutMs uint64) *Raft {
    ...
    heartbeatTimer:   time.NewTimer(time.Millisecond * time.Duration(heartbeatTimeOutMs)),
    electionTimer:    time.NewTimer(time.Millisecond * time.Duration(MakeAnRandomElectionTimeout(int(baseElectionTimeOutMs)))),
    ...
}
```

Listing 8.9 Timeout starts election

```
//
// Heartbeat timeout triggers an election: change own status and initiate the election
//
func (rf *Raft) Tick() {
  for !rf.IsKilled() {
    select {
    case <-rf.electionTimer.C:
      {
        rf.SwitchRaftNodeRole(NodeRoleCandidate)
        rf.IncrCurrentTerm()
        rf.Election()
        rf.electionTimer.Reset(time.Millisecond * time.Duration(MakeAnRandomElectionTimeout(int(rf.baseElecTimeout))))
      }
     ...
  }
}
```

2. After reaching the election timeout, Node C first changes its status to Candidate, then increases its term number, and starts the election, as shown in Listing 8.9.
3. Listing 8.10 contains the core logic for initiating an election. As shown in Fig. 8.3, Node C first calls IncrGrantedVotes to vote for itself, setting votedFor to its own ID, making Node C's vote count 1. Then, it constructs the RequestVoteRequest RPC, including its current term number, its own ID as CandidateId, the index of the last log entry, and the term of the last log entry. After persisting the current Raft state, it sends RequestVote requests in parallel to the other nodes in the cluster. In Fig. 8.3, the black dots represent these vote requests.

Listing 8.10 Election logic

```
//
// Election initiates a new election
//
func (rf *Raft) Election() {
  fmt.Printf("%d start election \n", rf.me_)
  rf.IncrGrantedVotes()
  rf.votedFor = int64(rf.me_)
  voteReq := &pb.RequestVoteRequest{
    Term:         rf.curTerm,
    CandidateId:  int64(rf.me_),
    LastLogIndex: int64(rf.logs.GetLast().Index),
    LastLogTerm:  int64(rf.logs.GetLast().Term),
  }
  rf.PersistRaftState()
  for _, peer := range rf.peers {
    if int(peer.id) == rf.me_ {
      continue
    }
    go func(peer *RaftClientEnd) {
      PrintDebugLog(fmt.Sprintf("send request vote to %s %s\n", peer.addr, voteReq.String()))

      requestVoteResp, err := (*peer.raftServiceCli).RequestVote(context.Background(), voteReq)
      if err != nil {
        PrintDebugLog(fmt.Sprintf("send request vote to %s failed %v\n", peer.addr, err.Error()))
      }
      ...
    }(peer)
  }
}

```

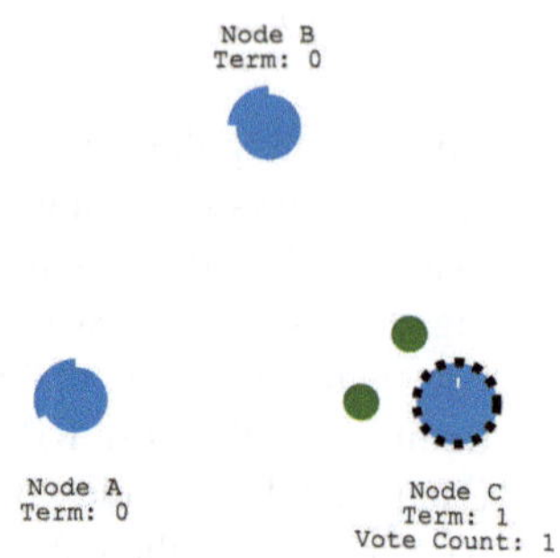

Fig. 8.3 Election process 2

If Nodes A and B receive the request before they have cast their votes (because they have not yet reached their election timeout), they will vote for Candidate Node C and reset their own election timeout timers.

Listing 8.11 Handle voting requests

```
//
// HandleRequestVote handles election voting requests from other nodes
//
func (rf *Raft) HandleRequestVote(req *pb.RequestVoteRequest, resp *pb.RequestVoteResponse) {
  rf.mu.Lock()
  defer rf.mu.Unlock()
  defer rf.PersistRaftState()

  // 1
  if req.Term < rf.curTerm || (req.Term == rf.curTerm && rf.votedFor != -1 && rf.votedFor != req.CandidateId) {
    resp.Term, resp.VoteGranted = rf.curTerm, false
    return
  }

  // 2
  if req.Term > rf.curTerm {
    rf.SwitchRaftNodeRole(NodeRoleFollower)
    rf.curTerm, rf.votedFor = req.Term, -1
  }

  ...

  rf.votedFor = req.CandidateId

  // 3
  rf.electionTimer.Reset(time.Millisecond * time.Duration(MakeAnRandomElectionTimeout(int(rf.baseElecTimeout))))
  resp.Term, resp.VoteGranted = rf.curTerm, true
}
```

4. Listing 8.11 is the detailed code for Raft to handle voting requests. We will analyze the logic below in conjunction with the example in Fig. 8.4. Assuming node A is handling a vote request from node C, if node C's term number is greater than node A's, the if branch in Section 1 of the code will not be executed. In Section 2 of the code, if node A discovers that the term number in the vote request message from node C is greater than its own, it will call SwitchRaftNodeRole to become a Follower node. Before responding to node C, in Section 3 of the code, node A calls electionTimer.Reset to reset its election timeout timer, and finally casts its vote for node C.

   As shown in Fig. 8.5, node C receives votes from a majority, that is, votes from either node A or node B in addition to its own vote. Its role will then change from Candidate to Leader, and it will stop its election timeout timer.

Fig. 8.4 Election process 3

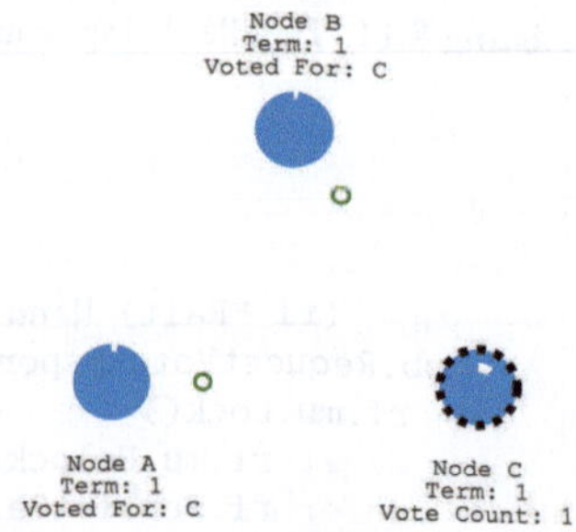

Fig. 8.5 Election process 4

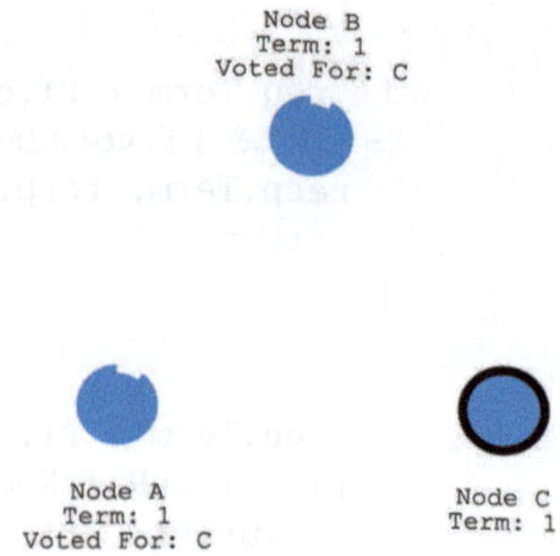

5. The response code for counting vote requests is as shown in Listing 8.12. Note that this code includes a lock because multiple goroutines may attempt to modify nonatomic variables in rf, and without a lock, this could lead to logical errors. In Section 1 of the code, if the received vote response VoteGranted is true, node C will call IncrGrantedVotes to increment its own vote count. Then, it checks if rf.grantedVotes > len(rf.peers)/2 to determine if it has received more than half of the votes. If so, it calls SwitchRaftNodeRole to switch its status to Leader. After that, as shown in Fig. 8.6, the Leader node broadcasts heartbeat messages (BroadcastHeartbeat) using red solid circles to represent the heartbeats, and resets its grantedVotes count to 0.

We know that during the previous handling of vote requests, A and B simply reset their election timeout timers. Then, if the timer reaches the timeout again, will they reinitiate an election, potentially leading to an election loop? The answer is no. We've only introduced the election timeout period earlier. There is also a heartbeat timeout, which is shorter than the election timeout, typically one-third of the election timeout. This means that before A and B reach the election timeout, the heartbeat timeout will occur first. If it's the Leader node, after reaching the heartbeat timeout, it will send heartbeat packets to other nodes in the cluster. When other nodes (A, B) receive the heartbeat packets, they will reset their own election timeout timers. In other words, as long as Leader C continues to operate normally and sends heartbeat packets, nodes A and B will not trigger an election. It is only when Leader C fails that nodes A and B will begin the next round of elections.

Listing 8.12 Counting vote requests

```
    if requestVoteResp != nil {
      rf.mu.Lock()
      defer rf.mu.Unlock()
      PrintDebugLog(fmt.Sprintf("send request vote to %s recive -> %s, curterm %d, req term %d", peer.addr, requestVoteResp.String(), rf.curTerm, voteReq.Term))
      if rf.curTerm == voteReq.Term && rf.role == NodeRoleCandidate {
          // 1
        if requestVoteResp.VoteGranted {
                // Vote succeeded
          PrintDebugLog("I grant vote")
          rf.IncrGrantedVotes()
          if rf.grantedVotes > len(rf.peers)/2 {
            PrintDebugLog(fmt.Sprintf("node %d get majority votes int term %d ", rf.me_, rf.curTerm))
            rf.SwitchRaftNodeRole(NodeRoleLeader)
            rf.BroadcastHeartbeat()
            rf.grantedVotes = 0
          }
          // 2
        } else if requestVoteResp.Term > rf.curTerm {
          // The initiated election failed
          rf.SwitchRaftNodeRole(NodeRoleFollower)
          rf.curTerm, rf.votedFor = requestVoteResp.Term, -1
          rf.PersistRaftState()
        }
      }
    }
```

Fig. 8.6 Election process 5

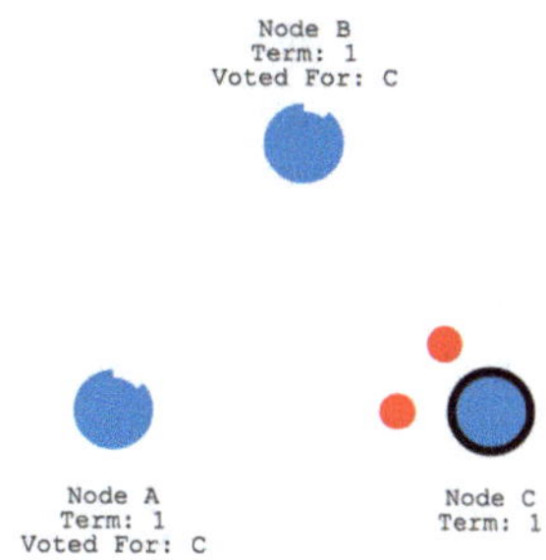

8.5 Log Replication Implementation Analysis

After the aforementioned election process, we now have a system with a Leader node and multiple Follower nodes. The Leader node will continuously send heartbeat messages to the Follower nodes. Now we need to consider how to handle client requests: When a client sends an operation, how does our system process it?

Listing 8.13 Counting the vote requests

```
//
// The Propose function handles proposals, which are processed only by
    the Leader
//

func (rf *Raft) Propose(payload []byte) (int, int, bool) {
  rf.mu.Lock()
  defer rf.mu.Unlock()
  if rf.role != NodeRoleLeader {
    return -1, -1, false
  }
  newLog := rf.Append(payload)
  rf.BroadcastAppend()
  return int(newLog.Index), int(newLog.Term), true
}
```

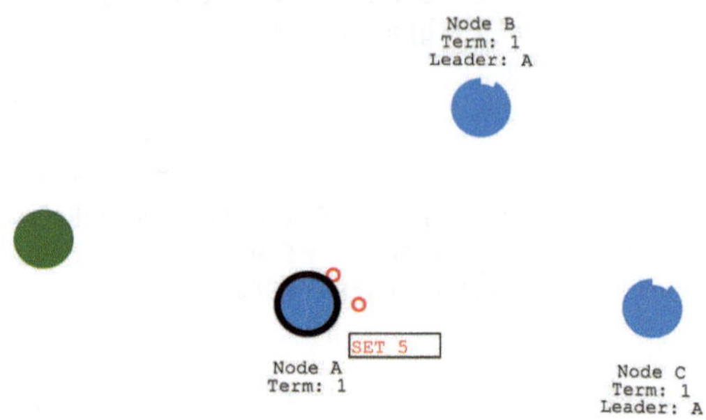

Fig. 8.7 Log replication 1

First, Raft stipulates that only the Leader node can handle write requests. Requests sent by clients will first arrive at the Leader node.

In the eraft library, the entry point function for user requests to interact with Raft is Propose, as shown in Listing 8.13. The Propose function first checks the current node's state. Only the Leader node can handle proposals. The Leader node serializes the user operation and then calls the Append function to append the operation to its own log. After that, BroadcastAppend is used to send the log contents to the Follower nodes in the cluster.

As shown in Fig. 8.7, the leftmost circle represents the client node, which sends a SET 5 request. Node A, as the current Leader in the cluster, will first encapsulate this SET 5 operation into a log entry, write it into its own log storage structure, and then include this log entry in the next heartbeat message sent to the Follower nodes, as depicted in Fig. 8.8.

In the eraft implementation, we have a dedicated set of Goroutines that handle log replication-related operations. When a user proposal reaches the Leader, the Leader node invokes the BroadcastAppend function to awaken the Goroutines that are responsible for log replication. We use a replicatorCond semaphore to achieve synchronization between the Goroutines, as shown in Listing 8.14.

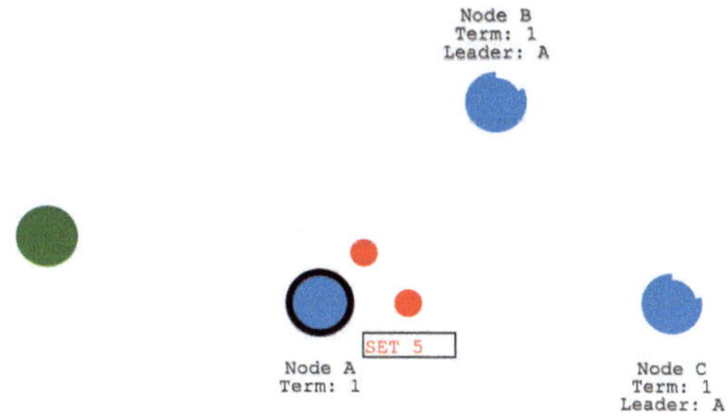

Fig. 8.8 Log replication 2

Listing 8.14 Synchronization

```
func (rf *Raft) BroadcastAppend() {
  for _, peer := range rf.peers {
    if peer.id == uint64(rf.me_) {
      continue
    }
    rf.replicatorCond[peer.id].Signal()
  }
}
```

Listing 8.15 Data replication

```
//
// Replicator manages data replication
//
func (rf *Raft) Replicator(peer *RaftClientEnd) {
  rf.replicatorCond[peer.id].L.Lock()
  defer rf.replicatorCond[peer.id].L.Unlock()
  for !rf.IsKilled() {
    PrintDebugLog("peer id wait for replicating...")
    for !(rf.role == NodeRoleLeader && rf.matchIdx[peer.id] < int(rf.
     logs.GetLast().Index)) {
      rf.replicatorCond[peer.id].Wait()
    }
    rf.replicateOneRound(peer)
  }
}
```

The task function executed by the Goroutine for replication is Replicator. When the semaphore is awakened in BroadcastAppend through the Signal function, rf.replicatorCond[peer.id].Wait() will stop blocking and continue executing, calling replicateOneRound to perform data replication, as shown in Listing 8.15.

The replicateOneRound function will then package the log into an AppendEntriesRequest and send the message to the Follower nodes.

After a Follower receives an append request, it appends the log entries to its own log storage structure and then sends a successful append response to the Leader.

Fig. 8.9 Log replication 3

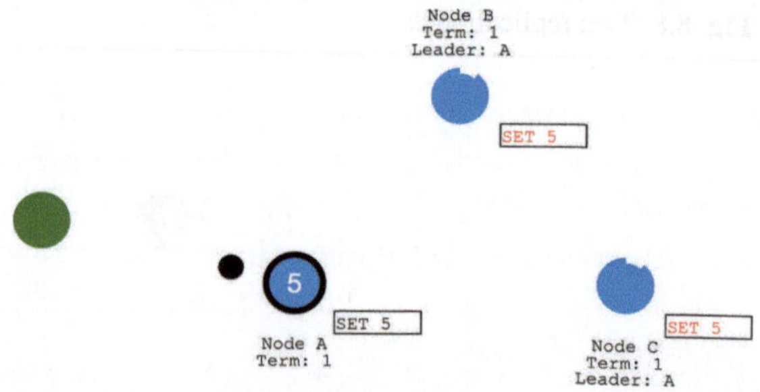

Once the Leader has received successful append acknowledgments from a majority of nodes in the cluster (including itself), it marks the log entry as committed and sends the result of the operation to the client, as shown in Fig. 8.9, where A sets SET 5.

After that, the information about the committed log will be sent to the Followers in the next heartbeat packet. Once the Followers receive the log commit information, they will also update their own log commit status.

After the log is committed, as shown in Fig. 8.9, the Apply goroutine will receive a notification and start to apply the committed log entries to the state machine. Once the logs are successfully applied, the Leader node will send a successful write response package to the client.

In the eraft implementation, after a log is committed, the Leader node will call the advanceCommitIndexForLeader function, as shown in Listing 8.16 This function calculates the current committed log index and compares it with the previously committed commitIdx. If the new index is greater, it will update commitIdx and simultaneously call rf.applyCond.Signal() to wake up the Goroutine that handles the Apply operation. The Applier function is the task function executed by the Apply Goroutine. It waits on the Wait applyCond semaphore. Once the Apply Goroutine is woken up, it copies the committed logs from the Leader node, packages them into an ApplyMsg, and sends this message to the applyCh channel to notify the application layer. The application layer, upon receiving the apply message, will update the state machine and return the data to the client.

8.6 Raft Snapshot Implementation Analysis

Listing 8.17 is the RPC definition for log snapshots

In the InstallSnapshotRequest, the term represents the term of the Leader that is sending the snapshot. The Follower compares it with its own term number to determine whether to accept the snapshot. The *leader_id* is the *id* of the current Leader. This allows clients to quickly identify the Leader information when they access a Follower node. Among them, *last_included_index* and *last_included_term*

Listing 8.16 Commit and apply

```
func (rf *Raft) advanceCommitIndexForLeader() {
  sort.Ints(rf.matchIdx)
  n := len(rf.matchIdx)
  newCommitIndex := rf.matchIdx[n-(n/2+1)]
  if newCommitIndex > int(rf.commitIdx) {
    if rf.MatchLog(rf.curTerm, int64(newCommitIndex)) {
      PrintDebugLog(fmt.Sprintf("peer %d advance commit index %d at term %d", rf.me_, rf.commitIdx, rf.curTerm))
      rf.commitIdx = int64(newCommitIndex)
      rf.applyCond.Signal()
    }
  }
}

//
// The Applier method writes the already committed //message to the applyCh channel and updates the //lastApplied value
//
func (rf *Raft) Applier() {
  for !rf.IsKilled() {
    rf.mu.Lock()
    for rf.lastApplied >= rf.commitIdx {
      PrintDebugLog("applier ...")
      rf.applyCond.Wait()
    }
    firstIndex, commitIndex, lastApplied := rf.logs.GetFirst().Index, rf.commitIdx, rf.lastApplied
    entries := make([]*pb.Entry, commitIndex-lastApplied)
    copy(entries, rf.logs.GetRange(lastApplied+1-int64(firstIndex), commitIndex+1-int64(firstIndex)))
    rf.mu.Unlock()

    PrintDebugLog(fmt.Sprintf("%d, applies entries %d-%d in term %d", rf.me_, rf.lastApplied, commitIndex, rf.curTerm))

    for _, entry := range entries {
      rf.applyCh <- &pb.ApplyMsg{
        CommandValid: true,
        Command:      entry.Data,
        CommandTerm:  int64(entry.Term),
        CommandIndex: int64(entry.Index),
      }
    }

    rf.mu.Lock()
    rf.lastApplied = int64(Max(int(rf.lastApplied), int(commitIndex)))
    rf.mu.Unlock()
  }
}
```

Listing 8.17 RPC definition for log snapshots.

```
message InstallSnapshotRequest {
    int64 term =                1;
    int64 leader_id =           2;
    int64 last_included_index = 3;
    int64 last_included_term  = 4;
    bytes data                = 5;
}

message InstallSnapshotResponse {
    int64 term = 1;
}

rpc Snapshot (InstallSnapshotRequest) returns (InstallSnapshotResponse) {}
```

record the index number and term number of the first log entry after the snapshot was taken, while data records the serialized data of the state machine.

When does Raft create snapshots? Raft will create snapshots when there are too many log entries. In eraft, this is achieved by calculating the number of log entries in the current phase using s.Rf.GetLogCount(). The entry function for creating a snapshot is takeSnapshot(index int), which takes in the *id* of the currently applied log entry. Then, it serializes the state machine data and calls the Raft layer's Snapshot function. This function performs log deletion operations using EraseBeforeWithDel and then persists the state data in the snapshot to the storage engine using PersisSnapshot, as shown in Listing 8.18.

When does a Leader send a snapshot? During replication, we check the peer's *prevLogIndex*. If it is smaller than the index of the first current log entry, it indicates that the Leader has already compacted that log entry into a snapshot. In this case, we construct an InstallSnapshotRequest and invoke the Snapshot RPC to send the snapshot data to the Follower node. Upon receiving a successful response, we update rf.matchIdx[peer.id] and rf.nextIdx[peer.id] to *LastIncludedIndex* and *LastIncludedIndex* + 1 respectively, reflecting the updated replication progress for the Follower node, as shown in Listing 8.19.

On the Follower side, the operation is relatively simple. It will call HandleInstallSnapshot to handle the snapshot data and construct a pb.ApplyMsg containing the snapshot data, which is then written to rf.applyCh. The Goroutine responsible for log application will call CondInstallSnapshot to install the snapshot. Finally, restoreSnapshot will parse the data from the snapshot and write it to its own state machine.

Listing 8.18 Snapshot persistence operation

```
//
// Execute and persist snapshot operation
//
func (rf *Raft) Snapshot(index int, snapshot []byte) {
  rf.mu.Lock()
  defer rf.mu.Unlock()
  rf.isSnapshoting = true
  snapshotIndex := rf.logs.GetFirstLogId()
  if index <= int(snapshotIndex) {
    rf.isSnapshoting = false
    PrintDebugLog("reject snapshot, current snapshotIndex is larger in
     cur term")
    return
  }
  rf.logs.EraseBeforeWithDel(int64(index) - int64(snapshotIndex))
  rf.logs.SetEntFirstData([]byte{}) // Set the first operation log
     number to null
  PrintDebugLog(fmt.Sprintf("del log entry before idx %d", index))
  rf.isSnapshoting = false
  rf.logs.PersisSnapshot(snapshot)
}
```

8.7 Summary

In this chapter, we have implemented the strongly consistent Raft algorithm library—eraft. We introduced the core data structures of eraft, such as the client-side data structure RaftClientEnd, the Raft algorithm struct, and the goroutine model of eraft. We presented an overview of the entire consensus algorithm's operational flow, as well as eraft's RPC definitions, including RequestVote for voting requests and AppendEntries for appending logs. Regarding the RPC definitions, since the original Raft paper has provided detailed descriptions of the fields within the RPCs, our implementation is also based on the original paper. In addition to the aforementioned content, this chapter also analyzes the Leader election algorithm, log replication implementation, and snapshot implementation of the Raft algorithm. This includes when elections occur, how to implement node heartbeat timeouts, compress logs with snapshots, the log replication process, and so on. While the Raft algorithm claims to be an easy-to-understand consensus algorithm, the core algorithmic processes still require careful comprehension and mastery by the reader.

Listing 8.19 The timing to send snapshot

```
if prevLogIndex < uint64(rf.logs.GetFirst().Index) {
  firstLog := rf.logs.GetFirst()
  snapShotReq := &pb.InstallSnapshotRequest{
    Term:              rf.curTerm,
    LeaderId:          int64(rf.me_),
    LastIncludedIndex: firstLog.Index,
    LastIncludedTerm:  int64(firstLog.Term),
    Data:              rf.ReadSnapshot(),
  }

  rf.mu.RUnlock()

  PrintDebugLog(fmt.Sprintf("send snapshot to %s with %s\n", peer.addr,
    snapShotReq.String()))

  snapShotResp, err := (*peer.raftServiceCli).Snapshot(context.
    Background(), snapShotReq)
  if err != nil {
    PrintDebugLog(fmt.Sprintf("send snapshot to %s failed %v\n", peer.
     addr, err.Error()))
  }

  rf.mu.Lock()
  PrintDebugLog(fmt.Sprintf("send snapshot to %s with resp %s\n", peer.
    addr, snapShotResp.String()))

  if snapShotResp != nil {
   if rf.role == NodeRoleLeader && rf.curTerm == snapShotReq.Term {
     if snapShotResp.Term > rf.curTerm {
       rf.SwitchRaftNodeRole(NodeRoleFollower)
       rf.curTerm = snapShotResp.Term
       rf.votedFor = -1
       rf.PersistRaftState()
     } else {
       PrintDebugLog(fmt.Sprintf("set peer %d matchIdx %d\n", peer.id,
    snapShotReq.LastIncludedIndex))
       rf.matchIdx[peer.id] = int(snapShotReq.LastIncludedIndex)
       rf.nextIdx[peer.id] = int(snapShotReq.LastIncludedIndex) + 1
     }
   }
  }
  rf.mu.Unlock()
}
```

Chapter 9
Building a Distributed Key-Value Storage System Based on a Strong Consistency Algorithm Library

In the preceding chapter, we implemented the Raft consensus algorithm using the Go programming language, resulting in our custom Raft library named "eraft." Building upon this foundation, this chapter focuses on constructing a distributed key-value store system called "eraftkv," which leverages our eraft library. We will begin by outlining the architecture of eraftkv, providing a comprehensive overview of its design principles and components. Subsequently, we will guide you through a hands-on deployment process, allowing you to experience firsthand the robustness and efficiency of a distributed key-value system. The chapter concludes with an in-depth exploration of eraftkv's implementation, dissecting its core functionalities and discussing key design decisions.

9.1 The Architecture and Operational Flow of eraftkv

Before running eraftkv, let's familiarize ourselves with its architecture, as depicted in Fig. 9.1. This will provide us with a clear understanding of the workflow of eraftkv. The diagram above is the architecture of eraftkv. As a distributed KV storage system, its core components are the bucket and config table. The node roles in eraftkv are divided into Client, ConfigServer, and ShardServer. Different core components are deployed on different nodes, and we determine the node's role based on the components deployed on it. Next, we will introduce the core components and node roles of eraftkv.

The Core Components

1. Bucket—It is the logical unit responsible for data management within the cluster.
2. Config table—Cluster configuration table. It primarily maintains the mapping between cluster service groups and buckets.

Z. Xu, *Distributed Data Services*, https://doi.org/10.1007/978-981-96-9392-4_9

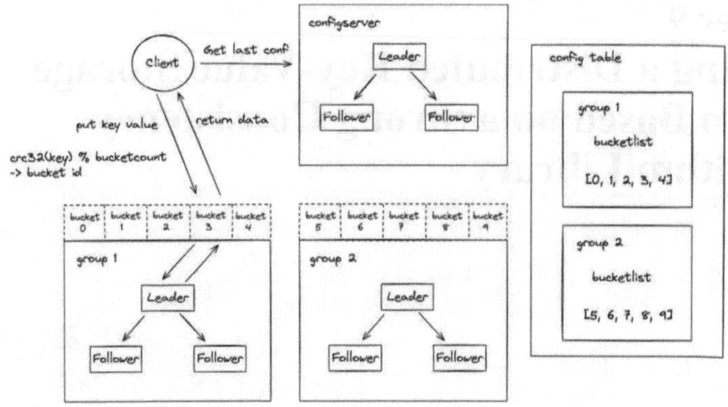

Fig. 9.1 Architecture diagram of eraftkv

The Three Roles in the System

1. Client—The client interface through which users access eraftkv.
2. ConfigServer—The configuration server acts as the central management hub for system configurations. It hosts the config table component and stores routing configuration information for the cluster. Before accessing data, clients query the ConfigServer to obtain the list of service groups where buckets reside.
3. ShardServer—The data server in the system responsible for storing user data. ShardServer deploys the bucket component and manages data for multiple buckets.

Request Processing Flow

After becoming familiar with the architecture of eraftkv, we will analyze a specific request example to understand the workflow of eraftkv. The client sends a "put <testkey, testvalue>" request to the eraftkv cluster:

1. Upon startup, the client retrieves the latest routing information table and cluster configuration table from the ConfigServer.
2. The client computes the CRC32 hash value of the key "testkey" and then calculates the modulo operation with the number of buckets in the cluster to determine which bucket the key "testkey" will map to.
3. The client packages the put request into an RPC request packet and sends it to the ShardServer service group responsible for the corresponding bucket, based on the information from the cluster configuration table.
4. Upon receiving the RPC request, the leader ShardServer node does not immediately write the data to the storage engine. Instead, it constructs a Raft proposal and submits it to the Raft state machine. When a majority of nodes in the ShardServer group agree on this operation, the leader ShardServer node can then respond to the client with a successful write confirmation.

9.2 Configuration of the eraftkv Environment

Installing the Go Compilation Environment First, we need to install the Go language compiler on your computer. You can download the installation package for your operating system version from the official website at [https://go.dev/dl/]. Follow the instructions at [https://go.dev/doc/]install to set up the Go language compilation environment.

Compiling and Building eraftkv Prior to compiling eraftkv, ensure that your development environment is properly configured. This involves not only installing the Go language compiler but also setting up essential development tools. Follow these steps to prepare your environment and compile eraftkv:

```
git clone https://github.com/eraft-io/eraft.git
cd eraft
make
```

9.3 Get the System Running

After building eraftkv, in the eraft directory, there is an output folder containing the necessary bin files that we need to run:

```
colin@B-M1-0045 eraft % ls -l output
total 124120
-rwxr-xr-x  1 colin  staff  12119296  5 25 20:40 bench_cli
-rwxr-xr-x  1 colin  staff  12114784  5 25 20:40 cfgcli
-rwxr-xr-x  1 colin  staff  13578848  5 25 20:40 cfgserver
-rwxr-xr-x  1 colin  staff  12127328  5 25 20:40 shardcli
-rwxr-xr-x  1 colin  staff  13600368  5 25 20:40 shardserver
```

Introduction to Executable Files

1. cfgserver: The executable file for ConfigServer serves as the central management hub for system configurations and needs to be started first.
2. cfgcli: A client tool for ConfigServer is used to interact with ConfigServer for managing cluster configurations.
3. shardserver: The executable file for ShardServer is responsible for storing user data.
4. shardcli: A client tool for ShardServer allows users to write data to the cluster.
5. bench_cli: A performance testing tool for the system.

Starting the Services

1. Starting the ConfigServer group:

```
    ./cfgserver 0 127.0.0.1:8088,127.0.0.1:8089,127.0.0.1:8090
    ./cfgserver 1 127.0.0.1:8088,127.0.0.1:8089,127.0.0.1:8090
    ./cfgserver 2 127.0.0.1:8088,127.0.0.1:8089,127.0.0.1:8090

```

2. Initializing cluster configuration:

```
    ./cfgcli 127.0.0.1:8088,127.0.0.1:8089,127.0.0.1:8090
    join 1 127.0.0.1:6088,127.0.0.1:6089,127.0.0.1:6090

    ./cfgcli 127.0.0.1:8088,127.0.0.1:8089,127.0.0.1:8090
    join 2  127.0.0.1:7088,127.0.0.1:7089,127.0.0.1:7090

    ./cfgcli 127.0.0.1:8088,127.0.0.1:8089,127.0.0.1:8090
    move 0-4 1

    ./cfgcli 127.0.0.1:8088,127.0.0.1:8089,127.0.0.1:8090
    move 5-9 2

```

3. Starting the data server group:

```
./shardserver 0 1 127.0.0.1:8088,127.0.0.1:8089,127.0.0.1:8090
 127.0.0.1:6088,127.0.0.1:6089,127.0.0.1:6090

./shardserver 1 1 127.0.0.1:8088,127.0.0.1:8089,127.0.0.1:8090
 127.0.0.1:6088,127.0.0.1:6089,127.0.0.1:6090

./shardserver 2 1 127.0.0.1:8088,127.0.0.1:8089,127.0.0.1:8090
 127.0.0.1:6088,127.0.0.1:6089,127.0.0.1:6090


./shardserver 0 2 127.0.0.1:8088,127.0.0.1:8089,127.0.0.1:8090
 127.0.0.1:7088,127.0.0.1:7089,127.0.0.1:7090

./shardserver 1 2 127.0.0.1:8088,127.0.0.1:8089,127.0.0.1:8090
 127.0.0.1:7088,127.0.0.1:7089,127.0.0.1:7090

./shardserver 2 2 127.0.0.1:8088,127.0.0.1:8089,127.0.0.1:8090
 127.0.0.1:7088,127.0.0.1:7089,127.0.0.1:7090
```

4. Reading and writing data:

```
./shardcli 127.0.0.1:8088,127.0.0.1:8089,127.0.0.1:8090 put
testkey testvalue

./shardcli 127.0.0.1:8088,127.0.0.1:8089,127.0.0.1:8090 get
testkey
```

5. Running benchmark tests:

```
./bench_cli 127.0.0.1:8088,127.0.0.1:8089,127.0.0.1:8090 100
put
```

9.4 External Interface Definition

Let's first define the interaction interface between the system and the client. The client can send Put and Get operations to write KV data into the system. We need to define these two operations as RPCs. We will combine them into one RPC request, defined as shown in Listing 9.1.

The OpType defines the supported operation types: Put, Append, Get. The content of the client request is defined in the CommandRequest. CommandRequest

Listing 9.1 The definition of the external interface

```
//
// Client operation types
//
enum OpType {
    OpPut = 0;
    OpAppend = 1;
    OpGet = 2;
}

//
// Client command request
//
message CommandRequest {
    string key = 1;
    string value = 2;
    OpType op_type = 3;
    int64  client_id = 4;
    int64  command_id = 5;
    bytes  context = 6;
}

//
// Client command response
//
message CommandResponse {
    string value = 1;
    int64  leader_id = 2;
    int64  err_code = 3;
}

rpc DoCommand (CommandRequest) returns (CommandResponse) {}
```

includes *key*, *value*, operation type, client *id*, command *id*, and a byte type context (which can hold additional uncertain content that we may need to attach).

The CommandResponse includes the returned *value*, *leader_id* field (indicating which node is currently the leader. Initially, when the client sends a request, it doesn't know which node has been elected as the leader. Through an initial probing request, it determines the leader node's *id* and subsequently sends future requests directly to the leader node), and an error code *err_code* (to record potential errors that may occur during command execution and provide detailed feedback to the client).

9.5 Core Server Implementation Analysis

The client sends a DoCommand RPC request to our distributed KV system. How does our system handle these requests? First, let's examine the server's structure definition based on the code snippet (referenced as Listing 9.2): "mu" is a read-write lock used to synchronize operations prone to concurrent conflicts. "dead" indicates the current liveness status of the service node, whether it is alive or not. "Rf" is particularly important, as it is a pointer to the Raft structure we previously constructed. "applyCh" is a channel used to receive applied logs from our algorithm library. Once the server receives these logs, it needs to apply them to the actual state machine ("stm"), representing our state machine. "notifyChans" is a map storing client response channels. After applying logs to the state machine, KvServer sends responses to clients via channels stored in this map. "stopApplyCh" is used to halt our application operations.As shown in Listing 9.2, once we have this structure, how do we apply the Raft algorithm library to achieve a highly available distributed KV system as depicted in the diagram?

Listing 9.2 Server-side structure

```
type KvServer struct {
  mu       sync.RWMutex
  dead     int32
  Rf       *raftcore.Raft
  applyCh chan *pb.ApplyMsg

  lastApplied int

  stm          StateMachine
  notifyChans map[int]chan *pb.CommandResponse
  stopApplyCh chan interface{}

  pb.UnimplementedRaftServiceServer
}
```

Listing 9.3 Constructing the server side

```
func MakeKvServer(nodeId int) *KvServer {
  clientEnds := []*raftcore.RaftClientEnd{}
  for id, addr := range PeersMap {
    newEnd := raftcore.MakeRaftClientEnd(addr, uint64(id))
    clientEnds = append(clientEnds, newEnd)
  }
  newApplyCh := make(chan *pb.ApplyMsg)

  logDbEng, err := storage_eng.MakeLevelDBKvStore("./data/kv_server" + "/node_" + strconv.Itoa(nodeId))
  if err != nil {
    raftcore.PrintDebugLog("boot storage engine err!")
    panic(err)
  }

    // Constructing the Raft structure, passing in clientEnds, the current node $id$, the database for log storage, the apply channel, heartbeat timeout, and election timeout.
    // For testing purposes and ease of observing election logs, we set the timeouts to 1s and 3s, which you can adjust to be shorter.
  newRf := raftcore.MakeRaft(clientEnds, nodeId, logDbEng, newApplyCh, 1000, 3000)
  kvSvr := &KvServer{Rf: newRf, applyCh: newApplyCh, dead: 0, lastApplied: 0, stm: NewMemKV(), notifyChans: make(map[int]chan *pb.CommandResponse)}
  kvSvr.stopApplyCh = make(chan interface{})

    // Starting apply Goruntine
  go kvSvr.ApplingToStm(kvSvr.stopApplyCh)

  return kvSvr
}
```

1. We need to construct an RPC client for each server.
2. Construct the "applyCh" channel and set up the log storage structure.
3. Invoke MakeRaft to construct the core structure of our Raft algorithm library.
4. Start the apply goroutine to listen for messages returned from the channel after passing through the Raft algorithm library.

The specific code is shown in Listing 9.3.

Upon the arrival of a client command, the initial call is to the "DoCommand" function, as shown in Listing 9.4, where we examine its operations: first, the "DoCommand" function invokes "Marshal" to serialize our "CommandResponse" into the "reqBytes" byte array. Then, it calls the "Propose" interface of the Raft library to submit the proposal to our consensus algorithm implementation. In Raft, only the leader can handle proposals. If the current node is not the leader, we immediately return the error code "ErrCodeWrongLeader" to the client.

Listing 9.4 Handling requests

```
func (s *KvServer) DoCommand(ctx context.Context, req *pb.CommandRequest
    ) (*pb.CommandResponse, error) {
  raftcore.PrintDebugLog(fmt.Sprintf("do cmd %s", req.String()))

  cmdResp := &pb.CommandResponse{}

  if req != nil {
    reqBytes, err := json.Marshal(req)
    if err != nil {
      return nil, err
    }
    idx, _, isLeader := s.Rf.Propose(reqBytes)
    if !isLeader {
      cmdResp.ErrCode = common.ErrCodeWrongLeader
      return cmdResp, nil
    }

    s.mu.Lock()
    ch := s.getNotifyChan(idx)
    s.mu.Unlock()

    select {
    case res := <-ch:
      cmdResp.Value = res.Value
    case <-time.After(ExecCmdTimeout):
      cmdResp.ErrCode = common.ErrCodeExecTimeout
      cmdResp.Value = "exec cmd timeout"
    }

    go func() {
      s.mu.Lock()
      delete(s.notifyChans, idx)
      s.mu.Unlock()
    }()

  }

  return cmdResp, nil
}
```

Next, we retrieve the notification channel for the current log *id* from "getNotifyChan." Only when this log notification occurs will the "select" statement proceed further, retrieving the value and placing it into "cmdResp.Value." If the operation exceeds the "ErrCodeExecTimeout" time, an error code indicating execution timeout will also be generated and returned to the client.

Finally, let's take a look at what the Apply goroutine does, as shown in Listing 9.5. It waits for messages to arrive on the s.applyCh channel. As mentioned earlier in the Raft library, applyCh is used to notify the application layer that logs

Listing 9.5 Apply request

```
func (s *KvServer) ApplingToStm(done <-chan interface{}) {
  for !s.IsKilled() {
    select {
    case <-done:
      return
    case appliedMsg := <-s.applyCh:
      req := &pb.CommandRequest{}
      if err := json.Unmarshal(appliedMsg.Command, req); err != nil {
        raftcore.PrintDebugLog("Unmarshal CommandRequest err")
        continue
      }
      s.lastApplied = int(appliedMsg.CommandIndex)

      var value string
      switch req.OpType {
      case pb.OpType_OpPut:
        s.stm.Put(req.Key, req.Value)
      case pb.OpType_OpAppend:
        s.stm.Append(req.Key, req.Value)
      case pb.OpType_OpGet:
        value, _ = s.stm.Get(req.Key)
      }

      cmdResp := &pb.CommandResponse{}
      cmdResp.Value = value
      ch := s.getNotifyChan(int(appliedMsg.CommandIndex))
      ch <- cmdResp
    }
  }
}
```

have been committed and can be applied to the state machine. When an appliedMsg arrives in applyCh, we update KvServer's *lastApplied* number. Then, based on the client's operation type, we perform different operations on our state machine. After completing these operations, we place the response into notifyChan, the channel that DoCommand is waiting on. With this, the entire request-handling process is concluded.

9.6 Summary

In this chapter, based on the eraft strong consistency algorithm library built in the previous chapter, we implemented a complete distributed key-value storage system called eraftkv using Raft. The strong consistency algorithm library serves as the core of a distributed storage system, providing high availability guarantees to handle various errors in distributed environments. However, relying solely on the strong consistency algorithm library is insufficient to provide external services. Therefore, in this chapter, we introduced several external interfaces based on the eraft library, such as Put and Get operations. These operations resemble typical read and write operations in a database. When combined with the strong consistency algorithm library, they enable the achievement of high availability in a distributed system.

Chapter 10
Optimized Design and Implementation of the Strong Consistency Algorithm Raft: Multi-Raft

10.1 Design Considerations

In the previous chapter, we applied Raft to implement a single-group Raft KV cluster. Client write requests were sent to the leader, which replicated the operations to follower nodes. In the event of leader failure, as described in Chap. 8 using the Raft algorithm library, a new election was conducted to elect a new leader to continue providing service. This setup ensured a highly available cluster.

Through the single-group KV cluster, we achieved high availability and partition tolerance. However, a distributed system also requires scalability, which means increasing throughput by adding more machines. A single-group cluster has only a single-leader node that handles client writes, so the throughput limit of a single-group cluster system depends on the performance of a single machine. How do we achieve distributed scalability?

Next, we will introduce Multi-Raft implementation, which addresses the scalability issues of a single-group cluster. The idea is this: since a single group has its limits, can we achieve scalability by using multiple Raft KV clusters? We need a configuration center to manage the address information of multiple-group servers. In a multi-group architecture, uniform distribution of user requests across the corresponding group servers is crucial. This can be achieved through consistent hashing algorithms, which compute hash values of user keys and map them to distinct service groups, ensuring optimal load balancing and minimizing redistribution during scaling operations.

Integrating the above ideas, we can outline the system architecture diagram as shown in Fig. 10.1.

In previous chapters, we covered foundational knowledge of the Go programming language, the Raft consensus algorithm library, and demonstrated its application. It is expected that readers now have a deeper understanding of the system.

Z. Xu, *Distributed Data Services*, https://doi.org/10.1007/978-981-96-9392-4_10

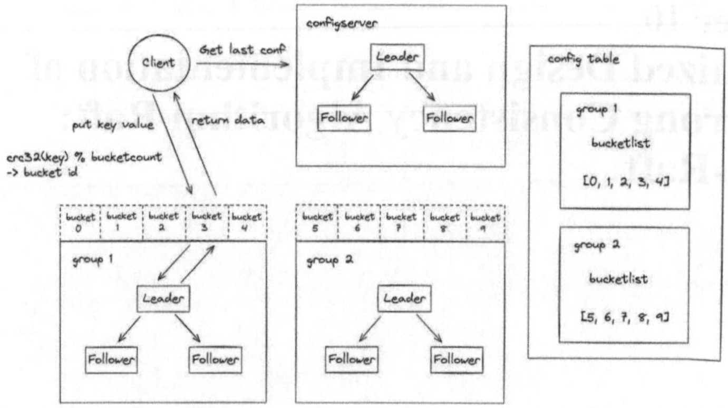

Fig. 10.1 Multi-Raft system architecture

Upon initialization, the client retrieves cluster partitioning metadata and corresponding data shard (Bucket) allocations from the configuration server (ConfigServer), caching this information in its local environment. When the client sends requests, it calculates the hash value of the key to determine the appropriate bucket and the server group (ShardServer) responsible for that bucket. The operation is then sent to the respective server group for processing. Both the ConfigServer and ShardServer server groups are designed to be highly available, with multiple ShardServer instances enhancing system scalability.

10.2 Configuration Server Implementation Analysis

The implementation of the configuration service is located in the "eraft/configserver" directory. According to the architecture diagram above, we can roughly determine what information the configuration server needs to store: (1) server addresses for each service group and (2) information about the data buckets managed by each service group. The structure definition is shown in Listing 10.1 as follows.

The "Version" represents the current version number of the configuration. "Buckets" store information about the buckets managed by service groups. "NBuckets" is

Listing 10.1 Config structure

```
type Config struct {
  Version int
  Buckets [common.NBuckets]int
  Groups  map[int][]string
}
```

Listing 10.2 Operations interface

```go
type ConfigStm interface {
  Join(groups map[int][]string) error
  Leave(gids []int) error
  Move(bucket_id, gid int) error
  Query(num int) (Config, error)
}
```

a constant indicating the maximum number of buckets in the system, defaulting to 10 in this case. For large-scale distributed systems, "Buckets" can be set to a much larger value.

Here's an example configuration: Version 6, 10 buckets, and 1 service group. We set up three servers, 127.0.0.1:6088, 127.0.0.1:6089, and 127.0.0.1:6090, to form service group 1. Buckets 0–4 are managed by service group 1. Buckets 5–9 are assigned a value of 0, indicating that currently no service group is responsible for writing data to these buckets:

```
{"Version":6,"Buckets":[1,1,1,1,1,0,0,0,0,0],"Groups
    ":{"1":["127.0.0.1:6088","127.0.0.1:6089","127.0.0.1:6090"]}}
```

In eraft, the configuration information described above is stored in leveldb. Listing 10.2 defines several interface operations: Join adds servers of a service group to the cluster configuration; Leave removes server configuration for specified service groups; Move assigns a bucket to its corresponding responsible group; and Query retrieves configuration information. These operations modify the Config structure and persist data to leveldb. The implementation logic for these operations can be found in "eraft/configserver/config_stm.go."

The core logic of the configuration server is located in config_server.go. Its architecture closely parallels the implementation of the single-group key-value store (KVS) discussed in Chap. 4. The primary distinction lies in the replacement of Put and Get operations with Join, Leave, Move, and Query operations, which are specifically designed for managing cluster configurations. Each configuration operation is subject to a distributed consensus protocol, ensuring strong consistency across the triplicated service nodes. This consensus-driven approach guarantees that all nodes within the cluster maintain an identical view of the current configuration state. Consequently, in the event of a leader configuration node failure, the cluster can seamlessly retrieve up-to-date configuration information from the newly elected leader configuration server, maintaining system reliability and availability.

10.3 Shard Server Implementation Analysis

First, let's examine the definition of the Bucket structure, which is delineated in Listing 10.3.

A bucket has a unique identifier, or ID. As previously mentioned when introducing the configuration server, a service group is responsible for a portion of buckets' data. In the configuration server, the ID associated with a bucket corresponds to the index number in the configuration array. Additionally, each bucket is associated with an interface for KvStore. When writing data, the current server's data storage engine is passed in. Listing 10.4 is an operation on the data in Bucket, including Get, Put, and Append operations.

Next, let's take a look at the structure defined for ShardServer, as shown in Listing 10.5.

This structure closely resembles our implementation of a single-group KvServer using Raft. Here, the state machine is of type "map[int]*Bucket," representing the data buckets managed by the current server. As a sharded server architecture, it establishes communication with a configuration server to determine its assigned shard responsibilities. The "cvCli" object encapsulates a client-side interface for interacting with the configuration server, facilitating dynamic shard allocation and management. "lastConfig" and "curConfig" respectively record the previous and current versions of the cluster configuration table. Once the server knows this table, it understands which data shards it is responsible for. When there is a change in the cluster topology, the configuration table updates, allowing the sharded server to promptly perceive and apply the new configuration. The other structures are similar to those we discussed in the single-group KvServer.

Let's now examine the construction process of ShardServer and the differences compared to a single-group KvServer, as shown in Listing 10.6. Initially, when the server starts, we initialize two engines: one (logDbEng) for storing Raft logs and another (newdbEng) for storing actual data. "cvCli" is a client connection to the configuration server group. We use MakeCfgSvrClient to construct the client to the configuration server, passing in the list of addresses of the configuration server group.

Listing 10.3 Bucket structure

```
//
// A bucket in a distributed system is a logical partition with a
//  specific ID, a pointer to a storage engine, and a state.
//
type Bucket struct {
  ID     int
  KvDB   storage_eng.KvStore
  Status buketStatus
}
```

Listing 10.4 Operations on data within buckets

```
//
// To retrieve encodeKey from the storage engine
//
func (bu *Bucket) Get(key string) (string, error) {
  encodeKey := strconv.Itoa(bu.ID) + SPLIT + key
  v, err := bu.KvDB.Get(encodeKey)
  if err != nil {
    return "", err
  }
  return v, nil
}

//
// Writing <key, value> data to the storage engine
//
func (bu *Bucket) Put(key, value string) error {
  encodeKey := strconv.Itoa(bu.ID) + SPLIT + key
  return bu.KvDB.Put(encodeKey, value)
}

//
// To append a value to an existing key in a storage engine
//
func (bu *Bucket) Append(key, value string) error {
  oldV, err := bu.Get(key)
  if err != nil {
    return err
  }
  return bu.Put(key, oldV+value)
}
```

The initialization of each Bucket within the state machine is performed by the "initStm function." Subsequently, the system invokes "cvCli.Query(-1)" to retrieve the most recent configuration, which is then cached locally as curConfig. During the initial bootstrap process, lastConfig is synchronized with this configuration to establish a consistent baseline state.

Next, there's a Goroutine performing tasks labeled as ConfigAction, and its core logic is shown in Listing 10.7. Let's delve into what it does. The core logic runs in a loop with a 1-s interval. Initially, it attempts to fetch the next configuration version using "cvCli.Query."

In the absence of modifications to the current cluster configuration, the function returns nil and proceeds to the subsequent iteration, maintaining the system's current state. Upon detection of a new configuration change, such as the addition of a server group, the system performs a version comparison between the new configuration and the existing one. If version parity is confirmed and the current node holds leadership status, it propagates the configuration change information to the relevant

Listing 10.5 ShardKV structure

```
type ShardKV struct {
  mu       sync.RWMutex
  dead     int32
  rf       *raftcore.Raft
  applyCh chan *pb.ApplyMsg
  gid_     int
  cvCli    *configserver.CfgCli

  lastApplied int
  lastConfig  configserver.Config
  curConfig   configserver.Config

  stm map[int]*Bucket

  dbEng storage_eng.KvStore

  notifyChans map[int]chan *pb.CommandResponse

  stopApplyCh chan interface{}

  pb.UnimplementedRaftServiceServer
}
```

server group. This process ensures configuration consistency across all servers within the group. The propagation is achieved through the proposal of an operation of type "OpType_OpConfigChange" utilizing the Propose method, leveraging the underlying consensus protocol.

Finally, let's see how servers in the group apply this log, as shown in Listing 10.8.

Listing 10.8 will update the Server's lastConfig and curConfig configuration information.

10.4 Client Implementation Analysis

When a client writes a key into the system, the first step is to determine which shard server is responsible for that key. When constructing the client, we first cache the latest configuration information locally for this purpose, as shown in Listing 10.9.

The client-side API exposes two primary interface methods: Get(key string) and Put(key, value string). Both of these methods internally invoke a common Command method (illustrated in Listing 10.10), which orchestrates the communication with the distributed shard servers.

1. First, we use the "Key2BucketID" function to compute the CRC32 hash of the key, determining the ID of the bucket to which it should be assigned.

Listing 10.6 Creating ShardKV

```
//
// The MakeShardKVServer function creates a new sharded KV server.
// peerMaps: Initialize peer map in a Raft group.
// nodeId: The nodeId of a peer in the Raft group.
// gid: The Raft group ID that identifies the node's membership in the
    group.
// configServerAddr: The address of the configuration server.
//
func MakeShardKVServer(peerMaps map[int]string, nodeId int, gid int,
    configServerAddrs string) *ShardKV {
    ...

  logDbEng := storage_eng.EngineFactory("leveldb", "./log_data/shard_svr
    /group_"+strconv.Itoa(gid)+"/node_"+strconv.Itoa(nodeId))
  newRf := raftcore.MakeRaft(clientEnds, nodeId, logDbEng, newApplyCh,
    500, 1500)
  newdbEng := storage_eng.EngineFactory("leveldb", "./data/group_"+
    strconv.Itoa(gid)+"/node_"+strconv.Itoa(nodeId))

  shardKv := &ShardKV{
      ...
    cvCli:       configserver.MakeCfgSvrClient(common.UN_UNSED_TID,
    strings.Split(configServerAddrs, ",")),
    lastApplied: 0,
    curConfig:   configserver.DefaultConfig(),
    lastConfig:  configserver.DefaultConfig(),
    stm:         make(map[int]*Bucket),
    ...
  }

  shardKv.initStm(shardKv.dbEng)

  shardKv.curConfig = *shardKv.cvCli.Query(-1)
  shardKv.lastConfig = *shardKv.cvCli.Query(-1)
    ...

  go shardKv.ConfigAction()

  return shardKv
}
```

2. Next, we look up in the locally cached “kvCli.config” configuration to find the server group responsible for data of this bucket ID.
3. Once we have identified the server group, we send a DoCommand RPC to the first server in the group.
4. If this server is not the leader, it will return the ID of the leader. Then the client will resend the DoCommand RPC to the leader node.

Listing 10.7 Configuration changes

```
if _, isLeader := s.rf.GetState(); isLeader {
...
  nextConfig := s.cvCli.Query(int64(curConfVersion) + 1)
  if nextConfig == nil {
    continue
  }
  nextCfBytes, _ := json.Marshal(nextConfig)
  curCfBytes, _ := json.Marshal(s.curConfig)
  raftcore.PrintDebugLog("next config -> " + string(nextCfBytes))
  raftcore.PrintDebugLog("cur config -> " + string(curCfBytes))
  if nextConfig.Version == curConfVersion+1 {
    req := &pb.CommandRequest{}
    nextCfBytes, _ := json.Marshal(nextConfig)
    raftcore.PrintDebugLog("can perform next conf -> "
        + string(nextCfBytes))
    req.Context = nextCfBytes
    req.OpType = pb.OpType_OpConfigChange
    reqBytes, _ := json.Marshal(req)
    idx, _, isLeader := s.rf.Propose(reqBytes)
    if !isLeader {
      return
    }

    ...
  }
}
```

Listing 10.8 Apply log

```
nextConfig := &configserver.Config{}
json.Unmarshal(req.Context, nextConfig)
if nextConfig.Version == s.curConfig.Version+1 {
    ...
  s.lastConfig = s.curConfig
  s.curConfig = *nextConfig
  cfBytes, _ := json.Marshal(s.curConfig)
  raftcore.PrintDebugLog("applied config to server -> "
      + string(cfBytes))
}
```

Listing 10.9 Creating a KvClient

```
//
// Creating a KvClient
//
func MakeKvClient(csAddrs string) *KvClient {
  ...
  kvCli.config = kvCli.csCli.Query(-1)
  return kvCli
}
```

10.5 Summary

In this chapter, we have reflected on and enhanced our previously implemented distributed key-value storage system, eraftkv. While benefiting from the strong consistency provided by the underlying algorithm, it also inherited inherent issues of traditional Raft, such as the bottleneck of a single leader for read and write operations. As the cluster scales, these deficiencies become increasingly apparent. Therefore, in this chapter, we continued to refine eraftkv by implementing a Multi-Raft version of the distributed key-value storage system. We organized the node cluster into groups, with each group responsible for ensuring consistency over a subset of data. Each group can be viewed as an independent Raft instance running in isolation. These optimizations are transparent to upper-layer applications, which require routing requests to the appropriate group for processing using hashing algorithms.

To achieve these goals, we designed a configuration server responsible for grouping data nodes. The configuration server stores information about server addresses and data bucket assignments for each service group. Buckets are organized in an array to record these groupings, where each bucket represents a set of data. Furthermore, we defined additional operations to manage data grouping: Join, Leave, Move, and Query. These operations collectively enable flexible and responsive cluster topology management. Additionally, we defined the shardKv responsible for processing key-value operations, which interacts with the configuration server to determine its data responsibilities. Using custom-defined functions, we mapped requests to the corresponding shard data, redirecting operations to the appropriate group for subsequent Raft algorithm processing. Ultimately, these efforts aim to achieve high availability and scalability in our distributed key-value storage system.

Listing 10.10 Command method

```
//
// The Command method
// Executing user commands
//
func (kvCli *KvClient) Command(req *pb.CommandRequest) (string, error) {
    bucket_id := common.Key2BucketID(req.Key)
    gid := kvCli.config.Buckets[bucket_id]
    if gid == 0 {
        return "", errors.New("there is no shard in charge of this
    bucket, please join the server group before")
    }
    if servers, ok := kvCli.config.Groups[gid]; ok {
        for _, svrAddr := range servers {
            if kvCli.GetConnFromCache(svrAddr) == nil {
                kvCli.rpcCli = raftcore.MakeRaftClientEnd(svrAddr,
    common.UN_UNSED_TID)
            } else {
                kvCli.rpcCli = kvCli.GetConnFromCache(svrAddr)
            }
            resp, err := (*kvCli.rpcCli.GetRaftServiceCli()).DoCommand(
    context.Background(), req)
            if err != nil {
                // node down
                raftcore.PrintDebugLog("there is a node down is cluster,
     but we can continue with outher node")
                continue
            }
            switch resp.ErrCode {
            case common.ErrCodeNoErr:
                kvCli.commandId++
                return resp.Value, nil
            case common.ErrCodeWrongGroup:
                kvCli.config = kvCli.csCli.Query(-1)
                return "", errors.New("WrongGroup")
            case common.ErrCodeWrongLeader:
                kvCli.rpcCli = raftcore.MakeRaftClientEnd(servers[resp.
    LeaderId], common.UN_UNSED_TID)
                resp, err := (*kvCli.rpcCli.GetRaftServiceCli()).
    DoCommand(context.Background(), req)
                if err != nil {
                    fmt.Printf("err %s", err.Error())
                    panic(err)
                }
                if resp.ErrCode == common.ErrCodeNoErr {
                    kvCli.commandId++
                    return resp.Value, nil
                }
            default:
                return "", errors.New("unknow code")
            }
        }
    } else {
        return "", errors.New("please join the server group first")
    }
    return "", errors.New("unknow code")
}
```

Bibliography

1. Abadi, D.J., Madden, S.R.: REED: Robust, Efficient Filtering and Event Detection in Sensor Networks. Technical Report MIT-CSAIL-TR-2004-011. MIT Computer Science and Artificial Intelligence Laboratory (2004)
2. Adya, A., Liskov, B., O'Neil, P.: Generalized isolation level definitions. In: Proceedings of 16th International Conference on Data Engineering (Cat. No. 00CB37073), pp. 67–78. IEEE, Piscataway (2000)
3. Ahamad, M., et al.: Causal memory: definitions, implementation, and programming. Distrib. Comput. **9**(1), 37–49 (1995)
4. ANSI SQL-92. https://www.contrib.andrew.cmu.edu/~shadow/sql/sql1992.txt. Accessed 1992
5. Ateniese, G., et al.: Improved proxy re-encryption schemes with applications to secure distributed storage. ACM Trans. Inf. Syst. Secur. (TISSEC) **9**(1), 1–30 (2006)
6. Attiya, H., Welch, J.L.: Sequential consistency versus linearizability. ACM Trans. Comput. Syst. (TOCS) **12**(2), 91–122 (1994)
7. Bailis, P., et al.: {HAT}, Not {CAP}: towards highly available transactions. In: 14th Workshop on Hot Topics in Operating Systems (HotOS XIV) (2013)
8. Bailis, P., et al.: Highly available transactions: virtues and limitations. Proc. VLDB Endowment **7**(3), 181–192 (2013)
9. Banker, K., et al.: MongoDB in Action: Covers MongoDB Version 3.0. Simon and Schuster, New York (2016)
10. Berenson, H., et al.: A critique of ANSI SQL isolation levels. ACM SIGMOD Rec. **24**(2), 1–10 (1995)
11. Bernstein, P.A., Goodman, N.: Concurrency control in distributed database systems. ACM Comput. Surv. (CSUR) **13**(2), 185–221 (1981)
12. Bernstein, P.A., Hadzilacos, V., Goodman, N.: Concurrency Control and Recovery in Database Systems, vol. 370. Addison-Wesley, Reading (1987)
13. Bian, N., Liu, X.: Implementation of non-blocking distributed transaction commit protocol (Chinese). Comput. Appl. Softw. **31**(7), 89–92 (2014)
14. Bloom, B.H.: Space/time trade-offs in hash coding with allowable errors. Commun. ACM **13**(7), 422–426 (1970)
15. Boru, D., et al.: Energy-efficient data replication in cloud computing datacenters. Clust. Comput. **18**(1), 385–402 (2015)
16. Brzezinski, J., Sobaniec, C., Wawrzyniak, D.: Session guarantees to achieve PRAM consistency of replicated shared objects. In: International Conference on Parallel Processing and Applied Mathematics, pp. 1–8. Springer (2003)

Z. Xu, *Distributed Data Services*, https://doi.org/10.1007/978-981-96-9392-4

17. Burckhardt, S., et al.: Principles of eventual consistency. Found. Trends®Program. Lang. **1**(1–2), 1–150 (2014)
18. ByteTCC. https://github.com/liuyangming/ByteTCC
19. Cahill, M.J., Röhm, U., Fekete, A.D.: Serializable isolation for snapshot databases. ACM Trans. Database Syst. (TODS) **34**(4), 1–42 (2009)
20. Cai, H., et al.: IoT-based big data storage systems in cloud computing: perspectives and challenges. IEEE Internet Things J. **4**(1), 75–87 (2016)
21. Cassandra. https://cassandra.apache.org/_/index.html
22. Cassandra Documentation. https://cassandra.apache.org/doc/latest/
23. Castro, M., Liskov, B., et al.: Practical byzantine fault tolerance. OsDI **99**, 173–186 (1999)
24. Chang, F., et al.: Bigtable: a distributed storage system for structured data. ACM Trans. Comput. Syst. (TOCS) **26**(2), 1–26 (2008)
25. Chen, M., Zhong, Z.: Block nested join and sort merge join algorithms: an empirical evaluation. In: International Conference on Advanced Data Mining and Applications, pp. 705–715. Springer, Berlin (2014)
26. Chervyakov, N., et al.: AR-RRNS: configurable reliable distributed data storage systems for Internet of Things to ensure security. Future Gener. Comput. Syst. **92**, 1080–1092 (2019)
27. Chung, S.M., Chatterjee, A.: Performance analysis of a parallel distributive join algorithm on the Intel Paragon. In: Proceedings 1997 International Conference on Parallel and Distributed Systems, pp. 714–721. IEEE, Piscataway (1997)
28. Chung, S.M., Chatterjee, A.: An adaptive parallel distributive join algorithm on a cluster of workstations. J. Supercomput. **21**(1), 5–35 (2002)
29. Chung, S.M., Yang, J.: Distributive join algorithm for cube-connected multiprocessors. In: Database Systems for Advanced Applications' 93, pp. 253–260. World Scientific, Singapore (1993)
30. CMU 15-445 MVCC. https://15445.courses.cs.cmu.edu/fall2019/slides/19-multiversioning.pdf
31. Codd, E.F.: A relational model of data for large shared data banks. Commun. ACM **13**(6), 377–387 (1970)
32. Coulouris, G.F., Dollimore, J., Kindberg, T.: Distributed Systems: Concepts and Design. Pearson Education, Edinburgh (2005)
33. Dageville, B., et al.: The snowflake elastic data warehouse. In: Proceedings of the 2016 International Conference on Management of Data, pp. 215–226 (2016)
34. DeCandia, G., et al.: Dynamo: Amazon's highly available key-value store. ACM SIGOPS Oper. Syst. Rev. **41**(6), 205–220 (2007)
35. DeWitt, D.J., Gerber, R.H.: Multiprocessor hash-based join algorithms. In: VLDB (1985)
36. Dimakis, A.G., et al.: Network coding for distributed storage systems. IEEE Trans. Inf. Theory **56**(9), 4539–4551 (2010)
37. Dimakis, A.G., et al.: A survey on network codes for distributed storage. Proc. IEEE **99**(3), 476–489 (2011)
38. dynamodb. https://aws.amazon.com/cn/dynamodb/
39. Egenhofer, M.J.: Spatial SQL: a query and presentation language. IEEE Trans. Knowl. Data Eng. **6**(1), 86–95 (1994)
40. etcd. en. https://etcd.io/. Visited on 10/12/2024
41. Eventual Consistency. https://en.wikipedia.org/wiki/Eventual_consistency
42. Fang, Y., Zhu, Y., Gong, X., et al.: Distributed transaction processing in microservice architecture (Chinese). Comput. Appl. Softw. **36**(1), 152–158 (2019)
43. Faroult, S., Robson, P.: The Art of SQL. O'Reilly Media, Inc., Sebastopol (2006)
44. Fischer, M.J., Lynch, N.A., Paterson, M.S.: Impossibility of distributed consensus with one faulty process. J. ACM (JACM) **32**(2), 374–382 (1985)
45. Fox, A., et al.: Cluster-based scalable network services. In: Proceedings of the Sixteenth ACM Symposium on Operating Systems Principles, pp. 78–91 (1997)
46. Garcia-Molina, H., Salem, K.: Sagas. ACM Sigmod Rec. **16**(3), 249–259 (1987)

47. Gilbert, S., Lynch, N.: Brewer's conjecture and the feasibility of consistent, available, partition-tolerant web services. SIGACT News **33**(2), 51–59 (2002). ISSN: 0163-5700. https://doi.org/10.1145/564585.564601
48. Gohil, J.A., Dolia, P.M.: Study and comparative analysis of basic pessimistic and optimistic concurrency control methods for database management system. Int. J. Adv. Res. Comput. Commun. Eng. **5**(1), 178–186 (2016)
49. Graefe, G.: The cascades framework for query optimization. IEEE Data Eng. Bull. **18**(3), 19–29 (1995)
50. Graefe, G.: Efficient columnar storage in b-trees. ACM SIGMOD Rec. **36**(1), 3–6 (2007)
51. Graefe, G.: A generalized join algorithm. In: Datenbanksysteme für Business, Technologie und Web (BTW) (2011)
52. Gray, J., Lorie, R., Putzolu, G.F., Traiger, I.L.: Granularity of locks and degrees of consistency. In: Nijssen, G.M. (ed.) Modeling in Data Base Management Systems. North Holland Publishing, Amsterdam (1976)
53. Gulutzan, P., Pelzer, T.: SQL Performance Tuning. Addison-Wesley Professional, Boston (2003)
54. Han, J., et al.: Survey on NoSQL database. In: 2011 6th International Conference on Pervasive Computing and Applications, pp. 363–366. IEEE, Piscataway (2011)
55. Helland, P.: Life beyond distributed transactions: an apostate's opinion. Queue **14**(5), 69–98 (2016)
56. Herlihy, M.P., Wing, J.M.: Linearizability: a correctness condition for concurrent objects. ACM Trans. Program. Lang. Syst. (TOPLAS) **12**(3), 463–492 (1990)
57. Hunt, P., et al.: {ZooKeeper}: wait-free coordination for Internet-scale systems. In: 2010 USENIX Annual Technical Conference (USENIX ATC 10) (2010)
58. Jain, H.K.: A comprehensive model for the storage structure design of CODASYL databases. Inf. Syst. **9**(3–4), 217–230 (1984)
59. Jalote, P.: Fault Tolerance in Distributed Systems. Prentice-Hall, Inc., Upper Saddle River (1994)
60. Jepsen. https://jepsen.io/consistency
61. Junqueira, F.P., Reed, B.C., Serafini, M.: Zab: high-performance broadcast for primary-backup systems. In: 2011 IEEE/IFIP 41st International Conference on Dependable Systems & Networks (DSN), pp. 245–256. IEEE, Piscataway (2011)
62. Kallman, R., et al.: H-store: a high-performance, distributed main memory transaction processing system. Proc. VLDB Endowment **1**(2), 1496–1499 (2008)
63. Karvounarakis, G., et al.: RQL: a declarative query language for RDF. In: Proceedings of the 11th International Conference on World Wide Web, pp. 592–603 (2002)
64. Kleppmann, M.: Designing Data-Intensive Application. O'Reilly Media, Inc., Sebastopol (2017)
65. Kossmann, D.: The state of the art in distributed query processing. ACM Comput. Surv. (CSUR) **32**(4), 422–469 (2000)
66. Koutris, P.: Bloom Filters in Distributed Query Execution. University of Washington, Seattle (2011)
67. Kung, H.-T., Robinson, J.T.: On optimistic methods for concurrency control. ACM Trans. Database Syst. (TODS) **6**(2), 213–226 (1981)
68. Lamport, L.: Proving the correctness of multiprocess programs. IEEE Trans. Software Eng. **SE-3**(2), 125–143 (1977). https://doi.org/10.1109/TSE.1977.229904
69. Lamport, L.: How to make a multiprocessor computer that correctly executes multiprocess programs. IEEE Trans. Comput. **c-28**(9), 690–691 (1979)
70. Lamport, L.: Paxos made simple. ACM SIGACT News (Distrib. Comput. Col.) **32**(4) (Whole Number 121, December 2001), 51–58 (2001)
71. Lamport, L.: The part-time parliament. In: Concurrency: The Works of Leslie Lamport, pp. 277–317 (2019)
72. Lamport, L.: Time, clocks, and the ordering of events in a distributed system. In: Concurrency: The Works of Leslie Lamport, pp. 179–196 (2019)

73. Lamport, L., Shostak, R., Pease, M.: The Byzantine generals problem. In: Concurrency: The Works of Leslie Lamport, pp. 203–226 (2019)
74. Lampson, B.W., Sturgis, H.E.: Crash Recovery in a Distributed Data Storage System. Technical Report. Xerox Palo Alto Research Center (1979)
75. Larson, P., Hanson, E.N., Price, S.L.: Columnar storage in SQL server 2012. IEEE Data Eng. Bull. **35**(1), 15–20 (2012)
76. Lee, T., Kim, K., Kim, H.-J.: Join processing using bloom filter in mapreduce. In: Proceedings of the 2012 ACM Research in Applied Computation Symposium, pp. 100–105 (2012)
77. Li, R., et al.: Blockchain for large-scale internet of things data storage and protection. IEEE Trans. Serv. Comput. **12**(5), 762–771 (2018)
78. Li, S., et al.: Distributed storage strategy for spatial big data based on NoSQL database (Chinese). Geomatics Inf. Sci. Wuhan Univ. **42**(2), 163–169 (2017)
79. Li, X., et al.: Study of query of distributed database based on relation semi join. In: 2010 International Conference on Computer Design and Applications, vol. 1, pp. V1–134. IEEE, Piscataway (2010)
80. Li, Y., et al.: Intelligent cryptography approach for secure distributed big data storage in cloud computing. Inf. Sci. **387**, 103–115 (2017)
81. Lin, Y., et al.: Llama: leveraging columnar storage for scalable join processing in the mapreduce framework. In: Proceedings of the 2011 ACM SIGMOD International Conference on Management of Data, pp. 961–972 (2011)
82. Liskov, B., Cowling, J.: Viewstamped Replication Revisited. Technical Report MIT-CSAIL-TR-2012-021. MIT Computer Science and Artificial Intelligence Laboratory (2012)
83. Liu, B., Rundensteiner, E.A.: Revisiting pipelined parallelism in multi-join query processing. In: Proceedings of the 31st International Conference on Very Large Data Bases, pp. 829–840. Citeseer (2005)
84. Lloyd, W., et al.: Don't settle for eventual: scalable causal consistency for wide-area storage with COPS. In: Proceedings of the Twenty-Third ACM Symposium on Operating Systems Principles, pp. 401–416 (2011)
85. Long, R., et al.: IMS primer. In: IBM International Technical Support Organisation (2000)
86. Melton, J.: Database language SQL. In: Handbook on Architectures of Information Systems, pp. 105–132. Springer, Berlin/Heidelberg (1998)
87. Melton, J., Simon, A.R.: SQL: 1999: Understanding Relational Language Components. Elsevier, Burlington (2001)
88. Miller, J.J.: Graph database applications and concepts with Neo4j. In: Proceedings of the Southern Association for Information Systems Conference, Atlanta, GA, vol. 2324, no. 36 (2013)
89. Mishra, P., Eich, M.H.: Join processing in relational databases. ACM Comput. Surv. (CSUR) **24**(1), 63–113 (1992)
90. Mohan, C., Lindsay, B., Obermarck, R.: Transaction management in the R* distributed database management system. ACM Trans. Database Syst. (TODS) **11**(4), 378–396 (1986)
91. Molinaro, A.: SQL Cookbook: Query Solutions and Techniques for Database Developers. O'Reilly Media, Inc., Sebastopol (2005)
92. MongoDB Documentation. https://www.mongodb.com/docs/
93. Mullin, J.K.: Optimal semijoins for distributed database systems. IEEE Trans. Software Eng. **16**(5), 558–560 (1990)
94. Nakamoto, S.: Bitcoin: a peer-to-peer electronic cash system. Decent. Bus. Rev. 21260 (2008). https://doi.org/10.2139/ssrn.3440802
95. Negri, M., Pelagatti, G.: Distributive join: a new algorithm for joining relations. ACM Trans. Database Syst. (TODS) **16**(4), 655–669 (1991)
96. Neo4j Documentation. https://neo4j.com/docs/
97. Oki, B.M., Liskov, B.H.: Viewstamped replication: a new primary copy method to support highly-available distributed systems. In: Proceedings of the Seventh Annual ACM Symposium on Principles of Distributed Computing, pp. 8–17 (1988)

98. Ongaro, D., Ousterhout, J.: In search of an understandable consensus algorithm. In: 2014 USENIX Annual Technical Conference (Usenix ATC 14), pp. 305–319 (2014)
99. Overview of the X/Open DTP model. https://infocenter.sybase.com/help/index.jsp?topic=/com.sybase.dc36123_1251/html/xainterf/X60196.htm
100. Özsu, M.T., Valduriez, P.: Principles of Distributed Database Systems, vol. 2. Springer, Upper Saddle River (1999)
101. Pawar, S., Rouayheb, S.E., Ramchandran, K.: On secure distributed data storage under repair dynamics. In: 2010 IEEE International Symposium on Information Theory, pp. 2543–2547. IEEE, Piscataway (2010)
102. Peng, D., Dabek, F.: Large-scale incremental processing using distributed transactions and notifications. In: 9th USENIX Symposium on Operating Systems Design and Implementation (OSDI 10) (2010)
103. Pérez, J., Arenas, M., Gutierrez, C.: Semantics and complexity of SPARQL. ACM Trans. Database Syst. (TODS) **34**(3), 1–45 (2009)
104. Ports, D.R.K., Grittner, K.: Serializable snapshot isolation in PostgreSQL (2012). arXiv preprint arXiv:1208.4179
105. PRAM. https://www.cs.princeton.edu/techreports/1988/180.pdf
106. Pritchett, D.: BASE: an acid alternative: in partitioned databases, trading some consistency for availability can lead to dramatic improvements in scalability. Queue **6**(3), 48–55 (2008)
107. Qi, Y., et al.: Distributed transaction processing technology and its models (Chinese). Comput. Eng. Appl. **37**(9), 60–62 (2001)
108. raft. http://thesecretlivesofdata.com/raft/
109. Redis. https://redis.io/
110. Redis Documentation. https://redis.io/docs/
111. Reed, D.P.: Naming and synchronization in a decentralized computer system. PhD thesis, Massachusetts Institute of Technology (1978)
112. Reinsel, D., Gantz, J., Rydning, J.: Data age 2025: the evolution of data to life-critical. In: Don't Focus on Big Data, vol. 2 (2017)
113. Riak. https://riak.com/
114. Semi-synchronous Replication at Facebook. http://yoshinorimatsunobu.blogspot.com/2014/04/semi-synchronous-replication-at-facebook.html
115. Shafagh, H., et al.: Towards blockchain-based auditable storage and sharing of IoT data. In: Proceedings of the 2017 on Cloud Computing Security Workshop, pp. 45–50 (2017)
116. Shahabi, C., Khan, L., McLeod, D.: A probe-based technique to optimize join queries in distributed internet databases. Knowl. Inf. Syst. **2**(3), 373–385 (2000)
117. Shao, P.: Distributed Database Systems and Their Applications, vol. 6 (Chinese). Science Press, Alexandria (2000)
118. Shapiro, M., et al.: Conflict-free replicated data types. In: Symposium on Self-Stabilizing Systems, pp. 386–400. Springer, Berlin (2011)
119. Skeen, D.: Nonblocking commit protocols. In: Proceedings of the 1981 ACM SIGMOD International Conference on Management of Data, pp. 133–142 (1981)
120. Stocker, K., et al.: Integrating semi-join-reducers into state-of-the-art query processors. In: Proceedings 17th International Conference on Data Engineering, pp. 575–584. IEEE, Piscataway (2001)
121. Stonebraker, M.: SQL databases v. NoSQL databases. Commun. ACM **53**(4), 10–11 (2010)
122. Stonebraker, M., et al.: Mariposa: a new architecture for distributed data. In: Proceedings of 1994 IEEE 10th International Conference on Data Engineering, pp. 54–65. IEEE, Piscataway (1994)
123. Taylor, R.W., Frank, R.L.: CODASYL data-base management systems. ACM Comput. Surv. (CSUR) **8**(1), 67–103 (1976)
124. TCC-TRANSACTION. https://github.com/changmingxie/tcc-transaction
125. Terry, D.B., et al.: Session guarantees for weakly consistent replicated data. In: Proceedings of 3rd International Conference on Parallel and Distributed Information Systems, pp. 140–149. IEEE, Piscataway (1994)

126. The Open Group. https://www.opengroup.org/
127. TiKV. en-us. https://tikv.org/. Visited on 10/12/2024
128. Tiwari, S. Professional NoSQL. Wiley, Indianapolis (2011)
129. Vaish, G.: Getting Started with NoSQL, vol. 2103. Packt Publishing, Birmingham (2013)
130. Van Steen, M., Tanenbaum, A.: Distributed systems principles and paradigms. Network **2**, 28 (2002)
131. Viotti, P., Vukolić, M.: Consistency in non-transactional distributed storage systems. ACM Comput. Surv. (CSUR) **49**(1), 1–34 (2016)
132. Wang, L.: Join Query Optimization for Distributed Databases (Chinese). MA thesis, East China Normal University (2017)
133. Wang, Y., Qin, Y., Liu, W.: Progress in data query optimization algorithms for distributed databases (Chinese). J. Bohai Univ. (Nat. Sci. Ed.) **28**(1), 73–76 (2007)
134. Wilschut, A.N., Flokstra, J., Apers, P.M.G.: Parallel evaluation of multi-join queries. In: Proceedings of the 1995 ACM SIGMOD International Conference on Management of Data, pp. 115–126 (1995)
135. Wolfson, O., Jajodia, S., Huang, Y.: An adaptive data replication algorithm. ACM Trans. Database Syst. (TODS) **22**(2), 255–314 (1997)
136. Xu, Z., et al.: Edge AI-driven neural network predictions for replica sync optimization. Appl. Soft Comput. **165**, 112083 (2024)
137. Yu, C.T., Chang, C.C.: Distributed query processing. ACM Comput. Surv. (CSUR) **16**(4), 399–433 (1984)
138. Yu, C.T., Chang, C.C.: On the design of a query processing strategy in a distributed database environment. ACM SIGMOD Rec. **13**(4), 30–39 (1983)
139. Zhang, P., et al.: A survey on distributed stream processing technologies for big data (Chinese). J. Comput. Res. Dev. **3**(002), 1–9 (2014)
140. Zhao, G., Hu, Y.: Three-phase commit protocol based on heartbeat technology (Chinese). Comput. Eng. Appl. **40**(11), 177–179 (2004)
141. Zhou, L., et al.: The semi-join query optimization in distributed database system. In: 2012 National Conference on Information Technology and Computer Science, pp. 913–916. Atlantis Press, Dordrecht (2012)
142. Zhu, T., et al.: Relationship between consistency and availability in distributed databases (Chinese). J. Softw. **29**(1), 131–149 (2018)
143. Zou, J., Sheng, G., Li, D.: Research and implementation of transactional workflow system for long transactions (Chinese). J. Comput. Appl. **23**(2), 59–61 (2003)

MIX
Papier aus verantwortungsvollen Quellen
Paper from responsible sources
FSC® C105338

If you have any concerns about our products,
you can contact us on
ProductSafety@springernature.com

In case Publisher is established outside the EU,
the EU authorized representative is:
Springer Nature Customer Service Center GmbH
Europaplatz 3, 69115 Heidelberg, Germany

Printed by Libri Plureos GmbH
in Hamburg, Germany